D1756782

033996

Insect and Bird Interactions

Insect and Bird Interactions

Editors:

HELMUT F. VAN EMDEN

Department of Agriculture, The University of Reading, Berkshire, UK

MIRIAM ROTHSCHILD

Ashton Wold, Peterborough, UK

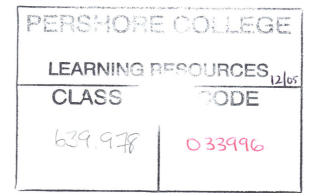

Intercept
Scientific Technical Medical Publishers

British Library Cataloguing in Publication Data
Insect and Bird Interactions.—

A CIP catalogue record for this book is available from the British Library
ISBN 1–898298–92–0

Published in January 2004 by Intercept Limited,
PO Box 716, Andover, Hampshire, SP10 1YG, UK.
Email: intercept@andover.co.uk
Website: www.intercept.co.uk

Typeset in Times by
Ann Buchan (Typesetters), Shepperton, Middlesex.
Printed by Athenaeum Press Ltd, Gateshead, Tyne & Wear.

Foreword

DAME MIRIAM ROTHSCHILD, F.R.S.

Ashton Wold, Peterborough, PE8 5LZ, UK

We have been told by Chris Mead (1983)[1] that 220 million swallows cross the equator to pass the winter in a salubrious climate, favouring a suitable supply of insect food.

Thousands of birds are dependent on this fragile prey ranging from clumsy grasshoppers to minute Diptera. We have no finer demonstration of the dependence of vertebrates on free living invertebrates than in this dramatic association.

At the end of August, my son, when four years old, remarked with childhood insight that the swallows 'looked sad'. There they were, beading the telegraph wires outside our house, unusually glum, motionless, and silent. Will they survive their migratory experience associated with the timely and regular emergence of the insect prey? Can they anticipate the dangers, bitter hardships, and agonising fatigue facing them, or do they enjoy a vision of a happy arrival, although over 2,000 miles away? Have they a sense of future events that we lack? Astonishingly, a pair of these birds will return unerringly next spring to their nest on the vertical side of our window sill.

Warblers attract less attention on their northward migration in the spring from Africa, although this shifting concourse of over 1,000 million birds outnumbers the swallows. When the warblers find their nesting sites, they advertise their survival with enchanting melodious tunes, both joyful and heart-rending – unconsciously producing a vision of their future brood and, simultaneously, a funeral dirge for both their many missing companions, and the freshly emerged insects they are now consuming. The migratory power of these mouse-like little birds gives us a glimpse of a world which exists beyond our comprehension. Is one of the clues which spurs them on an ultraviolet shimmer from the veins glittering in the wings of the flies, which we cannot appreciate, for we lack the ability to respond to this wavelength?

Bird migration is one of the most remarkable manifestations of natural selection. In the following pages, we also touch upon the problem of combining additional food production for ourselves with the complexity of the natural world, but we are still profoundly ignorant of the latter, and only learning slowly about the former. Thus, for example, we are still removing our protective hedgerows from our fields, and needlessly destroying the useful wild plants in field margins and ditches.

We have also commented on birds' vision, including ultraviolet appreciation, and sound reception, aposematic complexes, including warning signals, taste, memory,

---

[1]Mead, C. (1983) *Bird migration.* Country Life Books, Feltham, UK. 224 pp.

and resulting behaviour, breeding performance, and diet. Birds' foraging behaviour and our general farming practices have also been noted.

It is impossible for us to imagine the roar of the Dinosaur; will future generations understand what they have also lost, lacking the vanished liquid notes of the nightingale, the jubilant dawn chorus accompanied by the modest hum of the now dwindling bumble bees?

Miriam Rothschild

Preface

This book has its roots in a conference run by the Entomological Club at The University of Reading in September, 1997.

The Entomological Club was founded in 1826, and so is the oldest extant entomological society in the world. Founded as a dining club, membership is limited to just eight by constitution – there can be no doubt that the Club is exclusive, but it has no pretensions to being elitist.

As the result of a bequest from one of the members, George Verrall, an annual dinner (The Verrall Supper) was founded in 1887 at which other entomologists could join the eight Club members; today, the event runs at about 200 diners.

This history has relevance here in that the Club decided to use the centenary of The Verrall Supper in 1987 to organise the first conference in its 160-year history. Discussing the problems arising from the clandestine, and therefore undocumented, release in the UK of foreign specimens of rare or extinct insects in the name of 'conservation', the Club decided that a conference on *The Ecology of Insect Introductions* was needed. A successful Verrall Centenary Conference on the topic was held at The University of Reading in September, 1987.

As the ten-year anniversary of that first Club conference approached, the Club felt it was time to consider a second conference. One of us (MR) suggested that the variety of interactions that occurred between insects and birds would make a novel and interesting topic. The more the Club discussed it, the more interactions could be identified, and it was not a difficult decision to make *Insects and Birds* the second Entomological Club Conference run, again at The University of Reading, in September, 1997.

Fortunately, most of the speakers we invited to participate agreed enthusiastically, and some 80 participants enjoyed an excellent conference programme. There was considerable international interest, and there was great support for producing the first book exploring the diversity of interactions between insects and birds.

It is important to point out that this volume is not a conference proceedings, although most of the contributors to the conference have contributed to it. New information has appeared since 1997, and co-authors were recruited *de novo* to increase the coverage presented in what are now reviews, often with a broader title than was appropriate at the conference. As the topic of the conference was novel, so is the material presented here, which makes a first, and thus unique, compilation of information on insect/bird interactions.

The chapters are grouped into four sections. We put population management issues first. Here, the management of arable farmland is the principal topic, since today this has become such a topical issue as people worry about the possible indirect effect on farmland birds of the planting of large areas of genetically modified crops. Also, incentive schemes such as 'set-aside' and 'countryside stewardship' are offering new opportunities for increasing the biodiversity in our farming landscape, and birds are the organisms the schemes are especially designed to benefit. However, there is another aspect to population management – not of managing bird populations by providing insects, but of managing insects by exploiting birds as biological control agents. We believe there are few alternative sources of information on this subject to the chapter in this book.

The section on the effect of insecticides on bird populations follows on well after birds in agriculture. The contribution here from the RSPB presents an interesting approach to risk assessment in the absence of hard experimental data, and we follow this with an unusual example of how in a tourist island (one of the Seychelles) conservation of a declining bird population has had to rely on finding alternatives to neurotoxic pesticides used to achieve the level of control of domestic and urban pests required by the tourist population. Readers may then feel that exhuming the DDT–raptor decline story next is a bit *passé* in 2003, but the chapter reveals a surprising interpretation on what really happened to affect the raptors half a century ago.

The book then moves on to a different topic, the foraging behaviour of birds on insects. The first chapter in this section is another of the novelties the reader will find. It deals with colour vision in birds, but written by an ophthalmologist rather than an ornithologist. The specific area of ultraviolet vision in birds is then explored in relation to insect coloration, which takes us smoothly into warning colours and chemical defence of insects against predation by birds. More general aspects of selective foraging by birds are exemplified by studies of wryneck and great tit.

The final section deals largely with insects as parasites of birds, and begins with a review of the ecology and evolution of the co-adaptation of insect ectoparasites with birds. This is followed by another of the very novel aspects addressed in the book, one to which most readers will never have given any thought – how are ectoparasites transmitted between generations in the unusual life-style of the cuckoo? Many insects feed on detritus in birds' nests, and the final chapter addresses this with reference to moth species.

We need to add a special comment concerning the first chapter, separated out from the other chapters and identified as a 'case-study'. This chapter does not deal with insects, but with shellfish as food for shorebirds. It is a classic and detailed study of foraging by birds and how it is not density of prey, but its quality and availability, which is important in the population dynamics of the birds. We felt it had such relevance to foraging on insects by birds that it was very valuable to include it. We hope, having read this introductory chapter on shorebirds, that our readers will agree.

Our plans to produce this book were considerably delayed when our initial publisher was taken over by another company primarily interested in journals. We are therefore particularly grateful to Intercept for their interest in helping us achieve publication, and we are also most grateful to our authors for their patience, and for being so ready to revisit their subject some years after the conference to include more recent information. We believe their willingness to do this reflects a general conviction that the assembled material is well worth publishing. The chapter on transmission of ectoparasites between cuckoo generations was already published, and Cambridge University Press very kindly allowed us to reproduce the published version re-formatted to the pattern of our other chapters.

Additionally, we would like to thank our colleagues in the Entomological Club for taking on the bulk of the refereeing involved, and Mrs. Thelma Wise for voluntary editorial assistance. We are also most grateful to Ms. Berit Pedersen of the Royal Entomological Society for help with the references.

Helmut van Emden
Miriam Rothschild

Contents

PART 4: ECTOFAUNA

Contributors

AVERY, MARK I., *Royal Society for the Protection of Birds, The Lodge, Sandy, Bedfordshire, SG19 2DL, UK*

BARBA, EMILIO, *Instituto 'Cavanilles' de Biodiversidad y Biología Evolutiva & Unidad de Ecología, Universidad de Valencia, E-46100 Burjassot, Valencia, Spain*

BARKER, ALISON M., *The Game Conservancy Trust, Fordingbridge, Hampshire, SP6 1EF, UK (Present address: CABI-Bioscience Switzerland Centre, Rue des Grillons 1, CH-2800 Delémont, Switzerland)*

BARNEA, ANAT, *Department of Natural and Life Sciences, The Open University of Israel, Ramat-Aviv, Tel Aviv 61392, Israel*

BENNETT, ANDREW T. D., *School of Biological Sciences, University of Bristol, Woodland Road, Bristol, BS8 1UG, UK*

BIGNAL, ERIC M., *European Forum on Nature Conservation and Pastoralism, Kindrochaid, Gruinart, Bridgend, Isle of Islay, Argyll, PA44 7PT, UK*

BLAKE, SHONA, *Research Division, Scottish Agricultural College, Auchincruive, Ayr, KA6 5HW, UK*

BOWMAKER, JAMES K., *Department of Visual Science, Institute of Ophthalmology, University College London, Bath Street, London, EC1V 9EL, UK*

BROOKE, MICHAEL DE L., *Department of Zoology, University of Cambridge, Downing Street, Cambridge, CB2 3EJ, UK*

CAMPBELL, LENNOX H., *Royal Society for the Protection of Birds, The Lodge, Sandy, Bedfordshire, SG19 2DL, UK*

CHURCH, STUART C., *School of Biological Sciences, University of Bristol, Woodland Road, Bristol, BS8 1UG, UK*

CLAYTON, DALE H., *Department of Biology, University of Utah, Salt Lake City, UT 84112, USA*

CUTHILL, INNES C., *School of Biological Sciences, University of Bristol, Woodland Road, Bristol, BS8 1UG, UK*

EDWARDS, JOHN P., *Central Science Laboratory, Department for Environment, Food and Rural Affairs, Sand Hutton, York, YO41 1LZ, UK*

EVANS, ANDY D., *Royal Society for the Protection of Birds, The Lodge, Sandy, Bedfordshire, SG19 2DL, UK*

FOSTER, GARTH N., *Research Division, Scottish Agricultural College, Auchincruive, Ayr, KA6 5HW, UK*

FREITAG, ANNE, *Museum of Zoology, PO Box 448, CH-1000 Lausanne 17, Switzerland*

GIL-DELGADO, JOSÉ A., *Instituto 'Cavanilles' de Biodiversidad y Biología Evolutiva & Unidad de Ecología, Universidad de Valencia, E-46100 Burjassot, Valencia, Spain*

GLEN, DAVID M., *Cardiff University, School of Biosciences (Present address: Styloma Research & Consulting, Phoebe, The Lippiatt, Cheddar, Somerset, BS27 3QP, UK)*

GOSS-CUSTARD, JOHN D., *Natural Environment Research Council, Centre for*

Ecology and Hydrology, Winfrith Technology Centre, Dorchester, Dorset, DT2 8ZD, UK

GVARYAHU, GADI, *Department of Animal Sciences, Hebrew University of Jerusalem, Rehovot 76100, Israel*

HOLLAND, JOHN M., *The Game Conservancy Trust, Fordingbridge, Hampshire, SP6 1EF, UK*

MARPLES, NICOLA M., *Department of Zoology, Trinity College, Dublin D2, Ireland*

McCRACKEN, DAVID I., *Research Division, Scottish Agricultural College, Auchincruive, Ayr, KA6 5HW, UK*

MONRÓS, JUAN S., *Instituto 'Cavanilles' de Biodiversidad y Biología Evolutiva & Unidad de Ecología, Universidad de Valencia, E-46100 Burjassot, Valencia, Spain*

MOREBY, STEPHEN J., *The Game Conservancy Trust, Fordingbridge, Hampshire, SP6 1EF, UK*

MOYER, BRETT R., *Department of Biology, University of Utah, Salt Lake City, UT 84112, USA*

NAKAMURA, HIROSHI, *Faculty of Education, Shinshu University, Nishinagano, Nagano 380, Japan*

PARTRIDGE, JULIAN C., *School of Biological Sciences, University of Bristol, Woodland Road, Bristol, BS8 1UG, UK*

ROBINSON, GADEN S., *Department of Entomology, The Natural History Museum, Cromwell Road, London, SW7 5BD, UK*

ROPER, TIM J., *School of Biological Sciences, Sussex University, Brighton, East Sussex, BN1 9QG, UK*

ROTHSCHILD, MIRIAM, *Ashton Wold, Peterborough, PE8 5LZ, UK*

WALKER, COLIN H., *Cissbury, Hillhead, Colyton, Devon, EX24 6NJ, UK*

WEST, ANDY D., *Natural Environment Research Council, Centre for Ecology and Hydrology, Winfrith Technology Centre, Dorchester, Dorset, DT2 8ZD, UK*

PART 1

Population Management Issues

1

Case Study: An Example of the Kind of Study Now Needed for Insectivorous Birds – Population Level Implications of Variations in Prey Availability and Quality in Shorebirds: What We See is Not Necessarily What They Get

JOHN D. GOSS-CUSTARD AND ANDY D. WEST

Natural Environment Research Council, Centre for Ecology and Hydrology, Winfrith Technology Centre, Dorchester, Dorset, DT2 8ZD, UK

Introduction

Estuaries are highly productive biological systems, their productivity often matching that of the best arable land. This arises largely because organic material and nutrients generated outside the estuary itself are transported into it by river and tide. In northwest Europe, the resulting dense mats of vegetation, such as seagrass (*Zostera* spp.), provide rich feeding grounds for several species of herbivorous wildfowl (*Anatidae*). Macro-invertebrates reach very high numerical and biomass densities and provide productive feeding areas for carnivorous wildfowl and waders (*Charadrii*). The main invertebrate taxa are deposit-feeding and suspension-feeding bivalve molluscs, polychaete worms, and Crustacea that either live on or beneath the surface of the soft, muddy, or sandy sediments.

The herbivorous and carnivorous wildfowl and waders that exploit these feeding grounds do so mainly on the intertidal flats when these are exposed over the low water period. After their breeding season, most species arrive on the estuaries of the British Isles in late summer and autumn, at the end of the main growing and reproductive season for the vegetation and invertebrates, and so when the standing crop biomass of the food supply is at its highest. The generally large tidal range around the coasts of Britain encourages wide flats to develop, so that the feeding grounds are not only rich but also extensive. Because of the warming influence of the Gulf Stream, and in contrast to many intertidal areas further to the east in Europe, intertidal flats in the British Isles are seldom frozen over and thus rendered unuseable by birds in winter. British estuaries therefore provide relatively secure, dependable, and potentially rich

Insect and Bird Interactions
© Intercept Ltd., PO Box 716, Andover, Hampshire, SP10 1YG, UK.

feeding grounds through autumn, winter, and spring for many hundreds of thousands of migrant coastal birds. As a consequence, Britain has signed international agreements to protect and maintain these non-breeding season feeding habitats. This, in turn, has stimulated much research into the foraging and population ecology of shorebirds.

This research has now reached the stage at which predictive models can be built and used for predicting how bird populations would be affected by a whole range of human activities, referred to here as 'habitat change'. The models vary in complexity and in the amount of biological detail they include, depending on the nature of the predictions required. At one extreme, simple 'top-down', phenomenological and regression models are suitable for predicting how a change, for example, from sandy to muddy sediments, or *vice versa*, would affect the species composition and densities of waders on a shore (Yates *et al.*, 1996). Such models predict the distribution of a given number of birds, but do not forecast how that number of birds would itself be affected by habitat change (Goss-Custard & Durell, 1990). To do this, we need 'bottom-up' mechanistic models, coupled with demographic population models, which are able to predict how changes in the quality and quantity of the intertidal feeding grounds would affect overwinter survival rates, and thus population size (Goss-Custard *et al.*, 1995; Stillman *et al.*, 2000). This article briefly describes the main features of the foraging ecology of these populations which, at present, we think need to be incorporated into such mechanistic models if reliable predictions are to be obtained. This has relevance for future models of the foraging ecology of birds on insects, where the kind of data we present here do not yet seem to be available. The main message is that, even though food may be very abundant in productive areas (as here in the intertidal areas), temporal and spatial variations in its quality and in its availability to the birds have a profound influence on (i) their ability to exploit it, (ii) on how the carrying capacity of an area should be both measured and predicted, and (iii) on the year-round dynamics of the bird populations.

Seasonal variation in prey availability and quality: acute winter mortalities in waders

In general, the herbivores have easy access to the often abundant, surface-dwelling vegetation. This provides rich, but temporary, pickings before depletion by the birds themselves, or autumn gales, reduces its abundance to a thin and unprofitable smear on the sediment surface. The birds then either move on to other nearby areas, such as saltmarshes or fields, or move to other estuaries, further from the breeding areas, which have yet to be depleted.

For carnivores, much of the abundant food may be inaccessible because it is protected against carnivores (and dehydration) by various morphological and behavioural adaptations. Some species, such as the mussel (*Mytilus edulis*), live clearly visible on the sediment surface but have a thick shell which protects them from all but a few of their potential consumers. Others, such as the crab (*Carcinus maenas*), seek temporary shelter when the tide is out under algae or stones, or by immersing themselves under fluid surface sediments or in burrows. Most invertebrates, however, take advantage of the soft and penetrable sediments of intertidal flats by burrowing actively through, or into them, or by constructing either temporary or semi-permanent burrows. A vertical section through a typical mudflat reveals animals living at a range of depths, from the surface down to as deep as a metre in the case of the largest burrowing polychaetes, such as the lugworm (*Arenicola marina*).

These anti-predator 'hiding' adaptations must be circumvented by the birds if they are to exploit this rich food supply. In this they are aided not just by their own behavioural and morphological adaptations, but by two factors acting on the prey themselves and causing 'chinks in their anti-predator armoury'. One of these factors is parasites in the prey which, in order to secure their own onward transmission to an avian host, cause the behaviour of the prey to change in such a way that it becomes vulnerable to a foraging bird (e.g. Berthel & Holmes, 1973). The other, perhaps more widespread factor is the contradictory demands placed on the prey by the environment which forces them, from time to time, to 'break cover'. Most burrowing invertebrates, for example, must maintain some contact with the sediment surface to feed, to breathe, and to expel reproductive and waste products, and these contacts may make them vulnerable to birds on the surface.

An early and neat example of this 'prey dilemma' was demonstrated in bar-tailed godwits (*Limosa lapponica*) eating lugworms (Smith, 1975). These deep-burrowing polychaetes normally live well beyond the range of even the longest-billed wader, the curlew (*Numenius arquata*), whose bill can reach almost 17 cm in length. But Smith showed that, approximately every 40 minutes, lugworms back up their burrows to defecate and, in so doing, come within reach of even the shortest-billed waders, such as the grey plover (*Pluvialis squatarola*). Using both naturally occurring spatial and temporal variations in lugworm defecation rates, and field experiments, Smith showed that lugworms could be detected and caught by the godwits when they were at the very top of their burrows and forming their well-known worm castes at the surface. Only in these circumstances were the prey simultaneously detectable by the godwit sensory system and within reach of the bill, the two conditions required to make a prey available to the birds.

Other research has focused on the selection by shorebirds from the spectrum of prey available to them. Field studies on redshank (*Tringa totanus*) eating the polychaetes *Nereis diversicolor* and *Nephtys hombergii* demonstrated that, as predicted by simple rate-maximising foraging theory (see the review by Stephens & Krebs, 1986), waders may often select from the range of prey items available those most profitable size classes that maximise the bird's intake rate, measured as energy consumed per unit time spent foraging (Goss-Custard, 1977a). Thus, of the sometimes huge biomass of prey present in the sediment, waders can only exploit the fraction of it that is available (i.e. both detectable by its sense organs and obtainable by its feeding apparatus). From that fraction, they further select those prey items that are neither too small to be profitable nor too large to be swallowed, because gape-size is limiting. Zwarts & Wanink (1993) adopted the useful term 'harvestable biomass' to refer to the usable part of the prey biomass and, through field and laboratory studies, demonstrated that the harvestable biomass may, at any one time, only be a small fraction of the total prey biomass present in the sediment. Depletion by these carnivores tends to be a gradual, grazing-down process as the predators, in repeated passes over the same spot, remove the occasional, available prey. Whereas by early winter herbivores may rapidly remove most of the food supply that is not destroyed by other factors such as gales (Goss-Custard & Charman, 1976), the food supply of carnivores at present-day population densities typically lasts the whole non-breeding season, with usually less than 40% of the initial standing crop being removed by the birds themselves (Goss-Custard, 1980).

Studies over almost thirty years have demonstrated that the fraction of the prey that is available changes through autumn, winter, and spring. Sometimes, prey availability changes as prey respond immediately to changes in an environmental factor. Thus, lugworms become less active and move to the surface less often as the ambient temperature decreases towards 0°C, with a consequent reduction in the numbers detected and caught per unit time by godwits as the winter progresses (Smith, 1975). In other cases, prey may migrate deeper into the sediment through autumn and winter in response to environmental stimuli that vary less rapidly; indeed, the increasing numbers of waders in winter may themselves drive prey deeper into the sediment. Many bivalves burrow deeper as the day length shortens, putting an increasing proportion of them beyond the reach of waders (Zwarts & Wanink, 1993). Interestingly, and importantly for waders, it is generally the largest individual bivalves that burrow to the greatest depths and, within a size-class, those that are in the poorest condition remain closest to the surface (Zwarts & Wanink, 1991). The reason for this in bivalves, for example, is that the length of the siphon with which the animal maintains vital contact with the surface limits the depth to which the animal can burrow, and siphon length depends not only on the animal's size but also on its body condition. On top of this seasonal decline in the fraction of the prey that is available, the energy content of many intertidal invertebrates declines through autumn and spring as the food supply diminishes. Typically, an individual mussel, for example, may lose up to 40% of its flesh content, and thus energy value to the bird, between autumn and spring (Bayne & Worrall, 1980). In addition, depletion reduces prey abundance, especially of any fixed fraction that is accessible to the birds.

The decreasing abundance, the decreasing prey surface activity, the increasing burrowing depth, especially in the higher quality food items, and the declining food value of many prey through autumn and winter coincides with an increasing energy demand in the birds. The increasing energy demand is associated with decreasing ambient temperature and more frequent chilling high winds which, in the unsheltered intertidal habitats, is probably why shorebirds are amongst the most energy-demanding of birds (Kersten & Piersma, 1987). As winter progresses, the birds spend an increasing proportion of the period for which the intertidal flats are exposed, during both day and night, in feeding, and some may feed in terrestrial habitats over high water on supplementary foods, such as earthworms (*Lumbricidae*). Most waders lay down fat reserves through autumn and early winter, which help them survive periods of acute difficulty brought on, for example, by the freezing of parts of the intertidal zone and most of their terrestrial supplementary feeding areas during periods of low temperature. It is in these conditions that many starved waders can be found dead around our shores, and there is evidence from oystercatchers (*Haematopus ostralegus*) that it is the individuals that accumulated least fat reserves during autumn and early winter that are most at risk of starvation (Goss-Custard *et al.*, 1996b). At other times, however, waders are characterised by high overwinter survival rates. It seems that, in most circumstances, most individual waders of the species that regularly overwinter in the British Isles adapt to the generally deteriorating feeding conditions as the winter progresses. In spring, the feeding conditions improve as prey become more active, so that the birds are able to accumulate more easily the fat reserves needed to fuel their return to their often distant breeding grounds.

Prey availability and quality in relation to prey and predator densities: chronic winter mortality in waders

Although periodic mass mortalities of waders in severe weather have attracted much attention, the size of wader populations may be much more affected in the long term by the continuing low-level mortality rate in winter. In long-lived birds like these, with low *per capita* reproductive rates, equilibrium population size can be greatly affected by small changes in mortality rate (Goss-Custard *et al.*, 1996a), especially of the immature birds that are, in effect, the 'seed corn' of future generations (Goss-Custard & Durell, 1984). Importantly, this low-level mortality may also be density dependent, and thus contribute to the regulatory properties of the population.

The possibility that mortality is density dependent is also important for predicting the effect of anthropogenic factors on population size (Goss-Custard, 1977b). If birds compete for food at some stage during the non-breeding season, the competition will intensify if the food supply is diminished. This will happen even if, as in many circumstances, the diminution is only temporary. Many of the human activities that might affect wintering waders have the effect of reducing the extent, either temporarily or permanently, of their feeding areas, and thus increasing bird density. Competition will intensify if bird density increases, whether this is measured as the number of birds per unit area or per unit mass of food. For example, removing half the intertidal flats of an estuary would cause bird density in the remaining half to double, so that any competition would intensify. In the long run, bird density might decline as increased numbers of them were forced by the increased competition either to starve or to move to other estuaries, and this would lead to a new equilibrium population size being achieved as the new winter *per capita* mortality rate and summer *per capita* reproductive rate were brought into line (Goss-Custard, 1993). A person walking a dog on a sandflat temporarily prevents birds from feeding there, and so diminishes the area available to the birds and increases density in the remaining, disturbance-free areas. A person digging for lugworms not only drives birds away but also reduces the food abundance for the birds when they return after the bait digger leaves, so increasing density per unit of food. As the significance of density-related feedback processes in determining the response of populations to habitat loss and change has gradually been appreciated, increasing attention has been focused on identifying the mechanisms, and measuring the intensity, of competition for food amongst non-breeding shorebirds and wildfowl.

As in other systems, two broad classes of feedback mechanisms are involved in food competition in waders (Goss-Custard, 1980). As bird density increases, the removal (or 'depletion') rate of the prey increases. As prey abundance diminishes, the intake rate of individual birds starts to fall, once a threshold density of prey has been reached (the functional response). Any further increase in predation pressure arising from increased bird densities could therefore bring forward the stage in the winter when intake rates begin to fall because of depletion. The second feedback mechanism is interference, the immediate reduction in intake rate that occurs as bird density increases. One cause of interference, which may be particularly prevalent in

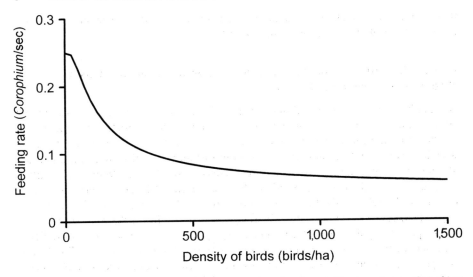

Figure 1.1. Interference in redshank (*Tringa totanus*) eating the amphipod crustacean *Corophium volutator*. The redshank do not take part in any noticeable social interactions, such as fighting, as bird density increases, and the proximity of a bird to its nearest neighbour does not affect its rate of feeding (Selman & Goss-Custard, 1988). But, as redshank density increases, the *Corophium* are thought increasingly to hide in their burrows, thus making them less available to the birds, whose rate of prey capture therefore falls. From Yates *et al.* (2000).

herbivorous wildfowl, is increased rates of social interactions between birds competing for food items or feeding micro-sites (Black *et al.*, 1991). But an additional cause in carnivores is likely to be 'prey depression', the increased frequency with which prey carry out anti-predator responses, such as retreating into the burrow, as bird density increases.

The decrease in prey availability as shorebird density increases has been suspected for many years (Goss-Custard, 1970); it makes sense that an anti-predator response by active prey involving withdrawal to added safety would reduce prey availability and thus depress predator intake rate as predator density increases. However, convincing field evidence has only emerged in recent years (*Figure 1.1*). Sutherland (1996) has suggested that prey availability may also vary with the density of prey, as well as that of predators, and affect the form of the functional response. If so, both feedback mechanisms – prey depletion and interference – could be influenced by variations in prey availability.

This possibility has been tested in a preliminary fashion in oystercatchers feeding on mussels (*Mytilus edulis*) (J. D. Goss-Custard, *unpublished*). Oystercatchers that open mussels by hammering a hole in the shell attack the mussels which have the thinnest shells. They may do this to minimise either the time taken to open their prey or the risk of damaging their bill. Shell thickness varies with animal size and with the length of time for which the mussel is exposed over low water, and therefore with its level on the shore (Goss-Custard *et al.*, 1993). With mussel size and shore level taken into account by multiple regression analysis, shell thickness does increase as mussel density increases (*Figure 1.2*). A greater proportion of the prey in low mussel density areas than in high density areas may therefore be

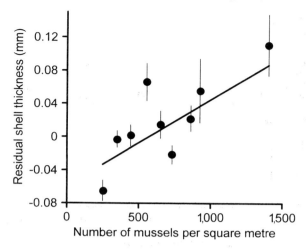

Figure 1.2.   The thickness of the shells of mussels on the ventral surface increases as the density of the large (>30 mm long) mussels consumed by oystercatchers increases on a mussel bed (numbered 26) of the Exe estuary. Shell thickness in mussels increases with the length of the mussel and decreases downshore (Goss-Custard *et al.*, 1993). Therefore, the shell thickness of the 1,074 mussels sampled in some 30 plots on bed 26 was related in a multiple regression analysis against shell length, the length of time for which the plot from which the mussel was taken is exposed on ordinary spring tides, and mussel density; the partial regression coefficients of all variables were highly significant (<0.001). The Figure shows the residual shell thickness once mussel length and exposure time has been taken into account. Data from individual mussels were grouped into mussel density classes, and the standard errors (vertical bars) and fitted line are shown. None of the oystercatchers on bed 26 open mussels by hammering them on the ventral surface (unpublished information), so these results are unaffected by selective predation by ventrally hammering birds.

available because a greater proportion of them can be opened, and are thus obtainable by oystercatchers.

We do not yet know in which direction the cause–effect relation goes in this association between mussel abundance and shell thickness. Mussels may be sparse in some areas because their shells are thin, so they are vulnerable to predators. Alternatively, low mussel density *per se* may cause shells to be thin, perhaps because mussel growth rates may be highest in low density areas and fast-growing mussels may have rather thin shells. Whatever the mechanism, the association between shell thickness and mussel density may affect the food intake rate of oystercatchers (that hammer mussels on the ventral surface). Intake rate may therefore remain high even when the prey numerical density, which is closely and positively correlated with prey biomass density (Goss-Custard *et al.*, 2001), is at a very low level (*Figure 1.3*). An alternative explanation of the flat functional response is that, with this surface-dwelling prey, oystercatchers can single out individual mussels for attack more readily where they are sparse than when they are tightly packed in dense clumps. There is evidence, for example, that oystercatchers prefer isolated to clumped prey (Coleman *et al.*, 1999). It may thus be increased detectability, and not just obtainability, of prey at low mussel densities that accounts for the flat functional response in hammering birds.

These findings do raise the possibility that prey availability may change not only with the density of the predator but also with that of the prey. Such variations in prey availability may contribute to the precise form of the feedback functions that relate

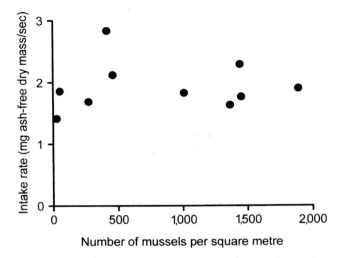

Figure 1.3.   The intake rate of oystercatchers that open mussels on the Exe estuary by hammering on the ventral surface in relation to the numerical density of the large mussels (>30 mm long) they eat. Each point represents the mean of 25–75 5 min observation periods on single birds, spread over two–three months, during which prey density would not have changed much. Data were collected from plots situated on six mussel beds. From Goss-Custard *et al.* (2001).

intake rate, and thus the risk of starvation, to bird density, either directly through increased rates of interference or indirectly through prey depletion over the non-breeding season. Since the dynamics of populations are so much affected by the form and parameter values of their feedback functions (May, 1981), variations in prey availability with predator and prey abundance are likely to have consequences at the population level.

Population consequences of variations in prey availability with prey and predator density

A PREDATOR–PREY MODEL OF SHOREBIRDS AND THEIR PREY

The possibility that a change in prey availability with the density of either predator or prey, or both, can be explored using a model of the interaction between oystercatcher and mussel populations developed and tested on the Exe estuary. At the outset, we should point out that interference in this system seems most likely to arise indirectly from increased rates of fighting for food as bird density increases (Ens & Cayford, 1996), rather than from changes in prey availability through prey depression. But this does not matter when exploring the population-level consequences of prey depression because the interference functions that arise from social interactions, and which are used in the model, have the same general form as those arising from prey depression (Stillman *et al.*, *in press*).

An individuals- and behaviour-based, game-theoretic model of the oystercatcher population predicts the response of individual oystercatchers to spatial and seasonal variations in the abundance and quality of mussels, and tracks the effects of the differing responses of each bird on its energy consumption rate, body condition, and survival chances (Goss-Custard *et al.*, 1995; Stillman *et al.*, 2000). Clarke & Goss-Custard (1996) give the mathematical formulation of the earlier version 2 of the model used here. Individual birds in the model choose each day to feed on the mussel bed on which they would maximise their intake rate. To do this, it is necessary to know the potential intake for each bird on each bed under the prevailing conditions, and this is calculated in two steps. First, the interference-free intake rate of an individual of average efficiency is calculated for each bed from the functional response relating intake rate to prey biomass density. This average intake rate is adjusted for each model bird for individual variations in foraging efficiency, which are important in this system (Caldow *et al.,*1999). Second, the reduction in intake rate arising from interference is calculated for each bird on each bed, given the density of its competitors and the fighting abilities, relative to its own, of the other individuals present on each bed. The potential intake rate is calculated by subtraction, and in the model, the bird spends the low water period where its intake rate is highest.

The ten main mussel beds of the estuary vary greatly in size and quality as feeding areas for oystercatchers. Each individual decides on which bed it should feed in every daily iteration from mid-September to mid-March. Birds continually change mussel bed as the relative quality of beds alters through oystercatcher depletion and other causes of mussel loss, such as gales. The energy content of individual mussel prey decreases over the winter as their flesh content declines; a seasonal decline in prey availability could also be incorporated, but is not done so as it is not thought to be important in this system. The daily energy requirements of individual oystercatchers are calculated from the air temperature and therefore increase as the winter progresses. Any daily consumption of food that is surplus to the current daily energy requirement is stored as fat, until an empirically-determined weight target has been reached. Birds failing to obtain enough energy by feeding on mussels over low tide feed upshore on supplementary prey, such as cockles *(Cerastoderma edule)*, as the tide advances and recedes and also, over high water, on earthworms in fields. The availability of worms decreases as the ambient temperature approaches 0°C, and this is included in the model. Fat reserves are utilised when an individual is unable to meet its current daily requirement; as in nature, this is most likely to happen in late winter when high bird energy demands coincide with reduced estuary food supplies and, in the fields, lower prey availability. Depending on the simulations required, a model bird either emigrates when its fat reserves fall to a given low point or dies if its reserves fall to zero at any point during the non-breeding season, which ends in mid-March.

MODEL PREDICTIONS FOR PRESENT-DAY MORTALITY AND SURVIVORSHIP FUNCTIONS AND CARRYING CAPACITY

The model was first run for the present-day autumn abundance of mussels on each of the ten mussel beds and for typical weather conditions. Simulations were run for a wide range of numbers of birds settling on the estuary in autumn. By plotting against initial population density the proportion that die of starvation before March, we produce density-dependent emigration/mortality curves and survival curves for the present-day climatic and food abundance conditions (dotted line in *Figure 1.4a*). The density on the mussel beds of the present-day wintering oystercatcher population on the Exe is *circa* 20–40 birds/ha, so the model predicts that winter mortality should

12   J. D. GOSS-CUSTARD AND A. D. WEST

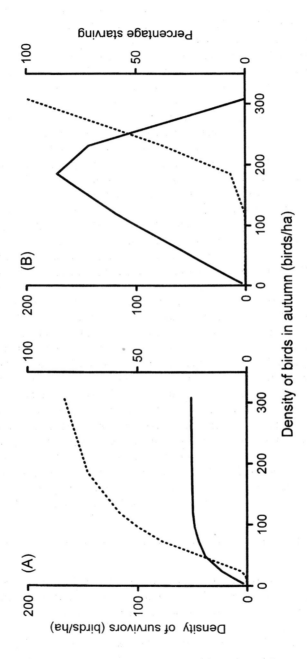

Figure 1.4. Predictions of the Exe estuary oystercatcher–mussel predator–prey model for the percentage of birds that, by the end of winter, survive (solid line) or starve (dotted line) or the density that survive (solid line) as a function of the numbers settling there during the previous September, according to the intensity of the interference between foraging birds. (A) Present-day levels of interference. (B) No interference. From Goss-Custard et al. (1996c).

now be density-dependent (*Figure 1.4a*), as has recently been confirmed in the field (Durell *et al.*, 2000). This stimulates some confidence that the model captures the most important features of the system.

As the initial numbers increase, the density of survivors more or less reaches a plateau at which one bird either emigrates or dies for every additional bird that arrives (solid line in *Figure 1.4a*). The density of survivors cannot increase any further, so the carrying capacity of the system for mussel-eating oystercatchers, as defined in terms of the number of survivors by Goss-Custard (1985), has been reached. This happens because, as bird densities increase, the two feedback processes of depletion and interference intensify (Goss-Custard *et al.*, 2001). The intake rates of an increasing proportion of birds are reduced to the point at which they are unable to accumulate sufficient fat reserves to survive the winter, and so either emigrate or, if they remain, die. By their departure, those that fail early in the winter reduce the subsequent rate of depletion of the food supply, allowing food density to be higher than it otherwise would have been for those that remain, whose survival chances are thereby enhanced.

EFFECT OF PREY AVAILABILITY CHANGING WITH PREDATOR DENSITY: IMPLICATIONS FOR CARRYING CAPACITY

The model was used to explore the influence of interference, the reduction in prey availability with increased predator density, on the form of the mortality and survival functions and on carrying capacity. Removing interference increased the slope of the density-dependent starvation function (dotted line in *Figure 1.4b*). The survivorship function changed from being a curve with a decelerating rise to a clearly identifiable plateau, and then to dome-shaped (solid line in *Figure 1.4b*).

No oystercatchers remained alive at the end of the winter after very large numbers had settled on the estuary the previous autumn, and densities on the mussel beds exceeded 300 birds/ha (*Figure 1.4b*). Without interference, large numbers of birds are able to exploit the food supply in autumn because their intake rates are not depressed through interference by the high densities of birds present. The food supply is rapidly reduced to such low levels that, by the end of the winter, only the most efficient foragers (if any) can feed at the rate required to survive. Like herbivores in which all the food is available at the sediment surface, interference-free carnivores are able rapidly to graze the food supply down to a low level. This happens even though, at any one time, only a fraction of prey biomass present is actually harvestable; a given patch of prey is passed over and exploited so frequently that depletion is rapid. This is to be expected because the removal of interference replaces contest competition with scramble competition, the demographic consequences of which were very well explored many years ago by entomologists (Varley *et al.*, 1973).

Without interference, carrying capacity can no longer be measured as the number of survivors because, at high initial population sizes, there are none. The usual procedure in such circumstances is to measure carrying capacity as the total number of bird-days over the winter (mean number of birds/day × number of days). This is done in *Figure 1.5*, with the curve resulting from using present-day levels of interference shown for comparison. Without interference, the relationship between bird-days and initial population size is still dome-shaped, although it declines much more slowly beyond the maximum capacity; compare the dotted line in *Figure 1.5*

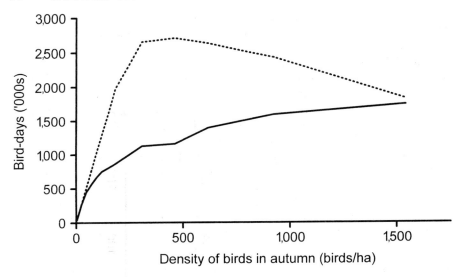

Figure 1.5.   Predictions of the Exe estuary oystercatcher–mussel predator–prey model for the total number of oystercatcher bird-days from October to March with either present-day levels of interference (solid line) or with no interference (dotted line). From Goss-Custard *et al.* (1996c).

with the solid line in *Figure 1.4b*. Once the initial population density exceeds 300 birds/ha, the density at which there are no survivors (*Figure 1.4b*), the predicted carrying capacity, measured as total bird-days per winter, depends on the numbers of birds that arrive in autumn (*Figure 1.5*). The non-linear functional response combined with the individual variation in bird foraging efficiency make the function shape difficult to anticipate without modelling (Goss-Custard *et al.*, 1997). In systems without interference, measuring the potential carrying capacity in bird-days per winter, as is often done, provides only an approximate measure of capacity, unless the number arriving in autumn is specified (Goss-Custard *et al.*, 1996c).

The solid line in *Figure 1.5* also shows the bird-days total for the simulations when there was interference between foraging birds. Measured in this way, carrying capacity continues to increase, although at a decelerating rate, as the initial densities in autumn rise to the incredibly high value of 1,500 birds/ha. This contrasts dramatically with carrying capacity measured as the number of survivors because this is attained at the much lower autumn density of *circa* 100 birds/ha (solid line in *Figure 1.4a*). Whether carrying capacity is likely ever to be reached in systems with interference thus depends critically on how it is measured. We conclude, therefore, that whether or not interference is present has a profound influence on how carrying capacity should be defined and measured, which in turn may affect how likely it is ever to be reached in the real world.

EFFECT OF PREY AVAILABILITY CHANGING WITH PREY DENSITY: IMPLICATIONS FOR CARRYING CAPACITY

The model was also used to explore how a change in prey availability with prey density, if confirmed, could affect the mortality and survivorship functions and the

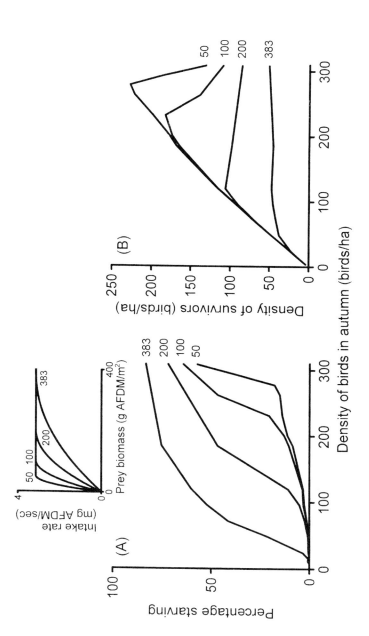

Figure 1.6. Predictions of the Exe estuary oystercatcher–mussel predator–prey model for the effect that changing the inflection point of the functional response has on (A) the percentage of birds that starve by the end of winter, or (B) the density that survive there during the previous September, with present-day levels of interference. The position of the inflection point along the x-axis of the functional response (prey biomass density, in g ash-free dry mass/m²) was varied as shown in the inset, and is indicated alongside each line in the two main plots.

measurement of carrying capacity. The very flat functional response in mussel-eating oystercatchers may partly arise because prey availability increases as prey density declines. Were this not to happen, the inflection point at which intake rate starts to decline might occur at a much higher density of prey, as illustrated in the inset to *Figure 1.6*.

To model the effect of lowering the prey density at which the inflection points in the functional response occur, the curves shown in the inset to *Figure 1.6* were used in simulations with present-day levels of interference. Moving the inflection point to the left causes the density-dependent starvation functions to be shifted to the right, so that a given percentage mortality occurs at ever higher bird densities (*Figure 1.6*). This happens because, when the inflection point of the functional response occurs at high prey densities, the inefficient and sub-dominant birds die early in the winter, when only small amounts of the prey have been removed. This reduces the subsequent overwinter rate of prey depletion which, in turn, allows a greater proportion of the remaining birds to survive than would otherwise be the case. Prey depletion starts to kill birds even when rather small numbers occupy the estuary in autumn, and the density-dependence therefore starts at a low population density and rises only gradually. In contrast, when the inflection point in the functional response occurs to the left, at low prey densities, depletion does not have much effect on survival unless very large numbers of birds settle in the autumn. At lower initial population sizes, the overwinter depletion is not so great as to cause many birds difficulty in obtaining their requirements. But with large numbers of birds, depletion does reach the point at which intake rate begins to fall before the end of winter. It does so very rapidly because there is a lag between the fall in intake rate and the starvation of the large numbers of birds necessary to stop the catastrophic decline in the food supply from continuing. The lag occurs because the poor-performing birds can live for some time on low intake rates by drawing on their fat reserves.

The sudden collapse in intake rates at low prey densities that is associated with a flat functional response, in effect, introduces scramble competition. The flat functional response overrides the potential for contest competition that would otherwise be present in a population in which competitive ability varies widely due to the considerable individual variation in foraging efficiency and susceptibility to interference. This, in turn, leads to a dome-shaped survival function (*Figure 1.6*), with the same implications for the measurement of carrying capacity as discussed above.

Discussion

This brief review shows how variations in prey availability with season, predator density, and prey density influence population-level phenomena in shorebirds, and probably other predators too. As has been realised for some time, seasonal changes in prey availability undoubtedly contribute to the difficulties shorebirds face in winter, especially during severe weather. The effect on the birds of the overall decrease in prey availability during the winter is increased by the tendency for the individual prey with the lowest flesh content to predominate amongst the decreased fraction that is available (Zwarts & Wanink, 1991). The modelling of shorebird predator–prey systems suggests that any tendency for prey availability to decrease as the densities of the birds and of their prey increase may have a significant influence on density-dependent mortality and survival functions, and on the measurement of carrying capacity. Any tendency for prey availability to be depressed more as bird density increases, in effect, introduces contest competition. Any tendency, if confirmed by future research, for prey availability to increase as prey density falls would tend to introduce scramble competition, even when the population is made up of individuals that vary greatly in competitive abilities. The interaction between contest and scramble competition arising from systematic variations in prey availability with prey and predator densities affects the density-dependent functions, and general theory tells us that this will affect the dynamics of the predator–prey interaction (May, 1981). Although these have not been explored here, we have illustrated the population-level effects of these sources of variation in prey availability using some preliminary model simulations, which show that they affect how carrying capacity should be measured. Whether comparable variations in prey availability occur in bird–insect predator–prey systems remains to be explored.

References

Bayne, B. L. & Worrall, C. M. (1980) Growth and production in mussels *Mytilus edulis* from two populations. *Marine Ecology Progress Series* **3**, 317–328.

Berthel, W. M. & Holmes, J. C. (1973) Altered evasive behaviour and responses to light in amphipods harbouring *Acanthocephalan cystacanths*. *Journal of Parasitology* **59**, 945–956.

Black, J. M., Carbonne, C., Wells, R. L. & Owen, M. (1991) Foraging dynamics in goose flocks: the cost of living on the edge. *Animal Behaviour* **44**, 41–50.

Caldow, R. W. G., Goss-Custard, J. D., Stillman, R. A., Durell, S. E. A. le V. dit, Swinfen, R. & Bregnballe, T. (1999) Individual variation in the competitive ability of interference-prone foragers: the relative importance of foraging efficiency and susceptibility to interference. *Journal of Animal Ecology* **68**, 869–878.

Clarke, R. T. & Goss-Custard, J. D. (1996) The Exe estuary oystercatcher–mussel model. In: *The oystercatcher: from individuals to populations* (ed. J. D. Goss-Custard), pp. 389–392. Oxford University Press, Oxford, UK.

Coleman, R. A., Goss-Custard, J. D., Durell, S. E. le V. dit & Hawkins, S .J. (1999) Limpet (*Patella* spp.) consumption by oystercatchers (*Haematopus ostralegus*): a preference for solitary prey items. *Marine Ecology Progress Series* **183**, 253–261.

Durell, S. E. A. le V. dit, Goss-Custard, J. D., Clarke, R. T. & McGrorty, S. (2000) Density-dependent mortality in oystercatchers *Haematopus ostralegus*. *Ibis* **142**, 132–138.

Ens, B. J. & Cayford, J. T. (1996) Feeding with other oystercatchers. In: *The oystercatcher: from individuals to populations* (ed. J. D. Goss-Custard), pp. 77–104. Oxford University Press, Oxford, UK.

Goss-Custard, J. D. (1970) Feeding dispersion in some overwintering wading birds. In: *Social behaviour in birds and mammals* (ed. J. H. Crook), pp. 3–35. Academic Press, London, UK.

Goss-Custard, J. D. (1977a) Optimal foraging and the size selection of worms by redshank, *Tringa totanus*, in the field. *Animal Behaviour* **25**, 10–29.

Goss-Custard, J. D. (1977b) The ecology of the Wash. III. Density-related behaviour and the possible effects of a loss of feeding grounds on wading birds (Charadrii). *Journal of Applied Ecology* **14**, 721–739.

Goss-Custard, J. D. (1980) Competition for food and interference among waders. *Ardea* **68**, 31–52.

Goss-Custard, J. D. (1985) Foraging behaviour of wading birds and the carrying capacity of estuaries. In: *Behavioural ecology: ecological consequences of adaptive behaviour* (eds. R. M. Sibly & R. H. Smith), pp. 169–188. Blackwell Scientific Publications, Oxford, UK.

Goss-Custard, J. D. (1993) The effect of migration and scale on the study of bird populations: 1991 Witherby lecture. *Bird Study* **40**, 81–96.

Goss-Custard, J. D. & Charman, K. (1976) Predicting how many wintering waterfowl an area can support. *Wildfowl* **27**, 157–158.

Goss-Custard, J. D. & Durell, S. E. A. le V. dit (1984) Feeding ecology, winter mortality and the population dynamics of oystercatchers, *Haematopus ostralegus*, on the Exe estuary. In: *Coastal waders and wildfowl in winter* (eds. P. R. Evans, J. D. Goss-Custard & W. G. Hale), pp. 190–208. Cambridge University Press, Cambridge, UK.

Goss-Custard, J. D. & Durell, S. E. A. le V. dit (1990) Bird behaviour and environmental planning: approaches in the study of wader populations. *Ibis* **132**, 273–289.

Goss-Custard, J. D., West, A. D. & Durell, S. E. A. le V. dit (1993) The availability and quality of the mussel prey (*Mytilus edulis*) of oystercatchers (*Haematopus ostralegus*). *Netherlands Journal of Sea Research* **31**, 419–439.

Goss-Custard, J. D., Caldow, R. W. G., Clarke, R. T., Durell, S. E. A. le V. dit, Urfi, A. J. & West, A. D. (1995) Consequences of habitat loss and change to populations of wintering migratory birds: predicting the local and global effects from studies of individuals. *Ibis* **137**, S56–66.

Goss-Custard, J. D., Durell, S. E. A. le V. dit, Clarke, R. T., Beintema, A. J., Caldow, R. W.G., Meininger, P. L. & Smit, C. (1996a) Population dynamics of the oystercatcher. In: *The oystercatcher: from individuals to populations* (ed. J. D. Goss-Custard), pp. 352–383. Oxford University Press, Oxford, UK.

Goss-Custard, J. D., Durell, S. E. A. le V. dit, Goater, C. P., Hulscher, J. B., Lambeck, R. H. D., Meininger, P. L. & Urfi, J. (1996b) How oystercatchers survive the winter. In: *The oystercatcher: from individuals to populations* (ed. J. D. Goss-Custard), pp. 133–154. Oxford University Press, Oxford, UK.

Goss-Custard, J. D., West, A. D., Caldow, R. W. G., Clarke, R. T. & Durell, S. E. A. le V. dit (1996c) The carrying capacity of coastal habitats for oystercatchers. In: *The oystercatcher: from individuals to populations* (ed. J. D. Goss-Custard), pp. 326–351. Oxford University Press, Oxford, UK.

Goss-Custard, J. D., West, A. D., Stillman, R. A., Durell, S. E. A. le V. dit, Caldow, R. W. G., McGrorty, S. & Nagarajan, R. (2001) Density-dependent starvation in a vertebrate without significant depletion. *Journal of Animal Ecology* **70**, 955–965.

Kersten, M. & Piersma, T. (1987) High levels of energy expenditure in shorebirds; metabolic adaptations to an energetically expensive way of life. *Ardea* **75**, 175–187.

May, R. M. (ed.) (1981) *Theoretical ecology: principles and applications (2nd edition)*. Blackwell Scientific Publications, Oxford, UK. 496 pp.

Selman, J. & Goss-Custard, J. D. (1988) Interference between foraging redshank, *Tringa totanus*. *Animal Behaviour* **36**, 1542–1545.

Smith, P. C. (1975) A study of the winter feeding ecology and behaviour of the bar-tailed godwit (*Limosa lapponica*). PhD thesis, University of Durham, UK.

Stephens, D. W. & Krebs, J. R. (1986) *Foraging theory*. Princeton University Press, New Jersey, USA. 247 pp.

Stillman, R. A., Goss-Custard, J. D., West, A. D., Durell, S. E. A. le V. dit, Caldow, R. W. G., McGrorty, S. & Clarke, R. T. (2000) Predicting to novel environments: tests and sensitivity of a behaviour-based population model. *Journal of Applied Ecology* **37**, 564–588.

Stillman, R. A., Goss-Custard, J. D. & Alexander, M. J. (*in press*). Predator search pattern and the strength of interference through prey repression. *Behavioural Ecology*.

Sutherland, W. J. (1996) *From individual behaviour to population ecology*. Oxford University Press, Oxford, UK. 224 pp.

Varley, G. C., Gradwell, G. R. & Hassell, M. P. (1973) *Insect population ecology*. Blackwell Scientific Publications, Oxford, UK. 212 pp.

Yates, M. G., Goss-Custard, J. D. & Rispin, W. E. (1996) Towards predicting the effect of loss of intertidal feeding areas on overwintering shorebirds (Charadrii) and shelduck (*Tadorna tadorna*); refinements and tests of a model developed for the Wash, east England. *Journal of Applied Ecology* **33**, 944–954.

Yates, M. G., Stillman, R. A. & Goss-Custard, J. D. (2000) Contrasting interference functions and foraging dispersion in two species of shorebirds Charadrii. *Journal of Animal Ecology* **69**, 314–322.

Zwarts, L. & Wanink, J. H. (1991) The macrobenthos fraction accessible to waders may represent marginal prey. *Oecologia* **87**, 581–587.

Zwarts, L. & Wanink, J. H. (1993) How the food supply harvestable by waders in the Wadden Sea depends on the variation in energy density, body weight, biomass, burying depth and behaviour of tidal-flat invertebrates. *Netherlands Journal of Sea Research* **31**, 441–476.

2
Birds of Lowland Arable Farmland: The Importance and Identification of Invertebrate Diversity in the Diet of Chicks

STEPHEN J. MOREBY

The Game Conservancy Trust, Fordingbridge, Hampshire, SP6 1EF, UK

Introduction

Farmland birds, as the name implies, have evolved to co-exist with agriculture, relying on the farmland habitat to provide food and nesting sites. Within the last few decades, however, farming practices and advances in mechanisation have started to alter radically the balance that had evolved between birds and man's use of the land. Between 1970 and 1990, 86% of the 28 species classified as farmland birds (Gibbons *et al.*, 1993; Fuller *et al.*, 1995) have declined, and of the 40 listed by Campbell *et al.* (1997), 78% have reduced breeding areas.

   Reduced food availability may be the major reason for many of these declines, with insects being a vital component in the chick diet of most species of farmland bird, as well as being important in the adult diet (*Table 2.1*). Adults can alter their feeding behaviour in response to changes in food availability, e.g. due to changes in cropping practices, and movement between suitable feeding areas usually posing no significant problems (O'Connor & Shrubb, 1986). However, when they have young to feed, any changes that affect the availability of food for the chicks can become critical (Potts, 1986). A good supply of insects in the diet is often vital for good chick survival, with research over 60 years ago highlighting the need for insect food (Ford *et al.*, 1938). Insects provide a high and easily assimilated source of protein, and are essential to chicks and nestlings as these lack the necessary gut flora for breaking down cellulose, are not able to utilise high fibre roughage as an energy source, and require ready-formed amino acids. Protein is also important for muscle development and feather growth, which are important for the early development of flying ability, which is essential for anti-predator avoidance (Potts, 1986). Insects can contain four times as much protein as plant food, with the protein digestibility being 70–90% compared to that of 20–80% in plant foods (Savory, 1989). Even the young of many vegetarian birds are known to require a good proportion of animal protein in their diet (Newton, 1967).

Insect and Bird Interactions
© Intercept Ltd., PO Box 716, Andover, Hampshire, SP10 1YG, UK.

Table 2.1. Farmland birds* commonly found breeding in southern England with the importance of invertebrates in relation to vegetation in the adult and chick diet**.

	Adult Diet	Chick Diet
Grey partridge (*Perdix perdix*)	Chiefly vegetation[a]	Chiefly invertebrates
Red-legged partridge (*Alectoris rufa*)	Chiefly vegetation[a]	Chiefly invertebrates
Pheasant (*Phasianus colchicus*)	Chiefly vegetation[a]	Chiefly invertebrates
Quail (*Coturnix coturnix*)	Omnivorous	Chiefly invertebrates
Lapwing (*Vanellus vanellus*)	Chiefly invertebrates	Chiefly invertebrates
Skylark (*Alauda arvensis*)	Omnivorous[a]	Chiefly invertebrates
Swallow (*Hirundo rustica*)	Invertebrates only	Invertebrates only
House martin (*Delichon urbica*)	Invertebrates only	Invertebrates only
Meadow pipit (*Anthus pratensis*)	Chiefly invertebrates	Chiefly invertebrates
Yellow wagtail (*Motacilla flava flavissima*)	Chiefly invertebrates	Chiefly invertebrates
Pied wagtail (*Motacilla alba yarrellii*)	Chiefly invertebrates	Chiefly invertebrates
Whitethroat (*Sylvia communis*)	Omnivorous	Chiefly invertebrates
Jackdaw (*Corvus monedula*)	Omnivorous	Omnivorous
Rook (*Corvus frugilegus*)	Omnivorous	Omnivorous
Starling (*Sturnus vulgaris*)	Omnivorous	Chiefly invertebrates
Tree sparrow (*Passer montanus*)	Omnivorous	Chiefly invertebrates
Greenfinch (*Carduelis chloris*)	Chiefly vegetation	Chiefly vegetation[b]
Goldfinch (*Carduelis carduelis*)	Chiefly vegetation	Chiefly vegetation[b]
Bullfinch (*Pyrrhula pyrrhula*)	Chiefly vegetation	Chiefly vegetation[b]
Linnet (*Carduelis cannabina*)	Chiefly vegetation	Chiefly vegetation[b]
Yellowhammer (*Emberiza citrinella*)	Chiefly vegetation	Chiefly invertebrates
Reed bunting (*Emberiza schoeniclus*)	Chiefly vegetation	Invertebrates only
Corn bunting (*Miliaria calandra*)	Chiefly vegetation	Chiefly invertebrates
Blackbird (*Turdus merula*)	Chiefly invertebrates	Invertebrates only
Song thrush (*Turdus philomelus*)	Chiefly invertebrates	Invertebrates only
Mistle thrush (*Turdus viscivorus*)	Chiefly invertebrates	Invertebrates only
Dunnock (*Prunella modularis*)	Chiefly invertebrates[c]	Chiefly invertebrates
Robin (*Erithacus rubecula*)	Chiefly invertebrates[c]	Chiefly invertebrates
Wren (*Troglodytes troglodytes*)	Chiefly invertebrates	Chiefly invertebrates
Spotted flycatcher (*Muscicapa striata*)	Chiefly invertebrates[c]	Invertebrates only
Blue tit (*Parus caeruleus*)	Chiefly invertebrates[c]	Invertebrates only
Great tit (*Parus major*)	Chiefly invertebrates[c]	Invertebrates only
Stock dove (*Columba oenas*)	Chiefly vegetation	Chiefly vegetation
Wood pigeon (*Columba palumbus*)	Chiefly vegetation	Chiefly vegetation
Turtle dove (*Streptopelia turtur*)	Chiefly vegetation	Chiefly vegetation
Collared dove (*Streptopelia decaocto*)	Chiefly vegetation	Chiefly vegetation

a – Invertebrates more important in summer
b – More invertebrates in diet compared to adult
c – Seeds, fruit and vegetation taken in winter
* Campbell *et al.* (1997) and Gibbons *et al.* (1993) **Cramp (1998)

A lack of invertebrate food during the breeding season has been suggested as contributing to these declines (Campbell *et al.*, 1997). This lack is particularly due to changes in cropping practices, such as a switch from spring- to winter-sown cereals, a widespread reduction in undersowing spring cereals with grass, and agricultural intensification. Modern pesticides perhaps have had the greatest impact on the food available to farmland birds by reducing both the arable flora and fauna. Apart from directly killing invertebrates (insecticides and some fungicides), fungicides remove the fungal food supply, and herbicides will both remove food plants of beneficial insects as well as plants and seeds that could have provided a replacement food source. Many studies have demonstrated direct detrimental effects of insecticides and fungicides on non-target invertebrate species within cereals (Vickerman & Sunder-

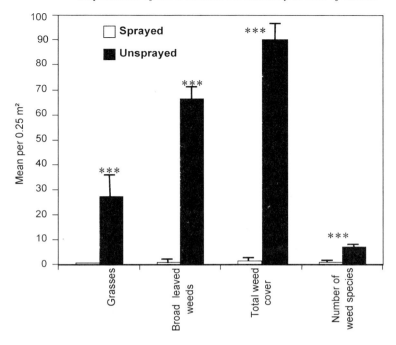

Figure 2.1. Mean per cent cover and number of weed species in winter wheat, found per 10 0.25 m² quadrats taken at 3 m into sprayed and unsprayed headlands.

land, 1977; Vickerman & Sotherton, 1983; Sotherton *et al.*, 1987; Theiling & Croft, 1988; Inglesfield, 1989; Moreby *et al.*, 1997), as well as from herbicide applications (Vickerman, 1974; Moreby & Southway, 1999).

The strongest evidence for a link between declines in arable invertebrate abundance and demographic changes in a declining bird population is for grey partridge (*Perdix perdix*) (Potts, 1986). However, recent research into skylarks (*Alauda arvensis*) (Poulsen *et al.*, 1998) and corn buntings (*Miliaria calandra*) (Ward & Aebischer, 1994) has identified an influence of invertebrate abundance on nestling survival and population density. While insect diversity on farmland is surprisingly large for such temporary habitats, attempts to increase nestling survival by increasing invertebrate abundance in arable habitats require an understanding of the invertebrate taxa most commonly represented in nestling diet. This need is particularly acute in relation to those species that may be affected by pesticide use (Aebischer, 1991; Burn & Cooke, 1995; Wilson *et al.*, 1997; Ewald & Aebischer, 1999). However, it is only recently, due to the studies into the decline of arable bird species, that it has been shown that insect food items for these birds have also often declined over large areas due to the impact of modern intensive agriculture (Aebischer, 1991). While reductions in certain key insect species may reduce the survival of chicks dramatically, many bird species can adapt to different food sources and it is often the general insect availability that is important rather than the availability of one or a few specific species (*Table 2.2*).

Research has found that the field headland, i.e. 6–10% of the field, has a greater floral and faunal diversity than the field centre, and a higher insect density (Marshall, 1985; Wilson, 1989; Moreby, 1995). The first few metres into a field are also the

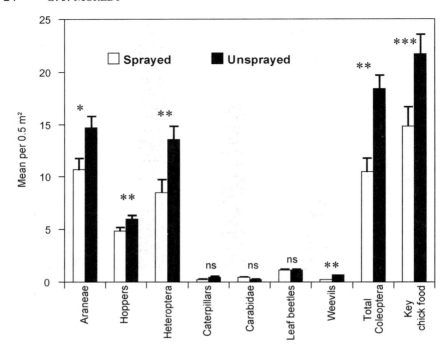

Figure 2.2.   Mean number of arthropods per 0.5 m^2 in sprayed and unsprayed winter wheat headlands.

preferred feeding area for game bird chicks (Green, 1984), as well as other ground-feeding arable birds (O'Connor & Shrubb, 1986). Any management to increase the availability of the invertebrate groups found in the diet of farmland birds would therefore most profitably be carried out within this headland area. To try and reverse this reduction in the arable flora and fauna, Conservation Headlands were developed. These, by the use of aphid specific insecticides and herbicides (Sotherton *et al.,* 1989; Sotherton, 1991), aimed to maximise the positive effects on the beneficial invertebrate groups in cereals without encouraging an increase in potential pest and weed species. Herbicides have a significant direct effect on the abundance and biodiversity of the arable flora (*Figure 2.1*). These arable plants provide food directly for farmland birds in the form of vegetation and seeds and, perhaps more importantly, indirectly by providing food for phytophagous insects which dominate in the diet of many birds (*Figures 2.3–2.6*). Coupled with reduced insecticide use, the increase in vegetation can also affect the microclimate and humidity at the crop base. This may directly increase invertebrates (*Figure 2.2*), and indirectly provide suitable conditions for the growth of micro-organisms, moulds and fungi, which are food for many invertebrates, including species of Coleoptera and Diptera eaten by birds. Positive effects have been demonstrated in studies where insect numbers were monitored and game bird brood survival was investigated by radio tracking (Rands, 1985, 1986).

   While the headland contains the highest floral diversity within the field, the uncropped field boundary contains an even greater species diversity of annual, and particularly of perennial species, most of which do not occur within the cultivated

Table 2.2.   Main invertebrate groups commonly found in the diet of young farmland birds up to 21 days old (identified from faecal analysis).

ARANEAE	– Spiders	
OPILIONIDAE	– Harvestmen	
HEMIPTERA	– Aphididae (aphids)	
	– Auchenorryncha (hoppers)	– Delphacidae
		– Cicadellidae
		– Cercopidae
	– Psyllidae (plant lice)	
	– Heteroptera (plant bugs)	
LEPIDOPTERA	– Larvae (caterpillars)	
	– Adult Moth	
HYMENOPTERA	– Adult Parasitic Wasps	
	– Formicidae (adult ants & pupae)	
	– Symphyta Larvae (sawflies)	
COLEOPTERA	The most common families eaten are:	
	– Carabidae (ground beetles) Adults & Larvae	
	– Chrysomelidae (leaf beetles) Adults & Larvae	
	– Staphylinidae (rove beetles) Adults & Larvae	
	– Scarabidae (dung beetles)	
	– Elateridae (click beetles)	
	– Curculionidae (weevils)	
	– Cryptophagidae (very small mould & vegetation feeders)	
	– Cantharidae (soldier beetles)	
	– Nitidulidae (pollen beetles)	

DIPTERA   Numerous Families which are generally unidentifiable due to their soft body, but may include:

– Tipulidae (craneflies)	– Lonchopteridae	– Syrphidae
– Chironomidae*	– Empididae	– Sphaeroceridae
– Mycetophilidae*	– Dolichopodidae	– Scatophagidae
– Bibionidae	– Psilidae*	– Muscidae
– Cecidomyiidae*	– Opomyzidae	– Simuliidae*
– Chloropidae	– Psychodidae*	– Pipunculidae

OTHERS
– Orthoptera (grasshoppers)
– Dermaptera (earwigs)
– Neuroptera (lacewings)
– Mollusca (snails)
– Collembola* (springtails)
– Lumbricidae (earthworms)

* = identified from neck collars (Moreby & Stoate, 1999)

area (Marshall, 1989). Careful pesticide applications in the field will give an additional benefit of helping to safeguard the biodiversity in field edges and boundaries, such as hedgerows. This type of habitat is also an important source for invertebrates eaten by birds (Moreby, 1994).

For birds reliant on insect food, therefore, the 50–100% increase in items available in the untreated weedy areas of a cereal field compared to the treated area (*Figure 2.2*) could potentially mean the difference between chicks fledging or not, especially over arable areas dominated by cereals. Environmentally sensitive management practices, such as conservation headlands and the planting of set-aside to increase invertebrate food generally (Holland, 2003), should therefore help improve the fledgling success of the many species of arable birds which have shown declines.

Determination of the invertebrate food

SAMPLE COLLECTION

Before recommending management changes on farmland to increase the availability of invertebrate food for birds, either directly by changes in the pesticide regime or indirectly by habitat improvement, the principal food groups in the birds' diet must first be identified (*Table 2.2*). Invertebrate diet determination can be achieved by a number of methods: direct observation of complete food items, use of neck collars to collect undigested items from the throat or crop, analysis of part-digested remains from the stomach, or identification of undigested fragments surviving in faecal samples.

Observation of food

Direct observation of feeding adult birds can often identify plant food, but invertebrate identification by this method is unlikely. Observation may allow identification of large invertebrates brought to nestlings when there is either a clear approach to the nest or photographic recording can be used within the nest (Kleintjes & Dahlsten, 1992). Photography has been used successfully for birds such as great tits (*Parus major*) (Royama, 1970) and blue tits (*Parus caeruleus*) (Banbura *et al.*, 1994*)*, where it has been used in conjunction with nest boxes, and also nests of black-throated blue warblers (*Dendroica caerulescens*) (Rodenhouse & Holmes, 1992). Monitoring equipment can, however, be prohibitively expensive, and photography will obviously prove difficult with species that nest in vegetation. Species such as tits, when foraging in wooded areas, are often able to feed their young on a small, highly selected range of food, such as a few species of caterpillars (Royama, 1970; Perrins, 1979), due to the woodland being able to support very high densities of such insects. For many other

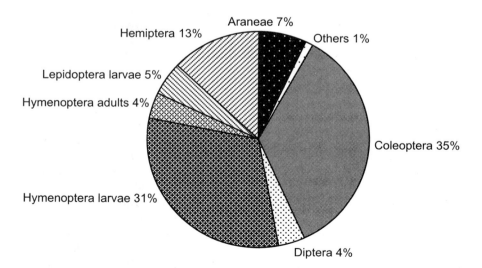

Figure 2.3. Diet of grey partridge chicks.

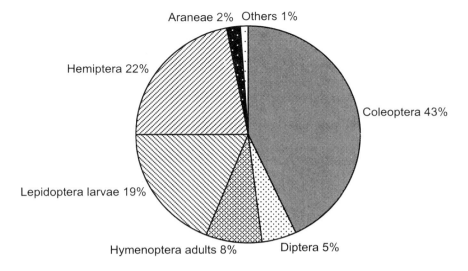

Figure 2.4.   Diet of pheasant chicks.

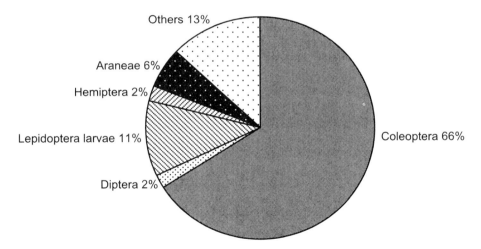

Figure 2.5.   Diet of corn bunting chicks.

species of nidicolous birds, such as yellowhammer (*Emberiza citrinella*), dunnock (*Prunella modularis*), whitethroat (*Sylvia communis*), or corn bunting, and for nidifugous species with mobile young, such as partridge and pheasant (*Phasianus colchicus*), the diet is generally very diverse (*Figures 2.3–2.6, Table 2.2*). Arable land also rarely contains numerous large-sized food items such as caterpillars and, as a result, the diet will normally comprise mainly small food items from a wide selection of invertebrate families in six or seven main Orders (*Table 2.4*). Since the accurate identification of small food items is often difficult (Read, 1994), observation is usually not a reliable method for bird species such as the above.

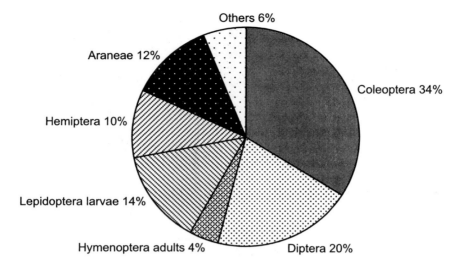

Figure 2.6.   Diet of yellowhammer chicks.

Neck collars

A neck collar or ligature round the throat to prevent the swallowing of food parcels brought to the nestling by the parent birds can be a very successful method, and allows the collection of undigested food with relatively easy identification of items. This method can, however, be used only for nidicolous chicks, and for many species of bird its use is often restricted to a short time window. For example, in skylarks, there is the possibility of injuring very young 1- to 3-day-old nestlings when placing a collar around their necks. Then, after 7–8 days, use is precluded by the possibility of chicks fledging with the collar attached (Poulsen & Aebischer, 1995). The collar must also apply sufficient pressure to prevent swallowing without choking the young bird or causing regurgitation of food; insufficient pressure may allow small food items to pass. To avoid excessive disturbance and stress, collars can only be used for short periods such as 30–60 minutes. This allows only the collection of a small number of food items. The number of samples obtained will be further reduced if adults spend time attempting to remove the collar (Robertson, 1973), or if the collar results in reduced begging by the chicks (Johnson *et al.,* 1980). However, despite problems associated with their use, collars have been used in many recent dietary studies (Jenny, 1990; Kleintjes & Dahlsten, 1992; Mellott & Woods, 1993; Kristin, 1995; Grim & Honza, 1996; Kristin & Patocka, 1997; Moreby & Stoate, 2000). In studies of how the parents prepare nestling food, such as by the removal of legs or wings from large insects (Barba *et al.,* 1996), this method is essential since it prevents any digestion.

Crop and stomach samples

Crop and stomach samples for adult or chick diet analysis can be collected from either living or dead birds. In older studies, food was normally obtained from stomachs of dead specimens, but with modern advances in field techniques, this method is generally inappropriate. Samples obtained from dead game birds can, however, be used to determine adult winter diet. For some birds, crop samples can be obtained using neck collars to prevent food from passing into the gut (Mellott & Woods, 1993), or small sub-samples can be obtained directly from the crop by use of a tube (Read, 1994), or an emetic can be administered to cause regurgitation of food items (Prys-Jones *et al.*, 1974; Valera *et al.*, 1997).

The above methods have advantages as well as disadvantages, but for most species of farmland birds the most appropriate method to determine the diet, depending on the type of nesting or time available during the field season, will be from faecal samples or collected by the use of neck collars.

Faecal analysis

Faecal analysis is perhaps the most widely used method of diet analysis for both adults and chicks (Davies, 1976; Calver & Wooller, 1982; Green, 1984; Bishton, 1985; Galbraith, 1989; Jenni *et al.*, 1990; Boddy, 1991; Galbraith *et al.*, 1993; Poulsen & Aebischer, 1995; Stoate *et al.*, 1998; Moreby & Stoate, 2001). It is a relatively non-invasive method for determining diet, and collection of samples by this method probably puts least stress on the birds, especially as they are handled only once. Samples from adult birds are usually best collected when the birds are caught by mist netting, either specifically for diet determination or for ringing. Nidicolous chicks usually provide samples automatically when handled. However, samples from nidifugous chicks prove more of a problem and are best collected from feeding areas or roost sites. Chance observation of broods, or searching in areas known to contain birds, may provide samples, but the most reliable method is to locate roost sites late in the evening by radio location (Hill, 1985; Rands, 1986) and collect samples the following morning, once the brood has moved on and there is no risk of disturbance.

INVERTEBRATE IDENTIFICATION

For invertebrate food items observed being brought to the nest, and for most of the items in crop/stomach and neck collar samples, identification of complete specimens is relatively easy to Family level using keys. Damaged or incomplete invertebrates that cannot be identified directly can usually be compared to complete specimens within the samples. For items which are too incomplete, and for diet determination using faecal analysis, invertebrates have to be identified from body structures.

Identification of prey items from faecal samples can be very time consuming because of the many small fragments present. Generally, however, all remains can be determined to the level of Order or sub-Order from the identification of characteristic structures such as mandibles, wings, or legs that have survived passage through the digestive system (*Table 2.2*). From key parts, an estimate of the number of items can be calculated (*Table 2.3*) (Moreby, 1988). Parts that do not correspond exactly to a particular Family can usually be identified to Order. At the beginning of a study, it is often helpful to familiarise oneself with structures and key parts by examining dissected complete specimens.

Table 2.3. Invertebrate structures commonly found in chick faecal samples of arable farmland birds that are useful in the general determination of the diet. Structures useful for calculating the number of individuals have the number of parts per one individual given in parentheses.

ARACHNIDA	– **Araneae**	*Fang*, may still be attached to chelicera (2); *legs* may be broken into many fragments.
	– **Opilionidae**	*Fang*, may be single or in a pair attached to sac-like body (4); *legs* always found as fragments.
INSECTA		
– **HEMIPTERA**		Due to the soft body of this group few parts survive from adults and very few from nymphs.
	–**Sternorryncha**	– **Aphididae** *Tibia* (6).
		– **Auchenorryncha** *Hind tibia* (2).
		– **Psyllidae** *Front wing* (2).
	– **Heteroptera**	*Tibia* (6); *Front tibia* in Nabidae (2); *Labium* (1); *Elytra* fragments.
– **LEPIDOPTERA**	– Larvae	– *mandibles* (2).
	– Adult	– *leg* fragments; *wing scales.*
– **HYMENOPTERA**	– Parasitic Wasps	– **Chalcidoidea/Proctotrupoidea** *Front wing* (2)
	– **Braconidae**	*Front wing* (2); *Femur & tibia.*
	– **Ichneumonidae**	*Petiole* (1); *Pterostigma* (2), may be detached from or attached to front wing.
	– **Formicidae**	– Adult – Numerous structures survive, head, antenna, petiole but the key parts to count are *mandibles* (2) and *tibia or femur* (6).
		– Pupae – tough fibrous sac.
	– **Symphyta**	– Adults – *Mandibles* (2); *Pterostigma* (2).
		– Larvae – *Mandibles* (2).
– **COLEOPTERA**		Numerous fragments from the head, leg, body and elytra survive. Coloured or striatted elytra fragments can be useful for relating to complete specimens from Families where no key structures occur.
		For specific Families however certain key parts are important.
	– **Carabidae**	– Adult – *Mandibles* (2); *Front tibia* (2).
		– Larvae – *Mandibles* (2).
	– **Curculionidae**	– *Femur or tibia* (6); *Head and snout* (1).
	– **Chrysomelidae**	– Adult – *Mandibles* (2); *Back femur* in Halticini (2).
		– Larvae – *Mandibles* (2).
	– **Staphylinidae**	– *Mandibles* (2); *Elytra* (2).
	– **Scarabidae**	– *Tibia* (6).
	– **Cantharidae**	– *Mandibles* (2).
	– **Elateridae**	– *Mandibles* (2); '*Spring peg*' (1).
	– **Nitidulidae**	– *Tibia* (6).
– **DIPTERA**		Due to the soft body very few identifiable parts survive in a form that allows numbers to be estimated. Fragmented legs will indicate dipteran presence. If large wing fragments are found, identification to Family is often possible.
	– **Empididae**	– *Front tibia* (2).
	– **Bibionidae**	– *Front tibia* (2).
– **ORTHOPTERA**	– **Acrididae**	– *Mandibles* (2).
– **DERMAPTERA**		– *Mandibles* (2); *Forceps* (2).
– **NEUROPTERA**	– Larvae	– *Mandibles* (4 parts).
MOLLUSCA		– *Shell fragments.*
LUMBRICIDAE		– *Chaetae.*

Table 2.4. Invertebrate groups comprising the diet of nestling dunnock and whitethroat on arable farmland over two years and identified by neck collar and faecal analysis (data summarised from Moreby & Stoate, 1999).

	Dunnock				Whitethroat			
	1996		1997		1996		1997	
	Neck Collar Per cent of diet	Faecal Sample Per cent of diet	Neck Collar Per cent of diet	Faecal Sample Per cent of diet	Neck Collar Per cent of diet	Faecal Sample Per cent of diet	Neck Collar Per cent of diet	Faecal Sample Per cent of diet
ARANEAE	3.1	3.6	7.1	7.2	15.9	13.0	16.0	14.3
OPILIONIDAE	0.3	1.2	1.9	1.8	0	3.0	0	2.3
Total HEMIPTERA	3.9	2.4	26.8	3.8	4.9	8.0	7.6	3.2
Aphididae	0.3	0.4	13.6	2.0	0	0.7	4.2	0.3
Psyllidae	1.2	0	7.2	0	0	0	0	0
Cicadellidae	0.9	0.6	5.3	0	0	0	0	0
Delphacidae	0.5	0	0.3	0	0	0.2	0	0
Cercopidae	0.3	0.4	0	0.2	3.7	6.9	2.5	2.7
Heteroptera	0.7	1.0	0.4	1.6	1.2	0.2	0.9	0.2
LEPIDOPTERA – Adults	8.9	2.8	1.4	0.9	32.9	9.4	25.2	10.3
– Larvae	3.6	4.5	7.5	15.1	32.9	28.6	29.4	28.8
Total HYMENOPTERA	1.2	5.4	1.7	7.1	8.6	9.4	7.6	6.7
– Sawfly Adults	0.3	0.2	0.4	1.4	0	0.7	2.5	2.9
– Sawfly Larvae	0	1.0	0.4	0.7	8.6	7.8	4.2	3.2
– Parasitica	0.9	4.2	0.9	5.0	0	0.9	0.9	0.6
Total COLEOPTERA	16.6	58.1	14.2	47.0	1.2	16.8	3.4	17.2
Carabidae – Adults	0.3	0.6	0	1.6	0	0	0	0
– Larvae	0.5	0.9	0	1.4	0	0	0	0
Staphylinidae	0.7	1.5	2.0	3.4	0	0.2	0	0
Cryptophagidae	0.3	0.4	0.4	1.6	0	0.2	0	0
Elateridae	0	1.5	0	2.0	0	0.2	0	0.8
Curculionidae	3.1	12.6	2.0	26.8	0	10.1	0	7.7
Chrysomelida – Adults	0	0.6	0.2	1.8	0	0.2	0	0.3
– Larvae	0	0	0	0	0	0.2	0	0.8
Cantharidae	0	1.3	0.5	2.7	1.2	1.2	3.4	4.5
Nitidulidae – Adults	7.9	32.0	0	2.5	0	0.4	0	0
– Larvae	2.6	0.4	8.9	0	0	0	0	0
Coccinellidae	1.2	0	0	0	0	0	0	0.3
Other Families	0	6.3	0.2	3.2	0	4.1	0	2.8
Total DIPTERA	33.6	22.2	26.4	15.4	2.4	11.2	5.8	12.6
Tipulidae	7.0	5.1	4.6	3.6	2.4	4.1	5.0	3.8
Bibionidae	9.3	8.9	1.8	1.1	0	0	0.8	1.9
Other Nematocera	9.6	1.4	11.3	0.5	0	1.1	0	0.7
Syrphidae Larvae	0	0	0.7	0	0	0	0	0
Empididae	1.4	1.9	2.0	3.6	0	0.4	0	0
Other Diptera	6.3	4.9	6.0	6.6	0	5.6	0	6.2
Other groups								
Mollusca	0.3	0	0.1	0.7	0	0.4	5.0	4.5
Collembola	21.6	0	12.1	0	0	0	0	0
Dermaptera	0.3	0	0.3	0.2	1.2	0	0	0
Neuroptera – Larvae	0.3	0.2	0.5	0.7	0	0	0	0
Orthoptera	0	0	0	0	0	0.2	0	0
Plecoptera	0.5	0	0	0	0	0	0	0
Isopoda	1.9	0	0	0	0	0	0	0
Lumbricidae	0.5	0	0	0	0	0	0	0

Faecal analysis does, however, have a potential inherent bias towards hard-bodied, less easily digested invertebrates, as opposed to soft-bodied ones such as Hemiptera and Collembola, as shown in the diet of dunnock chicks (*Table 2.4*), and this method may be better used in conjunction with limited use of neck collars (Moreby & Stoate, 2000).

While diet analysis will allow the determination of the main invertebrate groups eaten, variations in the relative use of groups may occur. This is illustrated by the diet of chicks studied on farmland over two breeding seasons in 1996 and 1997 (*Table 2.4*). In whitethroat chick diet, higher numbers of both Cercopidae and sawfly larvae occurred in 1996 compared to 1997, and higher numbers of Cantharidae and Mollusca occurred in 1997 compared to the previous year. While the relative use of the above groups in the overall diet was low, in the diet of dunnock chicks a large difference in the relative use of Nitidulidae and Curculionidae occurred between the two years. These differences may result from crop rotations, weather conditions, general agricultural practices, e.g. pesticide use, or be due to the ecology of a group or species. An example of this latter phenomenon is illustrated by sawfly larvae (*Dolerus* spp.), in which populations cycle, and while this might affect their distribution, they may be absent from an area due to specific habitat requirements (Barker, 2003). Therefore, for partridge and pheasant, two species that normally include similar invertebrate Orders in their diet, the relative use of sawfly larvae can be very different due to the two birds being studied in different areas (*Figures 2.3* and *2.4*).

Conclusion

To stop declines in the number of our arable birds, research is needed to determine at which stage in their life cycle they are most vulnerable. Analysis of the diet will be one of the main areas to study. While both summer and winter feeding habitats may be important, adult food alone is unlikely to be a limiting factor. However, for nestling survival, the supply of invertebrate food may be directly limiting. For a few farmland birds such as the tits, nestling diet can be identified by photography, but for most species faecal analysis and the use of neck collars will be the most appropriate methods. Once the important invertebrate groups comprising the diet are identified, sympathetic management can be targeted at the habitats where these invertebrates are found to increase the availability of food plants or provide suitable vegetation and structure in other areas.

References

Aebischer, N. J. (1991) Twenty years of monitoring invertebrates and weeds in cereal fields in Sussex. In: *The ecology of temperate cereal fields* (eds. L. G. Firbank, N. Carter, J. F. Darbyshire & G. R. Potts), pp. 305–331. Blackwell Scientific Publications, London, UK.
Banbura, J., Blondel, J., de Wilde-lambrechts, H., Galan, M.-J. & Maistre, M. (1994) Nestling diet variation in an insular Mediterranean population of blue tits *Parus caeruleus*: effects of years, territories and individuals. *Oecologia* **110**, 413–420.
Barba, E., Lopez, J. A. & Gildelgado, J. A. (1996) Prey preparation by adult great tits *Parus major* feeding nestlings. *Ibis* **138**, 532–538.
Barker, A. M. (2003) Insects as food for farmland birds – is there a problem? (*this volume*).
Bishton, G. (1985) The diet of nestling dunnocks *Prunella modularis*. *Bird Study* **32**, 113–115.
Boddy, M. (1991) Some aspects of frugivory by bird populations using coastal dune scrub in Lincolnshire. *Bird Study* **38**, 188–199.

Burn, A. J. & Cooke, A. S. (1995) *The role of pesticides in declines of seed-eating passerines* (eds. R. D. Fuller & J. D. Wilson), pp. 17–18. BTO Research Report, no. 149. BTO, Thetford, UK.

Calver, M. C. & Wooller, R. D. (1982) A technique for assessing the taxa, length, dry weight and energy content of the arthropod prey of birds. *Australian Wildlife Research* **9**, 293–301.

Campbell, L. H., Avery, M. I., Donald, P., Evans, A. D., Green, R. E. & Wilson, J. D. (1997) *A review of the indirect effects of pesticides on birds*. JNCC Report, no. 227. JNCC, Peterborough, UK. 148 pp.

Cramp, S. (ed.) (1998) *The complete birds of the Western Palearctic* [CD-Rom]. Oxford University Press, Oxford, UK.

Davies, N. B. (1976) Food, flocking and territorial behaviour of the pied wagtail (*Motacilla alba yarrellii* Gould) in winter. *Journal of Animal Ecology* **45**, 235–253.

Ewald, J. A. & Aebischer, N. J. (1999) *Pesticide use, avian food resources and bird densities in Sussex*. JNCC Report, no. 296. JNCC, Peterborough, UK. 103 pp.

Ford, J., Chitty, H. & Middleton, A. D. (1938) The food of partridge chicks (*Perdix perdix*) in Great Britain. *Journal of Animal Ecology* **7**, 251–265.

Fuller, R. J., Gregory, R. D., Gibbons, D. W., Marchant, J. H., Wilson, J. D., Baillie, S. R. & Carter, N. (1995) Population declines and range contractions among lowland farmland birds in Britain. *Conservation Biology* **9**, 1425–1441.

Galbraith, H. (1989) The diet of lapwings *Vanellus vanellus* chicks on Scottish farmland. *Ibis* **131**, 80–84.

Galbraith, H., Murray, S., Duncan, K., Smith, R., Whitfield, D. P. & Thompson, B. D. A. (1993) Diet and habitat use of the dotterel *Charadrius morinellus* in Scotland. *Ibis* **135**, 148–155.

Gibbons, D. W., Reid, J. B. & Chapman, R. A. (1993) *The new atlas of breeding birds in Britain and Ireland: 1988–1991*. Poyser, London, UK. 520 pp.

Green, R. E. (1984) The feeding ecology and survival of partridge chicks (*Alectoris rufa* and *Perdix perdix*) on arable farmland in East Anglia. *Journal of Applied Ecology* **21**, 817–830.

Grim, T. & Honza, M. (1996) Effect of habitat on the diet of reed warbler (*Acrocephalus scirpaceus*) nestlings. *Folia Zoologica* **45**, 31–34.

Hill, D. A. (1985) The feeding ecology and survival of pheasant chicks on arable farmland. *Journal of Applied Ecology* **22**, 645–654.

Holland, J. M. (2003) The impact of agriculture and some solutions for arthropods and birds (*this volume*).

Inglesfield, C. (1989) Pyrethroids and terrestrial non-target organisms. *Pesticide Science* **27**, 241–251.

Jenni, L., Reutimann, P. & Jenni-Eiermann, S. (1990) Recognizability of different food types in faeces and in alimentary flushes of *Sylvia* warblers. *Ibis* **132**, 445–453.

Jenny, M. (1990) Diet ecology of the skylark *Alauda arvensis* in a intensively cultivated agroecosystem in the Swiss midlands. *Der Ornithologische Beobachter* **87**, 31–53.

Johnson, E. J., Best, L. B. & Heagy, P. A. (1980) Food sampling biases associated with the 'ligature method'. *Condor* **82**, 186–192.

Kleintjes, P. K. & Dahlsten, D. L. (1992) A comparison of three techniques for analysing the arthropod diet of plain titmouse and chestnut-backed chickadee nestlings. *Journal of Field Ornithology* **63**, 276–285.

Kristin, A. (1995) The diet and foraging ecology of the penduline tit (*Remiz pendulinus*). *Folia Zoologica* **44**, 23–29.

Kristin, A. & Patocka, J. (1997) Birds as predators of Lepidoptera: selected examples. *Biologia* **52**, 319–326.

Marshall, E. J. P. (1985) Weed distribution associated with cereal field edges – some preliminary observations. *Aspects of Applied Biology* **9**, 49–58.

Marshall, E. J. P. (1989) Distribution patterns of plants associated with arable field edges. *Journal of Applied Ecology* **26**, 247–257.

Mellott, R. S. & Woods, P. E. (1993) An improved ligature technique for dietary sampling in nestling birds. *Journal of Field Ornithology* **64**, 205–210.

Moreby, S. J. (1988) An aid to the identification of arthropod fragments in the faeces of game bird chicks (Galliformes). *Ibis* **130**, 19–526.

Moreby, S. J. (1994) The influence of field boundary structure on heteropteran densities within adjacent cereal fields. In: *Field margins: integrating agriculture and conservation*, pp. 117–121. British Crop Protection Council, Farnham, UK.

Moreby, S. J. (1995) Heteroptera distribution and diversity within the cereal ecosystem. In: *Integrated crop protection: towards sustainability*, pp. 151–158. British Crop Protection Council, Farnham, UK.

Moreby, S. J. & Southway, S. E. (1999) Influence of autumn applied herbicides on summer and autumn food available to birds in winter wheat fields in southern England. *Agriculture, Ecosystems and Environment* **72**, 285–297.

Moreby, S. J. & Stoate, C. (2000) A quantitative comparison of neck collar and faecal analysis to determine passerine nestling diet. *Bird Study* **47**, 320–331.

Moreby, S. J. & Stoate, C. (2001) Relative abundance of invertebrate taxa in the nestling diet of three farmland passerine species, dunnock *Prunella modularis*, whitethroat *Sylvia communis* and yellowhammer *Emberiza citrinella*. *Agriculture, Ecosystems and Environment* **86**, 285–297.

Moreby, S. J., Sotherton, N. W. & Jepson, P. C. (1997) The effects of pesticides on species of non-target Heteroptera inhabiting cereal fields in southern England. *Pesticide Science* **51**, 39–48.

Newton, I. (1967) The adaptive radiation and feeding ecology of some British finches. *Ibis* **109**, 33–98.

O'Connor, R. J. & Shrubb, M. (1986) *Farming and birds.* Cambridge University Press, Cambridge, UK. 290 pp.

Perrins, C. (1979) *British tits.* Collins, London, UK. 304 pp.

Potts, G. R. (1986) *The partridge: pesticides, predation and conservation.* Collins, London, UK. 274 pp.

Poulsen, J. G. & Aebischer, N. J. (1995) Quantitative comparison of two methods of assessing diet of nestling skylarks (*Alauda arvensis*). *Auk* **112**, 1070–1073.

Poulsen, J. G., Sotherton, N. W. & Aebischer, N. J. (1998) Comparative nesting and feeding ecology of skylarks *Alauda arvensis* on arable farmland in southern England with special reference to set-aside. *Journal of Applied Ecology* **35**, 131–147.

Prys-Jones, R. P., Schifferli, L. & MacDonald, D. W. (1974) The use of an emetic in obtaining food samples from passerines. *Ibis* **116**, 90–94.

Rands, M. R. W. (1985) Pesticide use on cereals and the survival of grey partridge chicks: a field experiment. *Journal of Applied Ecology* **22**, 49–59.

Rands, M. R. W. (1986) The survival of game bird (Galliformes) chicks in relation to pesticide use on cereals. *Ibis* **128**, 57–64.

Read, J. L. (1994) The diet of three species of firetail finches in temperate south Australia. *Emu* **94**, 1–8.

Robertson, R. J. (1973) Optimal niche space of the red-winged blackbird. III. Growth rate and food of nestlings in marsh and upland habitats. *Wilson Bulletin* **85**, 209–222.

Rodenhouse, N. L. & Holmes, R. T. (1992) Results of experimental and natural food reductions for breeding black-throated blue warblers. *Ecology* **73**, 357–372.

Royama, T. (1970) Factors governing the hunting behaviour and selection of food by the great tit (*Parus major* L.). *Journal of Animal Ecology* **39**, 619–668.

Savory, C. J. (1989) The importance of invertebrates to chicks of gallinaceous species. *Proceedings of the Nutrition Society* **48**, 113–133.

Sotherton, N. W. (1991) Conservation headlands, a practical combination of intensive cereal farming and conservation. In: *The ecology of temperate cereal fields* (eds. L. G. Firbank, N. Carter, J. F. Darbyshire & G. R. Potts), pp. 373–397. Blackwell Scientific Publications, Oxford, UK.

Sotherton, N. W., Moreby, S. J. & Langley, M. G. (1987) The effects of the foliar fungicide pyrazophos on beneficial arthropods in barley fields. *Annals of Applied Biology* **111**, 75–87.

Sotherton, N. W., Boatman, N. D. & Rands, M. R. W. (1989) The 'Conservation Headland' experiment in cereal ecosystems. *Entomologist* **108**, 135–143.

Stoate, C., Moreby, S. J. & Szczur, J. (1998) Breeding ecology of farmland yellowhammers *Emberiza citrinella*. *Bird Study* **45**, 109–121.

Theiling, K. M. & Croft, B. A. (1988) Pesticide side-effects on arthropod natural enemies: a database summary. *Agriculture, Ecosystems and Environment* **21**, 191–218.

Valera, F., Gutierrez, J. E. & Barros, R. (1997) Effectiveness, biases and mortality in the use of apomorphine for determining the diet of granivorous passerines. *The Condor* **99**, 765–772.

Vickerman, G. P. (1974) Some effects of grass weed control on the arthropod fauna of cereals. *Proceedings of the 12th British Weed Control Conference* **3**, 929–939.

Vickerman, G. P. & Sotherton, N. W. (1983) Effects of foliar fungicides on the chrysomelid beetle *Gastrophysa polygoni* (L.). *Pesticide Science* **14**, 405–411.

Vickerman, G. P. & Sunderland, K. D. (1977) Some effects of dimethoate on arthropods in winter wheat. *Journal of Applied Ecology* **14**, 767–777.

Ward, R. S. & Aebischer, N. J. (1994) *Changes in corn bunting distribution on the South Downs in relation to agricultural land use and cereal invertebrates.* English Nature Research Report, no. 129. English Nature, Peterborough, UK. 78 pp.

Wilson, J. D., Evans, J., Browne, S. J. & King, J. R. (1997) Territory distribution and breeding success of skylarks *Alauda arvensis* on organic and intensive farmland in southern England. *Journal of Applied Ecology* **34**, 1462–1478.

Wilson, P. J. (1989) The distribution of arable weed seedbanks and the implications for the conservation of endangered species and communities. *Proceedings of the Brighton Crop Protection Conference, Weeds, 1989*, 1081–1086.

3

Insects as Food for Farmland Birds – Is There a Problem?

ALISON M. BARKER

The Game Conservancy Trust, Fordingbridge, Hampshire, SP6 1EF, UK (Present address: CABI-Bioscience Switzerland Centre, Rue des Grillons 1, CH-2800, Delémont, Switzerland)

Introduction

The decline of farmland birds in Britain is a topical subject. Much of the discussion has been centred on the deleterious effects of ever increasing pesticide use on birds (Campbell *et al.*, 1997; Ewald & Aebischer, 1999; Avery *et al.*, 2003). Rather than being directly toxic, like many of the early pesticides, modern pesticides are thought to influence bird survival indirectly by affecting the availability of food items in their diet. In fact, insects have been affected by a wide variety of changes in farming practice as agriculture has become more intensive (Potts, 1997; Sotherton, 1998; Barker *et al.*, 1999; Holland, 2003). This chapter discusses the extent to which insects figure in the diet of farmland birds, and identifies the insect groups of primary importance to birds. The evidence for declines in insect numbers in agricultural habitats will then be presented. Finally, the evidence linking these declines with reported declines in the abundance of farming birds will be considered.

The importance of insects in the diet of farmland birds

A recent major review of the indirect effects of pesticides on birds summarised the diet of 36 bird species known to depend wholly or partly on lowland farmland (Campbell *et al.*, 1997). Over three-quarters of these (*Figure 3.1*) had a significant element of invertebrate food in the diet. Some were obligate insectivores throughout the life cycle, such as lapwing (*Vanellus vanellus*), stone curlew (*Burhinus oedicnemus*), starling (*Sturnus vulgaris*) and yellow wagtail (*Motacilla flava flavissima*). Others, including the skylark (*Alauda arvensis*), corn bunting (*Miliaria calandra*) and grey partridge (*Perdix perdix*), were seasonal insectivores which consume insects in the breeding season, or whose chicks eat a diet initially rich in insect material. Less than one-quarter had little reliance on insect food at any point in their lives; the greenfinch (*Carduelis chloris*) and the turtle dove (*Streptopelia turtur*) are examples.

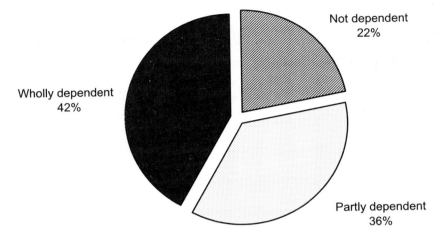

Figure 3.1.   Distribution of 36 farmland bird species into those with high, partial (e.g. seasonal), and low dependence on insects as a food source.

Detailed information is available on the particular insect groups selected by many species. One particularly well-studied example is the grey partridge. Grey partridges are birds of open landscapes, including cereal fields. The adults are largely herbivorous, feeding on seeds in winter and on growing shoots in the spring, but the chicks, for the first two weeks after hatching, feed mainly on insects. Without this important protein source, their feathers do not develop properly (Potts, 1986). Studies have looked at the diet of the chicks by a variety of techniques. These have involved dissection of the crop (Ford *et al.*, 1938; Southwood & Cross, 1969; Potts, 1970; Vickerman & O'Bryan, 1979), feeding trials (Vickerman & O'Bryan, 1979), analyses of chick faeces (Green, 1984), and modelling studies identifying insect groups significant in chick survival (Potts & Aebischer, 1991) (*Table 3.1*). Although many of these groups were taken by the birds in proportion to their abundance, others, such as sawfly and Lepidoptera larvae, seem to have been disproportionately selected; the significance of the relationships between chick survival and abundance of specific groups may indicate that these are of particular biological benefit. Most of these selected insects are relatively large with a high biomass. Also, they are often relatively inactive, which is likely to decrease handling problems.

Similar reviews of the diet of other farmland birds (Moreby, 2003) identify a core of insect groups that live in or around cropped land, and are frequently used as a summer food source by many insectivorous farmland birds: aphids (Hemiptera [Sternorrhyncha]: Aphididae), various beetle groups (Coleoptera: especially Chrysomelidae, Carabidae, Curculionidae), caterpillars (Lepidoptera and sawfly (Hymenoptera [Symphyta]: Tenthredinidae) larvae), plant bugs (Heteroptera and other Hemiptera excluding Aphididae), and grasshoppers (Orthoptera). Spiders (Araneae) and flies (Diptera) are also sometimes important. In winter, most of the above groups are in short supply. Although many bird species switch diet or migrate in winter, some remain dependent on invertebrates. Soil macro-invertebrates are important in the diet of these species during this period (Tucker, 1992); these include

Table 3.1.   Summary of studies into the diet of grey partridge chicks.

Study	Study type	Major insect food groups identified in order in numerical importance (* indicates most important in terms of biomass)
Ford *et al.* (1938)	Crop contents	Curculionidae*, Collembola (*Sminthurus viridis*), Aphididae, Cicadellidae, Lepidoptera larvae* Formicidae* (in older chicks)
Southwood & Cross (1969)	Crop contents	Hemiptera-Sternorryncha (mainly Aphididae), Collembola (*Sminthurus*), Coleoptera, Araneae
Potts (1970)	Crop contents	Aphididae, Coleoptera, Tenthredinidae larvae, Collembola (*Sminthurus*)
Vickerman & O'Bryan (1979)	Crop contents	Collembola (*Sminthurus*), Aphididae, Curculionidae*, Tenthredinidae larvae*, Chrysomelidae, Formicidae* (in older chicks)
Vickerman & O'Bryan (1979)	Feeding trials	Heteroptera, Tenthredinidae (larvae), Grasshoppers preferred to Formicidae and two Chrysomelid species
Green (1984)	Faecal analysis	Aphididae, Carabidae, Staphylinidae, Chrysomelidae
Green (1984)	Regression model	Significant groups for chick survival: Sawfly and Lepidoptera larvae, Chrysomelid larvae, Diptera, Aphids, other Hemiptera
Potts & Aebischer (1991)	Modelling study	Small Carabidae, larvae of Lepidoptera and Tenthredinidae, Chrysomelidae and Curculionidae, Heteroptera and Cicadellidae, Aphididae

insects such as larval Coleoptera and Diptera, as well as earthworms (Lumbricidae). For example, the winter diet of lapwing is based largely on earthworms and leatherjackets (the larvae of tipulid flies (Diptera)); in summer, surface-living beetles and spiders are added to the diet (Galbraith, 1989a,b; Baines, 1990).

Population trends in insect groups

Having established the importance of insects and other invertebrates as an element of the diet of farmland birds and identified the important groups, it is important to identify the trends in invertebrate numbers in farmland environments in the recent past. There are just two long-term datasets monitoring farmland insects in Britain. The Rothamsted Insect Survey, a network of light traps monitoring moth populations and suction traps monitoring aphid populations across Britain, was established in the early 1960s (Woiwod & Harrington, 1994), and The Game Conservancy Trust's Sussex Study, in which vacuum samples of insects have been taken in all cereal fields within a 62 km² area of farmland on the Sussex Downs, has been carried out annually every June since 1970 (Aebischer, 1991).

Aebischer (1991) found that total invertebrates (excluding Acari and Collembola) on the farms in The Game Conservancy Trust's Sussex study area declined by 4.2% per year. This meant that numbers had approximately halved over 20 years. When the data were split into smaller taxonomic units, many of the sub-groups were also found to be in decline. Of the potential chick-food insects considered, numbers of Araneae (spiders and harvestmen) declined by 4.1% per year, aphid numbers by 8.4% per year,

Table 3.2. Summary of reported trends in major chick-food insect groups. [n] indicates that trend is in insect numbers, [s] indicates trends in species diversity.

Invertebrate group	Trend
Total invertebrates in cereals	Down[n]
Araneae	Down[n]
Lepidoptera Butterflies	Down[ns]
Moths	Down[n]
Tenthredinidae	Down[n]
Carabids	? Down or stable
'Small beetles' *	Down[ns]
Orthoptera	Down[s]
Aphids	? Down[n] (locally) but Up[n] (nationally)
Chrysomelidae & Curculionidae	Down[n]
Non-aphid Hemiptera	Down[n]

*(Coleoptera excluding Carabidae and Elateridae)

and sawfly larvae by 4.4% per year. No decline was detected for carabid beetles. More recent analysis has also found declines in leaf-eating beetle groups (Chrysomelidae and Curculionidae) and in non-aphid Hemiptera (Ewald & Aebischer, 1999). The same dataset also reveals spatial changes in distribution, with contraction in the ranges occupied by spiders and harvestmen, caterpillars, and small beetles (Coleoptera excluding Carabidae and Elateridae) over the sampled farmland (Aebischer, 1995; Aebischer & Ward, 1997).

Data from the Rothamsted Insect Survey (Woiwod, 1991; Woiwod & Harrington, 1994) samples a much wider area, and is not limited to farmland habitats. The numbers of two of the most common aphid species (*Metopolophium festucae* and *Sitobion avenae*) sampled in this study increased, whilst the number of other species of cereal aphids showed no trends; this difference from the declines recorded in the Sussex study may be a result of different processes operating over different spatial scales. There was a large decline in both the numbers (55%) and diversity of macrolepidoptera over a 30-year period, and comparison with some data from earlier studies suggests that this decline had been occurring over even longer time scales. Use of subsets of data suggested a link with farming changes, as the biggest declines were obtained on cultivated areas, whilst woodland populations remained more stable.

Additional information is available from Europe on some key insect food groups. Data from Germany show declines in species richness of Coleoptera and Formicidae on arable land (Heydemann, 1983), and some groups of butterflies with larvae that feed in marginal arable habitats have shown declines over at least a five-year period (Dover *et al.*, 1990). One important group for which published data on long-term trends seems to be sparse is the grasshoppers (Orthoptera), which are insects of unimproved grassy areas rather than cereal fields, but there is, nevertheless, evidence that eutrophication from increased fertiliser load has reduced grasshopper diversity on agricultural grassland as slow-growing species are out-competed in high nitrogen environments (van Wingerden *et al.*, 1992). *Table 3.2* summarises the information known for the invertebrate groups of importance to birds, and suggests that, with the exception of aphids (the group providing the least biomass per individual), all important chick-food insect groups are in decline.

Why have these declines taken place? A suite of changes in farming practice has occurred in western Europe over the last 50 years. In Britain, for example, fields have been enlarged and hedgerows removed to accommodate mechanisation, cropping patterns have changed, with large changes from spring to winter cereals in many areas (Aebischer, 1991), and there has been a shift from mixed farming to wholly arable or livestock farming (Potts, 1997). At the same time, inputs to land have increased in order to boost productivity, with the introduction and increasing use of artificial fertilisers and pesticides – on average, UK cereal fields now receive over two applications of herbicides and fungicides and 0.7 applications of insecticides per year (data from the British Agricultural Association, quoted in Sotherton, 1998). All of these changes will have had effects on insects. For example, graminivorous sawflies, the larvae of which are an important chick-food item, appear to have been most affected by a combination of the loss of suitable host plants from the arable landscape through changes in crop rotation and increased use of graminicides, increased winter cultivation affecting the overwintering stages, increased grazing in stock farming, and increased use of insecticides (Vickerman, 1974; Aebischer, 1990; Barker *et al.*, 1999). Factors behind the decline of insects on farmland and suggested ways of reversing these declines are considered further by Holland (2003).

Decline in bird numbers

What evidence is there for a decline in bird numbers in farmland habitats? Data on temporal trends in farmland birds in Britain are available from survey work, principally the BTO's Common Bird Census (CBC). This survey estimates the numbers of birds holding territories in the breeding season on defined census areas, and uses these data to calculate an index of relative changes in population size over the full area surveyed. Separate surveys are carried out in woodland and farmland. There are limitations to the scope of the survey; for example, uncommon species tend not to be recorded sufficiently to estimate meaningful indices. Marchant *et al.* (1990) give details of the methodology and show the trends for many breeding bird species over the period 1962–1988 using these data. Campbell *et al.* (1997) give updated point values for the changes in CBC indices, where available, between 1969 and 1994 for a selection of 40 bird species that are dependent on the farmland habitat. Tucker & Heath (1994) provide estimates of European-wide declines for some of these species, plus other farmland species that are rare or non-breeders in Britain. Between them, these studies form a basis for identifying trends in the populations of farmland bird species. Using the list of farmland birds in Campbell *et al.* (1997) as a basis, and excluding predators of vertebrates, extensively released game birds, and species for which there are no data in the CBC surveys, leaves 28 species for which there are reliable data, including finches, thrushes, buntings, sparrows, wagtails, and pigeons. Trends for these in all three sources were assessed to obtain a general pattern. Following the classification in Tucker & Heath (1994), decreases of more than 50% were considered as large declines, –20 to –50% as moderate declines, changes of between –20 to +20% were considered stable, +20 to +50% moderate increases, and +50% as large increases. The species have been grouped into those with a European-wide decline, a British decline but no decline in Europe, no decline, or population increases (*Table 3.3*). The eleven species exhibiting large declines are marked; they

Table 3.3.   Trends in 28 British farmland bird species. Species are selected from those listed in Campbell *et al.* (1997) excluding vertebrate predators, released gamebirds and species with no CBC data. Trends are assessed from Marchant *et al.* (1990), Campbell *et al.* (1997), and Tucker & Heath (1994).

European decline 10 species (36%)	British decline 8 species (29%)	Stable 6 species (21%)	Increasing 4 species (14%)
Grey partridge[a] (*Perdix perdix*)	Lapwing[a] (*Vanellus vanellus*)	Pied wagtail (*Motacilla alba yarrellii*)	Stock dove (*Columba oenas*)
Turtle dove[a] (*Streptopelia turtur*)	Yellow wagtail (*Motacilla flava flavissima*)	Robin (*Erithacus rubecula*)	Wood pigeon (*Columba palumbus*)
Skylark[a] (*Alauda arvensis*)	Blackbird (*Turdus merula*)	Wren (*Troglodytes troglodytes*)	Collared dove (*Streptopelia decaocto*)
Swallow (*Hirundo rustica*	Dunnock (*Prunella modularis*	House sparrow (*Passer domesticus*)	Chaffinch (*Fringilla coelebs*)
Song thrush[a] (*Turdus philomelus*)	Starling (*Sturnus vulgaris*)	Greenfinch (*Carduelis chloris*)	
Mistle thrush (*Turdus viscivorus*)	Tree sparrow[a] (*Passer montanus*)	Goldfinch (*Carduelis carduelis*)	
Spotted flycatcher[a] (*Muscicapa striata*)	Bullfinch[a] (*Pyrrhula pyrrhula*)		
Linnet[a] (*Carduelis cannabina*)	Reed bunting[a] (*Emberiza schoeniclus*)		
Yellowhammer (*Emberiza citrinella*)			
Corn bunting[a] (*Miliaria calandra*)			

[a]Marks species which have declined by more than 50% on farmland in Britain (CBC Data 1969–1994; Campbell *et al.*, 1997).

include the grey partridge, the skylark, and the corn bunting. Additionally, the four British farmland corvids (rook (*Corvus frugilegus*), magpie (*Pica pica*), jackdaw (*Corvus monedula*), carrion crow (*Corvus corone*)) are generally increasing on farmland (Marchant *et al.*, 1990), and there is evidence for declines in the rare British breeding species quail (*Coturnix coturnix*), stone curlew, red-backed shrike (*Lanius collurio*), and cirl bunting (*Emberiza cirlus*), as well as many other European species of farmland habitats such as whitethroat (*Sylvia communis*) (Tucker & Heath, 1994; Campbell *et al.*, 1997).

In general then, survey results present a gloomy picture with nearly two-thirds of British farmland birds in decline. Many of these declines have been severe. According to the CBC farmland indices, eleven species have declined from 1969 to 1994 by over 50% and six species by more than 70%, including tree sparrow (*Passer montanus*) (–89%) and grey partridge (–82%). Few species have increased. Although three pigeon species have increased strongly in numbers, the fourth, turtle dove, has declined sharply. Increases in crow numbers are probably a result of reduced predator control by gamekeepers on farms. Rooks, which feed on a mixture of grain and grassland invertebrates, may, nevertheless, have been affected by agricultural intensification; numbers declined from the mid-1950s to the 1970s, although there is now a slow rise in numbers (Marchant *et al.*, 1990).

However, demonstrating causal links between trends in farmland bird numbers and trends in the abundance of their insect food is not easy, even though parallel declines have been demonstrated. Detailed autecological studies of individual species are required. One of the few studies where there is direct evidence for a link between declines in bird numbers and changes in the availability of insect food is the long-term study of the grey partridge carried out by The Game Conservancy Trust (Potts, 1986; Potts & Aebischer, 1991).

INSECTS AND THE DECLINE OF THE GREY PARTRIDGE

Declines in grey partridge numbers have occurred across the whole of its European range, with reductions of up to 90% in some countries (Aebischer & Potts, 1994). Potts (1997) estimated the overall decline in breeding stocks to be over 80%. Numbers in Britain have probably been declining since the end of the last century, but the decline accelerated in the 1950s (Potts, 1986). On The Game Conservancy Trust's Sussex Downs study area, where numbers have been intensively monitored since 1970, densities have dropped from 21 km^{-2} to 3 km^{-2} over a 25-year period.

The method used to monitor partridge numbers enables calculation of mortality rates for a series of sequential mortality factors, including loss of nests (usually from predation), chick losses to age six weeks, shooting losses, and winter losses between shooting and nesting, and identifies which of these contribute to the annual mortality (full methods given in Potts, 1986). Key factor analysis showed that, of all the mortalities measured, chick mortality was the key factor determining future population size (Blank *et al.*, 1967). Multiple regression analysis of probit chick survival showed that chick survival rates were significantly related to the availability of suitable insect food, which explained 52% of the variation in survival (Potts & Aebischer, 1991). Five groups (ground beetles, caterpillars, leaf beetles, heteropteran bugs, hoppers and aphids) were accepted by the stepwise regression as statistically significant.

The recorded declines in the densities of these insects in the study area (Aebischer, 1991) therefore appeared to be directly linked *via* chick mortality to the decline in partridges, with insufficient production to keep up bird population numbers. Evidence for the link was supported by computer modelling, which showed that changing the food supply to increase breeding success led to higher density breeding populations at equilibrium (Potts & Aebischer, 1991). On the same study area, patterns of change from mixed farming to monocultures of cereal or grassland have also been mirrored closely by changes in abundance of key chick-food insect groups such as sawflies (*Figure 3.2*) and by partridge numbers (*Figure 3.3*) (Aebischer, 1995; Aebischer & Ward, 1997; Aebischer & Potts, 1998). For example, partridge density is now highest on the one remaining traditional mixed farm in the area which still has relatively high caterpillar numbers (*Figures 3.2* and *3.3*, note bottom right hand corner of the study site).

An experimental demonstration of the link between chick survival and insect availability was conducted on a farm scale by leaving cereal headlands in selected areas of the farm unsprayed with herbicide. Overall, insect food availability in the unsprayed headlands was found to be over three times that of sprayed cereal headlands as insects colonised the weedy field edges (Rands, 1985). Chick survival in broods in the areas with unsprayed headlands was significantly higher than on

44   A. M. BARKER

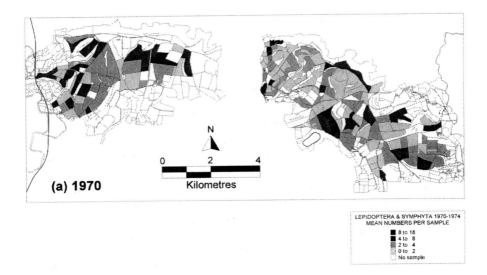

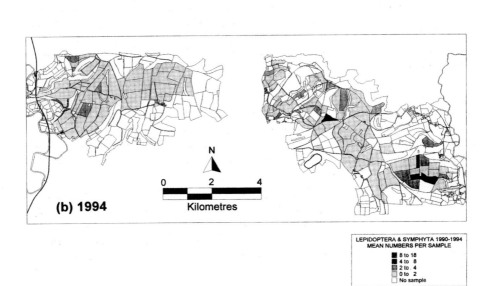

Figure 3.2.   Distribution of caterpillars (Lepidoptera and Symphyta larvae) across the Sussex study area in (a) 1970 and (b) 1994. For each field, the value is the mean of the samples that were collected over the five year period. From Aebischer & Ward, 1997.

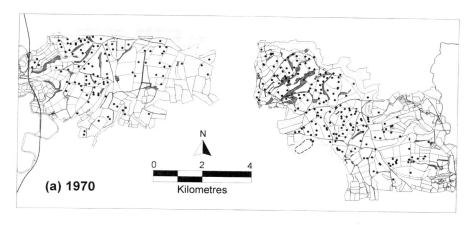

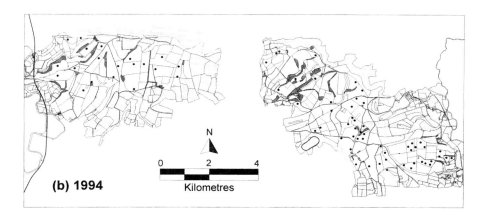

Figure 3.3.   Distribution of grey partridge (*Perdix perdix*) coveys in autumn on the Sussex study area (UK) in (a) 1970 and (b) 1994. Each dot denotes a covey, and the greyed area represents the overall study area outside the five main farms. From Aebischer & Potts, 1998.

matched headlands sprayed with normal herbicide applications (Rands, 1985, 1986). Similar benefits have been found for grey partridges in Sweden (Chiverton, 1999). Long-term results have been very positive, with chick survival rates on East Anglian farms on Conservation Headlands reaching the minimum of 30% needed for population recovery in five out of eight years, a level reached in only one of the eight years on fully sprayed areas (Sotherton, 1998). Additional benefits of conservation headlands are described by Holland (2003).

SUGGESTED REASONS FOR OTHER DECLINES

To what extent are the results from the grey partridge typical of farmland birds as a whole? Several aspects of the natural history of grey partridges would appear to make them especially vulnerable to declines in insect numbers on agricultural land. Firstly, their survival is highly dependent on insects in their diet during chick development (at a time when cereal fields are often sprayed with insecticides to control pests). Shortages will therefore have a critical impact. Secondly, farmland, in the shape of grassy banks for nesting and cereal field margins for feeding, is their primary habitat. As a result, birds cannot compensate for insect shortages by shifting to other habitats for breeding or by moving with their broods to forage elsewhere. It might be expected that other bird species which share these characteristics will be similarly affected, as will birds which rely extensively on invertebrate food from farmland in winter. Species with less reliance on insects, or which have alternative habitats such as woodland or parkland available, should be less affected by the decline in farmland insects. The fact that nine of the eleven most severely declining bird species mentioned above (*Table 3.1*: grey partridge, skylark, song thrush (*Turdus philomelus*), spotted flycatcher (*Muscicapa striata*), corn bunting, lapwing, tree sparrow, bullfinch (*Pyrrhula pyrrhula*), reed bunting (*Emberiza schoeniclus*)) are wholly or partly insectivorous, but that three (stock dove (*Columba oenas*), wood pigeon (*Columba palumbus*), collared dove (*Streptopelia decaocto*)) of the four species that have shown the most obvious increases are seed and plant feeders, supports the possibility of a link between invertebrate and bird declines.

Other detailed studies have begun to examine the role of changes in insect abundance in population trends in particular species of conservation concern. Three examples, the corn bunting, the lapwing and the stone curlew, illustrate respectively a species affected by declines in summer insect food, a species affected by availability of winter invertebrate food, and a species where other factors appear to have been the major cause of the decline in numbers.

Corn bunting

This species breeds and feeds mainly on lowland farmland and feeds its nidicolous chicks on invertebrate food. Loss of winter stubbles, leading to winter mortality through starvation, is considered to have been very important in leading to the decline in this species (e.g. Donald, 1997; Shrubb, 1997), but there is also evidence that the change away from spring tillage is associated with reductions in summer food availability, which limits chick production (Aebischer & Ward, 1997). Corn buntings on the Sussex Downs feed their chicks primarily on moth and sawfly larvae, spiders and harvestmen, and grasshoppers (Brickle & Harper, 1999). There is direct evidence for declines in abundance of three of these invertebrate groups over the same area (Aebischer, 1991), and for a direct relationship between the density of moth and sawfly larvae and corn bunting breeding density (Ewald & Aebischer, 1999), suggesting that reductions in chick food may have been involved in the decline and could hamper population recovery in this area.

Lapwing

Lapwings breed in rough upland pastures and spring-cultivated lowland agricultural areas; the nidifugous chicks and the adult birds feed on surface and soil invertebrates. Loss of unimproved grassland in upland habitats leads to low breeding success through increased predation and direct agricultural disturbance, but invertebrate food is not limiting (Baines, 1990). In lowland areas, loss of the required mosaic of habitats (spring-tilled land for nesting, grassland for chick rearing) and changes in management of grassland appears to have led to the species' decline (Marchant *et al.*, 1990). The trend to autumn rather than spring tillage reduces the availability of winter food as many soil invertebrates are negatively affected by cultivation (Edwards & Lofty, 1975). In winter, lapwings on farms are found in those fields which have the highest invertebrate densities (Tucker, 1992). It is therefore possible that winter mortality of this species, at least, is affected by declines in the availability of invertebrate food.

Stone curlew

Stone curlews nest on open ground on sandy soils. Adults feed themselves and their chicks on invertebrates such as large beetles and earthworms. Suitable habitats are very short, semi-natural grassland and sparse spring-sown crops such as sugar beet. Habitat loss, caused by changes in crop rotation and in grazing (including the myxomatosis-induced crash in rabbit numbers, as well as changes in stocking practice), appears to have been the main cause of decline in this species (Green, 1993). Increasing predation on grassland with reduction in gamekeeping may also have been a factor (Campbell *et al.*, 1997). There is little evidence that food availability has affected the survival of this species, although Green (1988) suggested that the reduction in mixed farming may have been deleterious because this system provided the birds with the combination of tilled land for nesting and ley pastures rich in invertebrate prey.

Conclusions

It has not been the purpose of this review to maintain that all declines in the abundance of farmland birds are a result of decreasing availability of insect food on farmland. Intensification of farming in the latter half of the twentieth century has encompassed a complex set of changes that have led to habitat loss and habitat degradation. In particular, changes in patterns of crop rotation have largely removed the mosaic of different habitats necessary to support the nesting and feeding requirements of many bird species through the course of the year, and loss of breeding habitats and of winter stubbles to provide seeds and grain for overwintering birds have frequently been connected to species declines (O'Connor & Shrubb, 1986; Marchant *et al.*, 1990; Fuller *et al.*, 1995; Shrubb, 1997; Gillings & Fuller, 1998). However, the role of invertebrate food has tended to be neglected in the past, and current studies are now beginning to recognise its importance. There is good evidence of widespread declines in the abundance of many of the key arthropod groups in bird diets over at least the last forty years, as a result of both changes in habitat quality and increased pesticide use. It is also clear that many bird species are reliant on insects and other invertebrates as a main food source during at least the breeding season, and they would appear to be vulnerable to the loss of this food source. The changes will not easily be reversed, but unless action is taken to improve the value of modern farmland as a habitat for invertebrates, their decline, and that of the bird species that depend on them, is likely to continue.

References

Aebischer, N. J. (1990) Assessing pesticide effects on non-target invertebrates using long-term monitoring and time-series modelling. *Functional Ecology* **4**, 369–373.

Aebischer, N. J. (1991) Twenty years of monitoring invertebrates and weeds in cereal fields in Sussex. In: *The ecology of temperate cereal fields* (eds. L. G. Firbank, N. Carter, J. F. Darbyshire & G. R. Potts), pp. 305–331. Blackwell Scientific Publications, Oxford, UK.

Aebischer, N. J. (1995) The changing pattern of farming and wildlife in Sussex 1970/1994. *The Game Conservancy Review of 1994*, 78–80.

Aebischer, N. J. & Potts, G. R. (1994) Partridge *Perdix perdix*. In: *Birds in Europe: their conservation status. Birdlife Conservation No. 3* (eds. G. M. Tucker & M. F. Heath), pp. 220–221. Birdlife International, Cambridge, UK.

Aebischer, N. J. & Potts, G. R. (1998) Spatial changes in grey partridge (*Perdix perdix*) distribution in relation to 25 years of changing agriculture in Sussex, UK. *Gibier Faune Sauvage* **15**, 293–308.

Aebischer, N. J. & Ward, R. S. (1997) The distribution of corn buntings *Miliaria calandra* in Sussex in relation to crop type and invertebrate abundance. In: *The ecology and conservation of corn buntings* Miliaria Calandra. *UK Nature Conservation No. 13* (eds. P. F. Donald & N. J. Aebischer), pp. 124–138. Joint Nature Conservation Committee, Peterborough, UK.

Avery, M. I., Evans, A. D. & Campbell, L. H. (2003) Can pesticides cause reductions in bird populations? (*this volume*).

Baines, D. (1990) The roles of predation, food and agricultural practice on determining the breeding success of the Lapwing (*Vanellus vanellus*) on upland grasslands. *Journal of Animal Ecology* **59**, 915–929.

Barker, A. M., Brown, N. J. & Reynolds, C. J. M. (1999) Do host-plant requirements and mortality from soil cultivation determine the distribution of graminivorous sawflies on farmland? *Journal of Applied Ecology* **36**, 271–282.

Blank, T. H., Southwood, T. R. E. & Cross, D. J. (1967) The ecology of the partridge. I. Outline of population processes with particular reference to chick mortality and nest density. *Journal of Animal Ecology* **36**, 549–556.

Brickle, N. W. & Harper, D. G. C. (1999) Diet of nestling corn buntings *Miliaria calandra* in southern England examined by compositional analysis of faeces. *Bird Study* **46**, 319–329.

Campbell, L. H., Avery, M. I., Donald, P., Evans, A. D., Green, R. E. & Wilson, J. D. (1997) *Review of the indirect effects of pesticides on birds.* JNCC Report no. 227, JNCC, Peterborough, UK. 190 pp.

Chiverton, P. A. (1999) The benefits of unsprayed cereal crop margins to grey partridges *Perdix perdix* and pheasants *Phasianus colchicus* headlands in Sweden. *Wildlife Biology* **5**, 83–92.

Donald, P. (1997) The corn bunting *Miliaria calandra* in Britain: a review of current status, patterns of decline and possible causes. In: *The ecology and conservation of corn buntings* Miliaria calandra. *UK Nature Conservation, No. 13* (eds. P. F. Donald & N. J. Aebischer), pp. 11–26. Joint Nature Conservation Committee, Peterborough, UK.

Dover, J. W., Sotherton, N. W. & Gobbett, K. (1990) Reduced pesticide inputs on cereal field margins: the effects on butterfly abundance. *Ecological Entomology* **15**, 17–24.

Edwards, C. A. & Lofty, J. R. (1975) The influence of soil cultivation on soil animal populations. In: *Progress in soil zoology* (ed. J. Vanek), pp. 399–407. Junk, The Hague, The Netherlands.

Ewald, J. A. & Aebischer, N. J. (1999) *Pesticide use, avian food resources and bird densities in Sussex.* JNCC Report no. 296, JNCC, Peterborough, UK. 103 pp.

Ford, J., Chitty, H. & Middleton, A. D. (1938) The food of partridge chicks (*Perdix perdix*) in Great Britain. *Journal of Animal Ecology* **7**, 251–265.

Fuller, R. J., Gregory, R. D., Gibbons, D. W., Marchant, J. H., Wilson, J. D., Baillie, S. R. & Carter, N. (1995) Population declines and range contractions among lowland farmland birds in Britain. *Conservation Biology* **9**, 1425–1441.

Galbraith, H. (1989a) The diet of lapwing (*Vanellus vanellus*) chicks on Scottish farmland. *Ibis* **131**, 80–84.

Galbraith, H. (1989b) Arrival and habitat use by lapwings *Vanellus vanellus* in the early breeding season. *Ibis* **131**, 377–388.

Gillings, S. & Fuller, R. J. (1998) Changes in bird populations on sample lowland English farms in relation to loss of hedgerows and other non-crop habitats. *Oecologia* **116**, 120–127.

Green, R. E. (1984) The feeding ecology and survival of partridge chicks (*Alectoris rufa* and *Perdix perdix*) on arable farmland in East Anglia. *Journal of Applied Ecology* **21**, 817–830.

Green, R. E. (1988) Stone curlew conservation. *RSPB Conservation Review* **2**, 30–33.

Green, R. E. (1993) Stone curlew *Burhinus oedicnemus*. In: *The new atlas of breeding birds in Britain and Ireland* (eds. D. W. Gibbons, J. B. Reid & R. A. Chapman), pp. 160–161. Poyser, London, UK.

Heydemann, B. (1983) Die Beurteilung von Zielkonflikten zwischen Landwirtschaft, Landschaftspflege und Naturschutz aus Sicht der Landschaftspflege und des Naturschutzes. *Schriftenreihe für ländliche Sozialfragen* **88**, 51–78.

Holland, J. M. (2003) The impact of agriculture and some solutions for arthropods and birds (*this volume*).

Marchant, J. H., Hudson, R., Carte, S. C. & Whittington, P. (1990) *Population trends in British breeding birds*. British Trust for Ornithology, Tring, Herts., UK. 300 pp.

Moreby, S. J. (2003) Birds of lowland arable farmland: the importance and identification of invertebrate diversity in the diet of chicks (*this volume*).

O'Connor, R.J. & Shrubb, M. (1986) *Farming and birds*. Cambridge University Press, Cambridge, UK. 290 pp.

Potts, G. R. (1970) Recent changes in the farmland fauna with special reference to the decline of the grey partridge (*Perdix perdix*). *Bird Study* **17**, 145–166.

Potts, G. R. (1986) *The partridge: pesticides, predation and conservation*. Collins, London, UK. 274 pp.

Potts, G. R. (1997) Cereal farming, pesticides and grey partridges. In: *Farming and birds in Europe* (eds. D. J. Pain & M. W. Pienkowski), pp. 150–177. Academic Press, London, UK.

Potts, G. R. & Aebischer, N. J. (1991) Modelling the population dynamics of the grey partridge: conservation and management. In: *Bird population studies: their relevance to conservation and management* (eds. C. M. Perrins, J.-D. Lebreton & G. J. M. Hirons), pp. 373–390. Oxford University Press, Oxford, UK.

Rands, M. R. W. (1985) Pesticide use on cereals and the survival of grey partridge chicks: a field experiment. *Journal of Applied Ecology* **22**, 49–54.

Rands, M. R W. (1986) The survival of game bird (Galliformes) chicks in relation to pesticide use on cereals. *Ibis* **128**, 57–64.

Shrubb, M. (1997) Historical trends in British and Irish corn bunting *Miliaria calandra* populations – evidence for the effects of agricultural change. In: *The ecology and conservation of corn buntings* Miliaria calandra. *UK Nature Conservation, No. 13* (eds. P. F. Donald & N. J. Aebischer), pp. 27–41. Joint Nature Conservation Committee, Peterborough, UK.

Sotherton, N. W. (1998) Land use changes and the decline of farmland wildlife: an appraisal of the set-aside approach. *Biological Conservation* **83**, 259–268.

Southwood, T. R. E. & Cross, D. J. (1969) The ecology of the partridge. III Breeding success and the abundance of insects in natural habitats. *Journal of Animal Ecology* **13**, 497–509.

Tucker, G. M. (1992) Effects of agricultural practices on field use by invertebrate-feeding birds in winter. *Journal of Applied Ecology* **29**, 779–790.

Tucker, G. M. & Heath, M. F. (1994) *Birds in Europe: their conservation status. Birdlife Conservation Series No. 3*. Birdlife International, Cambridge, UK.

van Wingerden, W. K. R. E., van Kreveld, A. R. & Bongers, W. (1992) Analysis of species composition and abundance of grasshoppers (Orth., Acrididae) in natural and fertilized grasslands. *Journal of Applied Entomology* **113**, 138–152.

Vickerman, G. P. (1974) Some effects of grass weed control on the arthropod fauna of cereals. *Proceedings of the 12th British Weed Control Conference* **3**, 929–939.

Vickerman, G. P. & O'Bryan, M. K. (1979) Partridge chick diet analysis from crop dissection. *The Game Conservancy Review of 1978*, pp. 35–43. The Game Conservancy Trust, Fordingbridge, UK.

Woiwod, I. P. (1991) The ecological importance of long-term synoptic monitoring. In: *The ecology of temperate cereal fields* (eds. L. G. Firbank, N. Carter, J. F. Darbyshire & G. R. Potts), pp. 275–304. Blackwell Scientific Publications, Oxford, UK.

Woiwod, I. P. & Harrington, R. (1994) Flying in the face of change: the Rothamsted Insect Survey. In: *Long-term experiments in agricultural and ecological science* (eds. R. A. Leigh & A. E. Johnston), pp. 321–342. CAB International, Wallingford, UK.

4

The Impact of Agriculture and Some Solutions for Arthropods and Birds

JOHN M. HOLLAND

The Game Conservancy Trust, Fordingbridge, Hampshire, SP6 1EF, UK

Introduction

Since the 1950s, many forms of wildlife associated with agricultural ecosystems, and particularly cereals, have declined throughout Europe. This has coincided with the intensification of agriculture driven by the EU policy for European self-sufficiency in food production, food security, and maintenance of farm incomes. Most affected have been the birds (Tucker & Heath, 1994), especially the grey partridge (*Perdix perdix*) (Potts, 1986), butterflies (Pollard & Yates, 1993), beneficial insects (Aebischer, 1991), and annual arable wildflowers (Schumacher, 1987). Many of these organisms are connected through the food chain, and therefore declines at the base of the chain have implications for the higher organisms. The most intensively studied relationship has been between invertebrates and the grey partridge, where chick survival is closely dependent on invertebrate abundance (Potts, 1986). However, arthropods are also essential for the survival of many other farmland birds (Campbell *et al.*, 1997), and this is examined in this volume by Moreby (2003) and Barker (2003). This review will consider the evidence for these declines and the way in which agricultural practices may be changed to encourage beneficial arthropods, and thereby the birds which are dependent on them.

Factors influencing the decline in arthropods and birds on farmland

PESTICIDES AND NITROGENOUS FERTILISER

Use of pesticides has risen steadily, doubling every 10 years since the early 1950s (Campbell *et al.*, 1997), despite the increasing awareness of the potential ecological damage which they may cause. The development of more efficient active ingredients has reduced the tonnage applied, but their greater effectiveness, along with an increase in the area treated, has led to a consistent decrease in farmland biodiversity. The finger has most often pointed at insecticides as being the main culprit, and this is still true, but herbicides, and to a lesser extent fungicides, have also been implicated

Insect and Bird Interactions
© Intercept Ltd., PO Box 716, Andover, Hampshire, SP10 1YG, UK.

in the decline of many arthropods because they reduce food availability for herbivorous and saprophytic groups, respectively (Aebischer, 1991; Ewald & Aebischer, 1999). Now, the high cost of developing new compounds is influencing the type of products being developed, and broad-spectrum compounds, which have a greater range of uses and consequently affect a higher number of non-target species, are more likely to reach the market than selective compounds. The environmental profile of many pesticides marketed in first world countries has improved since those used in the 1950–1970s, which were directly toxic, causing death or breeding failures of birds. However, these earlier products (e.g. organochlorine insecticides) are still widely used in developing countries, even when banned, and as a consequence can be detected in species feeding in agricultural areas, for example raptors and neotropical migrants. The persistence of organochlorines in the soil has also led them to accumulate in higher organisms, to cause secondary poisoning. The herbicides and fungicides available in first world countries are now regarded as being largely non-toxic to birds and arthropods. Unfortunately, many of the insecticides (e.g. organophosphates and synthetic pyrethroids) are still extremely toxic to a wide range of beneficial arthropod species (Theiling & Croft, 1988), and some (organophosphates and carbamates) are directly toxic to birds, especially when used as seed dressings (Newton, 1998). Moreover, it is now recognised that pesticides can have substantial indirect effects by reducing food supplies for both birds and arthropods.

The broad-spectrum organophosphorus insecticides are still used on arable crops, primarily because they are cheap and have longer persistence than the more recently developed synthetic pyrethroids. The latter are currently the most widely used insecticides, but they are still toxic to those arthropods which are important in the diet of farmland birds (Sotherton, 1990; Moreby et al., 1997). Their use in summer is most damaging to farmland birds because this is when many are feeding their chicks, and thus require a high abundance of arthropods. In addition, insecticides may cause taste aversion in birds, which of course has the same impact as invertebrate mortality of the invertebrate food (Nicolaus & Lee, 1999).

With these concerns in mind, a number of large-scale studies of farming systems has been carried out over the past 20 years in the UK to evaluate the environmental impact of pesticide use. The Boxworth project (1981–1988) was the first of these, and revealed reductions of 50% in the numbers of predators and phytophagous insects where pesticide use was highest (Vickerman, 1992). Birds were also monitored, and some were exposed to organophosphorus insecticides (Hart et al., 1992), but there was no conclusive evidence of an overall effect of the pesticide regimes on the breeding performance of the most common bird species (Fletcher et al., 1992). However, by the time the project had been completed, national pesticide use had declined, and the project did not reflect the then current practice (Greig-Smith & Griffin, 1992). To confirm the findings from Boxworth, a further study, the SCARAB project (1990–1996), was initiated, but this time at more sites and reflecting the current level of pesticide inputs (Cooper, 1990). Long-term effects of pesticides were only found for Collembola in just one field which had a grass–wheat rotation and, consequently, annual usage of an organophosphorus insecticide (Frampton, 2002). Short-term effects also occurred, but not always, and only for certain arthropods. Other, recently completed farming systems studies throughout Europe have also examined the impact of pesticides in modern production. Some of these studies have recorded increases in beneficial arthropods where pesticide use was reduced (Holland *et al.*, 1994a), but some only found short-term reductions of a few species following insecticide application (Büchs *et al.*, 1997; Holland *et al.*, 1998). The scale of these projects, although replicated in several fields, may be inadequate to reflect long-term changes occurring across the landscape because re-invasion can occur following a damaging pesticide application (Duffield & Aebischer, 1994; Denholm *et al.*, 1998; Holland *et al.*, 2000).

The potential direct and indirect impact of pesticides across the farmed landscape is now gaining recognition, and this is being addressed mainly through theoretical investigations (Aebischer, 1990; Sherratt & Jepson, 1993; Trumper & Holt, 1998) because of the practical difficulty in monitoring arthropods adequately across large areas. The only monitoring project which is of sufficient scale and duration to detect changes within the landscape is that conducted by The Game Conservancy Trust in the Sussex study. In this study, arthropod abundance and diversity, crop type, and weed abundance have been monitored in approximately 100 cereal fields covering 62 km^2 during the third week in June every year since 1970. Long-term declines have occurred during the study for many arthropod and weed species (Aebischer, 1991), and these were related to a number of factors, including the use of insecticides and herbicides, and the increase in area sprayed (Ewald & Aebischer, 1999, 2000). The importance of these factors is considered in more detail by Barker (2003). Analysis of data from the Sussex study has also made the link between the use of more efficient herbicides, which reduced within-field weed diversity, and the decline of some invertebrate species. This would be expected, given that the crop is unpalatable to many phytophagous insects, and they must rely on weeds if they are to survive within the crop. Indeed, four of the five most important partridge chick-food insects are dependent on broad-leaved weeds (Chiverton & Sotherton, 1991; Sotherton, 1991; Sotherton & Moreby, 1992). Weed cover may also alter the microhabitat and provide a source of pollen and nectar, thus increasing the number of habitat niches available. Modern fungicides, although relatively harmless to most arthropods, may also decrease food availability for saprophytic species, and this may explain the decline in the Sussex study of *Tachyporus* species, a group of fungal-feeding staphylinid beetles (Aebischer, 1991).

It is not only arthropods within the crop which are at risk from pesticides, but also those in the field boundary because pesticide contamination may occur if it is applied inaccurately, or allowed to drift. Insecticide drift poses the greatest risk, as these pesticides are typically applied in a finer droplet spectrum than herbicides and can cause considerable mortality or sub-lethal effects, even at low doses (Cuthbertson, 1988; Longley *et al.*, 1997). If herbicides reach the flora at the base of hedges, they may cause sufficient damage that the composition of plant species changes, and thereby the phytophagous insect community also. Alternatively, at lower concentrations, they may prevent flowering and deprive pollinators of food (Boatman *et al.*, 1994a). Similarly, misplaced nitrogen fertiliser may also alter the nutrient balance and allow nitrophilous annual weeds to flourish, which outcompete the perennial hedgerow species (Boatman *et al.*, 1994b). The physical structure of the hedgerow flora is also important, especially for overwintering species, and changes to this by herbicides or fertilisers may alter its suitability. Indeed, the decline in farmland butterflies in the UK (Pollard & Yates, 1993) may, in part, be attributed to pesticides.

This has been demonstrated experimentally where crop margins were selectively sprayed. More annual weeds then survived in the crop, whilst the field boundary was protected from spray drift (Rands & Sotherton, 1986; Dover *et al.*, 1990).

HABITAT CHANGES

Other changes which have occurred as agriculture has become more intensive include the removal of field boundaries and other non-crop areas to increase field size and allow the operation of larger farm machinery (Chapman & Sheail, 1994). Such areas provide a rich resource of arthropods. The herbaceous flora associated with a hedge base (Pollard, 1968), the woody structure of the hedge (Joyce *et al.*, 1997; Maudsley *et al.*, 1997), and grassy strips (Kromp & Steinberger, 1992) are all known to harbour high numbers and diversity of phytophagous, pollinating, and predatory insects. The high diversity of flora found in hedgerows is able to support many species during the summer (Joyce, 1998), whilst the hedge base is used as an overwintering site for beetles and spiders (Luff, 1966; Pollard, 1968; Desender *et al.*, 1981; Desender, 1982; Sotherton, 1984; Bayram & Luff, 1993). Many of the species overwintering in the boundary will then migrate in spring into the crop, where they play an important role in the regulation of crop pests (Coombes & Sotherton, 1986; Jensen *et al.*, 1989; Thomas *et al.*, 1991). Even so, many species remain associated with the boundary throughout the year, and overall invertebrate abundance is usually greatest near to the boundary (Holland *et al.*, 1999a). Non-crop areas also provide a resource from which re-invasion may occur following spraying (Duffield & Aebischer, 1994; Holland *et al.*, 1999b, 2000). Without the availability of this habitat, the effects of pesticides may be exacerbated for the less mobile species.

One of the greatest changes which has occurred across the British countryside is the polarisation into either arable or livestock farming, both within farms and nationally. Thus, there has been a decline in permanent grassland and rotational grassland in the eastern counties, and a nation-wide reduction in the practice of undersowing, which was often used to establish grass within an arable rotation (Potts, 1997). Undersowing supported additional insect species (e.g. sawflies and, to a lesser extent, leaf beetles, and weevils) within a cereal crop, and its demise has been implicated in the decline of sawflies within fields (Potts, 1986). Grassland and the following arable crop also supported a range of invertebrate pests such as leatherjackets, which are an important winter food for many birds, and consequently specialist arable areas are less attractive. As a result of these changes, field boundaries and woodland are often the only uncultivated areas remaining within arable farms and, therefore, they have gained in importance. The conversion from mixed to arable farming has also meant that animal manures are not returned to the land. Farmyard manure has been shown to benefit the epigeal arthropod fauna not only by increasing detrivores and their predators, but also through the modification of soil structure and the creation of extra habitat niches (Purvis & Curry, 1984; Hance & Gregoire-Wibo, 1987; Humphreys & Mowat, 1994).

Another major change which has occurred within arable areas is the switch to predominantly winter cropping, entailing cultivation in the autumn and a loss of overwinter stubbles. As a consequence, only those arthropod species survive which can tolerate cultivation at this time, or are able to escape the field and find alternative areas in which to overwinter. Moreover, stubbles were previously important winter feeding areas for many species, such as corn bunting (*Miliaria calandra*) (Donald & Evans, 1994) and cirl bunting (*Emberiza cirlus*) (Evans & Smith, 1994), although it is seeds these birds were seeking at this time of the year. The availability of arthropod food for birds within crops may also have changed; modern crops which receive inorganic fertilisers rapidly produce a dense ground cover, which is disliked by skylarks (*Alauda arvensis*) and partridges. Organic crops may be more favoured because they usually have lower ground cover and crop density is variable, creating a greater range of crop structures and opportunities for feeding

SOIL CULTIVATION

The type and timing of cultivation may affect arthropods. Ploughing is the most widely used method, and may reduce overwintering arthropods through physical destruction, desiccation, depletion of food, or increased exposure to frost and predators. There is some evidence that arthropod size influences their vulnerability to cultivation. Thus, smaller species, e.g. *Bembidion* species (Carabidae), have been shown to favour areas cultivated by minimum tillage (Baguette & Hance, 1997), although the opposite was found by Kendall *et al*. (1995). Such effects may be a result of differing dispersal ability; those which can escape quickly from the field following ploughing may be better able to survive (Luff & Sanderson, 1992). The time of cultivation is also important because different life stages may differ in their vulnerability (Tischler, 1955; Skuhravý, 1958). However, both carabid adults and larvae have been shown to tolerate cultivation (Hance & Gregoire-Wibo, 1987; Baguette & Hance, 1997). Indeed, Luff & Sanderson (1992) were unable to identify an effect of ploughing on species present as either larvae or adults.

Rove beetles (Coleoptera: Staphylinidae) are frequently found in arable farmland, and may also be sensitive to soil cultivations, though less is known of their ecology (Zimmerman & Büchs, 1996). Spiders are usually the most abundant arthropods within arable fields and some groups, e.g. wolf spiders (Lycosidae), are relatively sensitive to disturbance. The effect of cultivation on other arthropod groups has rarely been investigated. Arthropods important in the diet of pheasant chicks (*Phasianus colchicus*) were assessed in no-tillage and ploughed or disked fields, but there were few differences (Basore *et al*., 1987). Cultivation has been shown to reduce numbers of adult sawflies emerging from overwintering sites in the soil by up to 50% (Barker *et al*., 1999). Sawflies overwinter as pupae in the soil, and so are particularly vulnerable to disturbance. They are an important prey item for the chicks of farmland birds such as the grey partridge, pheasant, corn bunting, reed bunting (*Emberiza schoeniclus*), and skylark.

What measures are available to reverse these declines?

An extensive review identified three invertebrate groups as associated with declining farmland bird species. These are the Coleoptera (especially Carabidae), Orthoptera, and larvae of Lepidoptera (Campbell *et al*., 1997). The latter two are dependent on permanent grass and perennial herbs for their survival, and therefore provision of suitable non-crop areas which are protected from pesticide drift are needed. Carabidae are widely dispersed throughout the agroecosystem, with significant populations within fields. However, the only experimental evidence correlating bird survival rates with arthropod abundance is for grey partridge (Potts, 1986), corn bunting (Brickle, 1997), and skylark (Poulsen *et al.*, 1998). The insect species important in the diet of grey partridge chicks have been identified, and these are the Araneae, Auchenorrhyncha, Heteroptera, caterpillars (Symphyta and Lepidoptera larvae), Chrysomelidae, Curculionidae, and Carabidae (Potts, 1986). However, the food and habitat preferences of arthropods taken by farmland birds are not fully understood. The most studied arthropods are the polyphagous predators, notably Araneae and Carabidae (Luff, 1987; Ekschmitt *et al.*, 1997), because of their perceived value in pest control. Within these two groups, there is considerable diversity of species with a wide range of ecologies and, therefore, it is likely that some species will be able to exploit whatever habitat is provided. Less is known of the phytophagous arthropod groups. *Polygonum* species, such as knotgrass, are the host for the chrysomelid beetle *Gastrophysa polygoni* (Sotherton, 1982), and the graminivorous sawfly larvae (*Dolerus* spp. and *Pachynematus* spp.) preferred the grasses *Lolium perenne* and *Festuca rubra* compared to wheat (Barker & Maczka, 1996; Barker *et al.*, 1999). While a number of heteropteran species will be found in fully sprayed crops, numbers are greatly reduced compared to less intensively sprayed fields supporting a weed flora (Moreby, 1995).

Few techniques have been developed specifically to encourage arthropods as food for farmland birds, the exceptions being the measures advocated by The Game Conservancy Trust, where provision of food for game birds is the motivation. These techniques include Conservation Headlands in which only a selective range of pesticides are used within the outer 6 m of cereal fields to ensure broad-leaved arable weeds and arthropods are protected (Sotherton, 1991). The management of set-aside in the UK has also changed since its initiation, allowing a more desirable range and structure of vegetation to be created, and thereby encouraging arthropods and farmland birds (Sotherton, 1998a). Measures to increase numbers of predatory arthropod species, and thereby improve natural pest control as part of an integrated pest management programme, will also indirectly benefit a wider range of arthropod species.

CONSERVATION HEADLANDS

As mentioned above, the insect species important in the diet of game bird chicks have been identified as the Araneae, Auchenorrhyncha, Heteroptera, caterpillars, Chrysomelidae, Cuculionidae, and Carabidae (Potts, 1986). Five of these groups are composed of predominantly phytophagous species. Because grey partridge chicks forage mostly along the edges of cereal fields (Green, 1984), Conservation Headlands were devised as a means of improving food availability along the field margins. The practice involves only selectively spraying the field margin, usually the 6 m width of the outermost section of a spray boom. Thus, no broad-spectrum insecticides are allowed during the summer, and herbicides must be sufficiently selective to allow the survival of some broad-leaved weeds. The technique has been extensively tested within the UK, Sweden, Holland, and Germany, and the benefits have been shown for grey partridges (Rands, 1985a,b), wild pheasants (Sotherton *et al.*, 1993), arthropods important as chick food (Chiverton & Sotherton, 1991; Moreby, 1997), predatory insects (De Snoo *et al.*, 1995; De Snoo & De Leeuw, 1996), flower-visiting insects (Rands & Sotherton, 1986; Dover *et al.*, 1990; De Snoo & De Leeuw, 1996), and arable weeds (Sotherton, 1991; De Snoo, 1994; Kleijn & van der Voort, 1997). Other farmland birds, such as ground-feeding passerines, which feed at field edges (O'Connor & Shrubb, 1986) may also utilise the extra provision of insect and plant food provided by this technique. Furthermore, the headland location, chosen because of the foraging habits of game birds, was also fortuitously the area where the seed bank is most abundant and diverse (Wilson, 1995). Consequently, Conservation Headlands, especially if nitrogen is also withheld, have also proved to benefit rare arable weeds (Wilson, 1994), many of which are in decline throughout western Europe (Wiggington, 1999).

Conservation Headlands extend the range of habitats found within farmland, and consequently the species which can survive. They also have the added benefit of acting as a buffer zone for the field boundary, preventing spray drift (Cuthbertson & Jepson, 1988). But they primarily benefit those species, particularly grey partridges, which feed at the field edge and not those, such as the skylark and lapwing (*Vanellus vanellus*), which prefer field centres. Indeed, skylarks, which feed on plants as well as insects at ground level, avoided Conservation Headlands, probably because of the extensive weed growth (De Snoo *et al.*, 1994).

Moreover, the headlands are perceived by farmers as being difficult to manage, whilst causing an agronomic and economic burden. Consequently, the uptake of Conservation Headlands, despite their success in encouraging a wide range of species, has been relatively small compared to the number of field edges available (0.3%), and has declined since the introduction of compulsory set-aside, even where grant support is available (Potts, 1997).

SET-ASIDE

When set-aside was introduced in 1988, it was thought it might provide an improved habitat for the farmland birds by allowing broad-leaved plants to proliferate, and thereby improve the availability of arthropods and seeds. Unfortunately, the management options were too limited and, where natural regeneration was relied upon, a dense sward of annual and perennial grasses developed after several years (Fisher *et al.*, 1992; Wilson, 1992) with an arthropod fauna typical of naturally regenerating land (Hopper & Doberski, 1992). Even if a cover of sown grasses were chosen, this also resulted in limited botanical diversity and an arthropod fauna adapted to grassland. Rotational set-aside was also introduced in 1992 and, following recommendations by conservationists, there is now sufficient flexibility in both schemes to facilitate the creation of desirable habitats which have the potential to benefit farmland biodiversity. How this may be best achieved is described by Sotherton (1998a). The abundance of arthropods has been monitored on rotational set-aside and, generally, arthropods are more abundant compared to conventional cereals, but this is dependent on the vegetation which establishes (Moreby & Aebischer, 1992; Moreby & Sotherton, 1995). Insects may be better provided for if a mixture (known as 'brood-rearing cover') of cereals, brassicas, and red clover is sown, creating a habitat similar to Conservation Headlands (Sotherton, 1998b). Pollinating insects may also be attracted if dicotyledonous flowering species are included (Cowgill *et al.*,

1993; Dramstad & Fry, 1995; Dover, 1996), but this is an expensive option. Other set-aside options available in the UK, e.g. winter cover, may also encourage arthropods, but this has yet to be demonstrated.

HERBACEOUS STRIPS

Higher densities of arthropods are normally found within and close to field boundaries (Sotherton, 1985; Lagerlöf & Wallin, 1993). Recognising this, Nentwig (1989) created weedy strips across wheat fields in order to encourage a more even spread of predatory arthropods. Nentwig (1989) showed that the weed strips contained higher numbers of beneficial arthropods, and increased their density in the adjacent wheat crop. Further studies revealed higher activity densities and diversity of Carabidae in the weed strip, with a decline in densities into the adjacent crop (Lys et al., 1994; Frank, 1997). Mark–recapture studies revealed that some species were positively attracted to the strips because of the rich food supply available (Lys & Nentwig, 1992), which increased beetle reproductive potential (Zangger et al., 1994). Similarly, other beneficial arthropods were also attracted to the weed strips, including Araneae (Frank & Nentwig, 1995) and pollinators (wild bees and butterflies) which were attracted by the additional source of nectar and pollen (Frank, 1998). Adult hoverflies were also attracted by the flowers within the weed strips (Frank, 1998), but they did not remain within the strip because aphid densities were not higher there than in the adjacent crop (Salveter, 1998).

The strips are also suitable as overwintering sites for Carabidae, Staphylinidae, and Araneae (Lys & Nentwig, 1994). Thus, they are suitable habitats for encouraging beneficial species and may act as foraging sites for farmland birds, but their benefits for pest control have not been demonstrated (Jmhasly & Nentwig, 1995). This limits their economic appeal to farmers and wider uptake.

Using herbaceous strips to provide flowers rich in pollen or nectar for predatory arthropods has also been tested (Hickman & Wratten, 1996; Holland & Thomas, 1996). Several plant species were examined for their attractiveness to hoverflies, and Phacelia tanacetifolia was the most promising (Macleod, 1992; Lovei et al., 1993). This plant also supported a variety of other arthropods, in particular Miridae (Holland & Thomas, 1996), which are consumed by farmland birds.

BEETLE BANKS

In the UK, Beetle Banks were developed to provide additional overwintering sites for predators and, in effect, to reduce field size, allowing more rapid within-field dispersal of predatory arthropods in the spring (Thomas et al., 1991). They also provide additional nesting and foraging areas in the summer, encouraging gramini-vorous arthropods such as sawfly and lepidopteran larvae (A. M. Barker & C. J. M. Reynolds, unpublished). The uptake of Beetle Banks has been restricted; because the pest control benefits have not been demonstrated, farmers did not like reducing field size, and grant aid for their establishment and maintenance was not directly available until the implementation of Countryside Stewardship. They may also now be incor-porated within non-rotational set-aside (one of the options within 'Flexible set-aside'), but the minimum width of any such strip is 20 m (Potts, 1997). Grassy strips next to field boundaries or established as blocks of non-rotational set-aside may also provide similar benefits, provided a mixture of tussocky grasses is allowed to establish.

FARM-SCALE HABITAT MANAGEMENT

Providing specific habitats for beneficial arthropods may benefit some species locally, but many arthropod species are highly mobile, as are farmland birds. Therefore, if we are to encourage insects throughout the farming landscape, we have to look at whole farm practices. In addition, the management guidelines become unwieldy if the dietary and habitat requirements of individual farmland birds are addressed, as has occurred for the grey partridge. There is, however, evidence that when a combination of the above farm practices are implemented along with mammalian and avian predator control, then farmland bird populations can be substantially increased (Boatman & Stoate, 2000). This conservation management approach increased non-game bird populations by 42% between 1992 and 1998, and overall passerines increased relative to the national trend. Indeed, some species showed really substantial changes, such as the song thrush (*Turdus philomelus*), which increased by 243%. Similarly, populations of game birds were considerably increased. Whether additional benefits could be achieved by adopting farm husbandry aimed at further increasing arthropod and seed availability within the cropped areas has yet to be resolved.

Crop production methods to encourage arthropods

SOIL MANAGEMENT

The type and timing of soil cultivation is most likely to affect those species which overwinter within the soil, such as many of the Coleoptera and Diptera, which form a large proportion of the diet of farmland birds (Moreby, 2003). The effects on arthropods of adopting minimal tillage compared to ploughing have been rather variable. There have been a number of studies which have specifically examined the effect of ploughing compared to non-inversion or conservation tillage on soil macro-arthropods. Carabid beetles are the most frequently studied organisms in such studies because many species reside all year round within arable fields, and they are sensitive to the type and timing of cultivations. However, changes in the total number of arthropods following non-inversion tillage have been inconsistent, with increases (Blumberg & Crossley, 1983; Brust *et al.*, 1985; House & Parmelee, 1985; Stinner & House, 1990; Brust, 1994; Kendall *et al.*, 1995), decreases (Barney & Pass, 1986; Cárcamo, 1995), and no effect being recorded (Tyler & Ellis, 1979; Basore *et al.*, 1987; Huusela-Vesitola, 1996). Individual species may vary in their response, depending on their species-specific characteristics (Tyler & Ellis, 1979; Barney & Pass, 1986; Hance & Gregoire-Wibo, 1987; Weiss *et al.*, 1990; Clark *et al.*, 1993, 1997; Cárcamo, 1995; Kendall *et al.*, 1995). Results were not always consistent between sites (Hance *et al.*, 1990; Weiss *et al.*, 1990), and interactions often occurred with the cropping system (Clark *et al.*, 1997), the latter often exerting a greater effect than the tillage system (Weiss *et al.*, 1990). In a review of the literature, Holland & Luff (2000) found no clear benefit of minimum tillage for carabids, but rather a species-specific preference.

Some of the inconsistencies reported above may be because of inter-guild predation. Moreover, effects may not have been detected previously because in most studies, arthropod numbers are monitored well after emergence, when populations have had time to redistribute and predate each other. Detailed long-term studies to evaluate arthropod emergence are needed if minimal tillage is to be investigated unambiguously. There is evidence that emergence of sawflies is greater when the soil is left uncultivated until emergence has been completed (Barker *et al.*, 1999), and this was thought to be the reason why sawfly densities were high from grass leys established by undersowing spring cereals (Potts & Vickerman, 1974; Vickerman, 1978).

The application of organic manure also encourages arthropods through a richer supply of saprophagous mesofauna as a source of alternative prey (Purvis & Curry, 1984). Its use in organic fields was thought to be responsible for the higher carabid activity compared to conventional crops where mineral fertiliser was applied (Hokkanen & Holopainen, 1986). Whether the application of sewage slurry or treated sewage can have similar benefits has not been investigated, although compost generated from organic household rubbish did not increase the emergence of Carabidae and Araneae in the short term (Idinger *et al.*, 1996).

PESTICIDE INPUTS

Crop scouting for pests, use of economic thresholds, and pest resistant or tolerant crops are the first tools to protect beneficial arthropods from insecticide applications. If chemical intervention is required, then there are a number of measures which can be taken to reduce its impact. These were primarily developed for non-target predatory species, although other arthropod species may also benefit.

The first approach is to utilise the physiological selectivity of some insecticides by choosing one with the most restricted species-toxicity profile. Pirimicarb, a carbamate insecticide, is one of the few products available which is truly selective; it is marketed as an aphicide. It has been shown to be safe for many beneficial arthropods, including carabids and lycosids (Brown *et al.*, 1988; Dinter & Poehling, 1995; Cilgi *et al.*, 1996), sawfly larvae (Sotherton, 1990) and Heteroptera (Moreby *et al.*, 1997). The synthetic pyrethroids also vary, to some extent, in their spectrum of toxicity. Tau-fluvalinate was found to be the least toxic of the pyrethroids to insects important in the diet of farmland birds. Even greater selectivity may be achieved if lower dose rates are used, thereby exploiting differences in the dose response curves between pest and non-target arthropods. This is possible with pirimicarb, whilst still maintaining a dose rate adequate for aphid control, but is less achievable with the other insecticides (dimethoate, deltamethrin) commonly used in summer (Poehling, 1989; Mann *et al.*, 1991; Cilgi *et al.*, 1996). Some more selective inorganic insecticides, insect growth regulators, and products based on naturally occurring toxic organisms (microbials) have been developed, but fewer such products are now being developed because of the high cost of development and registration, combined with limited potential markets; as a consequence, they are not widely used (Theiling & Croft, 1988).

The second approach is to accept broad-spectrum insecticides and to employ ecological selectivity. This encompasses a range of techniques to reduce the exposure of the beneficial arthropods in space and/or time. Selectivity in space may be achieved by restricting the distribution of the insecticide to the areas of the crop where the pest is active. This may be most easily achieved for sucking insect pests where systemic insecticides are available. For example, in cereal crops a new systemic insecticide, imidacloprid, applied as a seed treatment, should eliminate the need for an insecticide spray in the autumn to control cereal aphids which may transmit barley yellow dwarf virus. However, because few arthropods are active at this time, the benefits for insectivorous birds will be limited. Better targeting of pests feeding on the upper parts of the canopy could also be achieved using spray application equipment which produces a narrower size range of droplets and would allow dose rate reductions (Holland *et al.*, 1997). Maximising deposition on the upper canopy reduces wastage and contamination of the ground, although it may prove more harmful to foliage-feeding and climbing predators. A hazard index was developed to assess the potential risk of an application of deltamethrin to a selection of beneficial arthropod species commonly found in cereal crops (Wiles & Jepson, 1994), but this did not take into account the distributions of insects within the canopy, or the difference in pesticide distribution through the canopy as a result of different applications systems. When these factors were taken into account, the overall risk was lower, and ground-active species were less susceptible when spray was better targeted at the upper part of the crop canopy (Alford *et al.*, 1998). The improvements in spraying technology also reduce the likelihood of drift because modern nozzles now produce fewer of the finer droplets most prone to wind dispersal, while air assistance can also be used to ensure that the spray is driven into the crop before drift can occur (Hislop, 1991). The introduction of compulsory unsprayed strips (buffer zones) next to vulnerable habitats may also provide further protection (Croxford, 1998).

Birds and arthropods may forage across several fields, and thus their chances of survival will be improved if some unsprayed refuges remain. Consequently, the spatial dynamics of insecticide application are important both within fields and at the landscape scale. The first step towards exploiting this is to avoid planting crops of the same type in adjacent fields, and thereby the need to treat large areas when an insecticide is needed. By creating a mosaic of crops, a greater range of habitats will also be available. This is especially important for breeding birds as it reduces the distance over which they need to forage. The adverse effects on breeding birds of wide-scale applications of broad-spectrum insecticides has been confirmed for finches (Simpson, 1984) and grey partridge (Potts, 1977, 1997).

Within annual crops, pests are unlikely to be homogeneously distributed because they must re-invade each year, and populations will spread from the sites of initial infestation. For example, cereal aphids were highest around the edge of a cereal field (Winder *et al.*, 1998). Similarly, seed weevil (*Ceutorhynchus assimilis*) and pollen beetle (*Meligethes* spp.) populations develop from the edges. Patch spraying may therefore be possible, although this relies on crop scouting because, at present, no automated systems are available to detect insect pests. The use of buffer zones to protect water courses from spray drift and Conservation Headlands also leads to a form of patch spraying, as they protect the habitat within the buffer (Boatman, 1998). As most re-invasion following an insecticide application occurs from the margins (Duffield & Aebischer, 1994), leaving the outer edge of the crop unsprayed will also help protect insects within the margins because spray drift will be reduced (Cuthbertson, 1988). Consequently, this resource of insects will be able to re-invade the crop. This was demonstrated when insect distributions were monitored across a whole field, with and without unsprayed margins (Holland et al., 1999b, 2000).

For those bird species requiring insects to rear their young, the timing of insecticide applications is crucial, because the birds only have a limited time available in which to gather sufficient insects. Avoiding insecticide applications during this period is, therefore, the priority. If an application is unavoidable, then the impact over time may be reduced by using chemicals with relatively short persistence, e.g. pirimicarb.

These principles can be applied broadly to herbicide use, the aim being to achieve sufficient weed control without compromising yield or harvesting efficiency, but allowing some weeds to survive as a food source within the crop. Mechanical weeding, which is used in organic production and integrated crop management, apart from the obvious risk to ground-nesting birds, may disturb the soil surface, creating harmful dust, and cause mechanical injury to arthropods. However, no deleterious effects were detected upon adult Carabidae, Staphylinidae, Araneae, or chick-food insects using mechanical weeding (J. M. Holland, *unpublished*).

The ecological impact of growing genetically modified (GM) crops resistant to broad-spectrum herbicides is at present unknown, although it is currently being evaluated in the UK. In crops drilled and herbicide treated in the autumn, weed numbers and diversity will be low at this time, but may allow more weeds to be tolerated in the spring. Furthermore, GM crops could lead to lower herbicide inputs because farmers may be more tolerant of weeds, especially grasses, in other parts of the rotation, knowing that effective control is available. Insect resistant GM crops will eliminate the need for insecticides, but may upset the ecosystem by depriving predatory species of crop-feeding prey. The scale at which GM crops are deployed across the landscape is likely to influence strongly their overall impact, and this must be considered in any appraisals.

CROP TYPE

The type of crop can also influence arthropod abundance and availability to farmland birds. A number of studies have revealed that the carabid fauna differs with crop type (Booij & Noorlander, 1992; Holland et al., 1994a). It is likely that this is a result of the husbandry practices associated with a particular crop, and especially type and timing of soil cultivation, rather than movement and active selection, although some species may prefer the microclimate provided by a particular crop. No one beneficial species has been linked with a particular crop plant, but there are some associations. The greatest differences occur between winter-sown crops (cereals and oilseed rape) and spring-sown root crops (potatoes, sugar beet, maize) (Kabacik-Wasylik, 1975). Spring root crops usually had lower abundance and diversity of beetles (Booij, 1994; Holland et al., 1994b) and spiders (Holland et al., 1998). It is probably the time of cultivation associated with a particular crop which is most influential, rather than factors such as microclimate and crop structure (Hance et al., 1990; Baguette & Hance, 1997). The microclimates provided by root and cereal crops, however, differs considerably, being much drier and warmer in cereals. Such a habitat is favoured by xerophilic or euryhygric species with a preference for warmth, as are most field-inhabiting carabid species (Thiele, 1977). Root crops provide an environment which is darker, damper, and colder.

Selection of crop type is not always an option for farmers, but providing a variety of crops and avoiding concentrations of one crop type will provide a range of different habitats to be exploited by birds and arthropods. This may be applied at the within-field scale by intercropping, where two or more crops are grown simultaneously on the same land area. Undersowing, in which a grass/clover mixture is established under a cereal crop prior to a ley, is one example; this was once practised on 25% of the European arable area, although now it is confined to mixed farms, which are becoming rarer. Arthropod diversity and density, especially Coleoptera, have been found to be greater within an undersown cereal crop and, because the ground is not cultivated after harvest, soil-dwelling organisms are better able to survive into the following year. Intercropping has not been applied to arable crops because of the potential yield loss due to crop competition between the main crop and intercrop, the availability of cheap agrochemicals, and management difficulties (Theunissen, 1997), although it has been shown to encourage arthropods in vegetable crops (McKinlay & McCreath, 1995; Booij *et al.*, 1997).

FARMING SYSTEMS APPROACH

A number of different farming systems have been advocated for their environmental benefits; indeed, arthropods are frequently monitored as bio-indicators. Of the various systems, organic farming is often regarded as having the greatest benefits. However, because of the relatively small area under organic management (0–8.4% in EU countries), its impact across the landscape can only be small in comparison to, for example, set-aside, although the area under organic farming is increasing in the EU. Interpreting its relative value is also confounded by the difference in farm structure between organic and conventional arable farming. Organic farms are usually mixed, produce different crops, and have smaller fields and a greater proportion of field boundaries. A number of studies have monitored arthropods important for farmland birds. Moreby *et al.* (1994) found there was no difference overall in total chick food. There were higher densities of Curculionidae, Araneae, Collembola, Auchenorrhyncha (plant hoppers), and Tenthredinidae larvae in the organic fields, but more crop pests and their predators in the conventional fields. Sampling was restricted to the field edge, where the differences were not so great as in the mid-field (Reddersen, 1997). The differences found were attributed to the greater density and diversity of weeds in the organic fields (Moreby & Sotherton, 1997; Reddersen, 1997), and Carabidae, especially seed-eating species from the genera *Amara* and *Harpalus*, were similarly encouraged (Kromp, 1989). In the UK, an intensive study of birds and insects (Anon., 1995) found for organic fields an increase in some insects (Curculionidae, small Staphylinidae, some Carabidae) and many farmland birds, with higher breeding success of yellowhammer (*Emberiza citrinella*) and skylark. Several other studies throughout western Europe have revealed higher farmland bird densities on organic, compared to conventional farms. The differences could not be attributed to reduction of agrochemicals, but were related to an increase in the application of organic manure, a more variable crop structure, which created greater habitat diversity, and smaller fields with wider boundaries. Organic farming is unlikely to replace conventional farming because, at present, profitability relies on achieving a premium for the produce. Integrated farming may, however, be more suitable for a greater number of farmers, and is seen as a way to improve the sustainability of arable production throughout Europe. This approach uses good husbandry practices to reduce the need for agrochemicals, whilst caring for the environment. Integrated farming incorporates a number of practices which, as discussed, may affect invertebrates and birds, and these include:

- A more diverse crop rotation.
- Closer monitoring and care in selection of agrochemicals whilst utilising resistant varieties to reduce the frequency and dose rate of pesticide.
- Nutrient management to optimise fertiliser inputs and increase organic content through incorporation of organic manures and crop residues.
- Soil management based on reduced cultivation intensity to retain soil structure and fertility, increase biodiversity, and prevent soil erosion.
- Protection and provision of non-crop areas (minimum of 5% recommended by IOBC, 1997)

Many projects throughout Europe have examined conventional and integrated farming (Holland *et al.*, 1994). Birds are rarely monitored, as the scale of experimental studies is usually too small, but invertebrates are often used as bio-indicators, although the emphasis is usually on predatory groups (Coleoptera and Araneae) which contribute to pest control. Overall, results have been variable, with some projects reporting no difference between integrated and conventional farming (Booij & Noorlander, 1992; Winstone *et al.*, 1996; Holland *et al.*, 1998), whilst others have reported greater numbers in the integrated system for some groups (El Titi, 1991; Büchs *et al.*, 1997). Often, differences were greater between crops, fields, geographic location and year, than between the farming systems (Booij & Noorlander, 1992; Holland *et al.*, 1998). When results for Carabidae alone were compared for integrated, organic, and biodynamic farming systems, the latter two encouraged the most number of species, while the number of species favoured by the integrated system was no greater than those favoured by conventional farming (Holland & Luff, 2000). There may be a number of explanations why integrated farming has not revealed many differences. Firstly, the effect is often dependent on the arthropod's life cycle; therefore, individual species must be examined, otherwise the effects may be masked within broad taxonomic groups (Büchs *et al.*, 1997). More practically, the temporal scale and spatial scale have usually been insufficient to prevent movement between systems. The differences in management practices implemented between systems may not be great enough to affect arthropods, especially if arthropods have already been reduced to such low numbers that recovery could not occur within the time scale of the studies. Finally, the sampling methods may not be sensitive enough to detect changes, or the species/families selected for monitoring may be insufficiently responsive. Non-crop areas are known to be important, but management of these has not usually been a component of the integrated studies because of scale limitations, although where such areas have been enhanced, arthropods were encouraged within fields (El Titi, 1991). Population models indicate that, where the proportion of non-crop habitats are increased within a landscape, this increases overall diversity of arthropods (Bhar and Fahrig, 1998). Evidence to support this was found by Ryszkowski *et al.* (1993) because a greater biomass and species diversity of above-ground insects were found in landscapes of greater complexity, compared to more uniform areas. It is also considered that field size and type of farming, in addition to the landscape structure, influence the overall biodiversity (Gliessman, 1990). This was tested in Poland by comparing the survival of grey partridge chicks in areas with small fields to that in areas with large fields. Although the abundance of one of the key food items (Heteroptera) was higher in the small arable fields, no difference in brood size was found between the small and large fields (Panek, 1997). At one site, brood size increased in the small fields with the amount of permanent cover.

Conclusions

There are many ways in which the number and diversity of arthropods could be increased in arable farmland to help the recovery of farmland birds, but they all require considerable management, and often entail some loss of farm income. Such practices are unlikely to be widely adopted unless there is: financial support or a statutory requirement, such as in the proposals of Potts (1997) to help revive grey partridge populations, or there is personal commitment by the landowner with, for example, shooting or specific conservation interests, or there is an agronomic and/or economic reward, as provided by improved pest control.

References

Aebischer, N. J. (1990) Assessing pesticide effects on non-target invertebrates using long-term monitoring and time-series modelling. *Journal of Functional Ecology* **4**, 369–373.

Aebischer, N. J. (1991) Twenty years of monitoring invertebrates and weeds in cereal fields in Sussex. In: *The ecology of temperate cereal fields* (eds. L. G. Firbank, N. Carter, J. F. Darbyshire & G. R. Potts), pp. 305–331. Blackwell Scientific Publications, Oxford, UK.

Alford, J., Miller, P. H. C., Goulson, D. & Holland, J. M. (1998) Predicting susceptibility of non-target species to different insecticide applications in winter wheat. *Proceedings of the Brighton Crop Protection Conference, Pests and Diseases, 1998* **2**, 599–606.

Anon. (1995) *The effect of organic farming regimes on breeding and winter bird populations. Parts I–IV.* BTO Research Report no. 154, Thetford, Norfolk, UK. 175 pp.

Baguette, M. & Hance, T. (1997) Carabid beetles and agricultural practices: influence of soil ploughing. *Biological Agriculture and Horticulture* **15**, 185–190.

Barker, A. M. (2003) Insects as food for farmland birds – is there a problem? (*this volume*).

Barker, A. M. & Maczka, C. J. M. (1996) The relationships between host selection and subsequent larval performance in three free-living graminivorous sawflies. *Ecological Entomology* **21**, 317–327.

Barker, A. M., Brown, N. J. & Reynolds, C. J. M. (1999) Do host-plant requirements and mortality from soil cultivation determine the distribution of graminivorous sawflies on farmland? *Journal of Applied Ecology* **36**, 271–282.

Barney, R. J. & Pass, B. C. (1986) Ground beetle (Coleoptera: Carabidae) populations in Kentucky alfalfa and influence of tillage. *Journal of Economic Entomology* **79**, 511–517.

Basore, N. S., Best, L. B. & Wooley, J. B. (1987) Arthropod availability to pheasant broods in no-tillage fields. *Wildlife Society Bulletin* **15**, 229–233.

Bayram, A. & Luff, M. L. (1993) Winter abundance and diversity of lycosids (Lycosidae, Araneae) and other spiders in grass tussocks in a field margin. *Pedobiologia* **37**, 357–364.

Bhar, R. & Fahrig, L. (1998) Local vs. landscape effects of woody field borders as barriers to crop pest movement. *Conservation Ecology* [online] **2** (2), 3.

Blumberg, A.Y. & Crossley, D. A. (1983) Comparison of soil surface arthropod populations in conventional tillage, no-tillage and old field systems. *Agro-Ecosystems* **8**, 247–253.

Boatman, N. D. (1998) The value of buffer zones for the conservation of biodiversity. *Proceeedings of the Brighton Crop Protection Conference, Pests and Diseases, 1998* **3**, 939–950.

Boatman, N. D. & Stoate, C. (2000) Integrating biodiversity conservation into arable agriculture. *Aspects of Applied Biology* **62**, 21–30.

Boatman, N. D., Blake, K. A., Aebischer, N. J. & Sotherton, N. W. (1994a) Factors affecting the herbaceous flora of hedgerows on farms and its value as wildlife habitat. In: *Hedgerow management and nature conservation* (eds. T. A. Watt & G. P. Buckley), pp. 33–46. Wye College Press, Wye, UK.

Boatman, N. D., Theaker, A. J., Froud-Williams, R. J. & Rew, L. J. (1994b) The impact of nitrogen fertiliser on field margin flora. In: *Field margins – integrating agriculture and conservation* (ed. N. D. Boatman), pp. 209–214. British Crop Protection Council, Farnham, UK.

Booij, C. J. H. & Noorlander, J. (1992) The impact of integrated farming on carabid beetles. *Proceedings of Experimental and Applied Entomology* **2**, 16–21.

Booij, C. J., Noorlander, J. & Theunissen, J. (1997) Intercropping cabbage with clover: effects on ground beetles. *Biological Agriculture and Horticulture* **15**, 261–268.

Booij, K. (1994) Diversity patterns in carabid assemblages in relation to crops and farming systems. In: *Carabid beetles: ecology and evolution* (ed. K. Desender), pp. 425–431. Kluwer, Dortrecht, The Netherlands.

Brickle, N. (1997) The use of game cover and game feeders by songbirds in winter. *Proceedings of the Brighton Crop Protection Conference, Weeds, 1997* **3**, 1185–1190.

Brown, R. A., McMullin, L. C., Jackson, D., Ryan, J. & Coulson, J. M. (1988) Beneficial arthropod toxicity assessments with three insecticides in laboratory, semi-field and field studies. *Proceedings of the Brighton Crop Protection Coference, Pests and Diseases, 1988* **1**, 527–534.

Brust, G. E. (1994) Natural enemies in straw-mulch reduce Colorado potato beetle populations and damage in potatoes. *Biological Control* **4**, 163–169.

Brust, G. E., Stinner, B. R. & McCartney, D. A. (1985) Tillage and soil insecticide effects on predator–black cutworm (Lepidoptera: Noctuidae) interactions in corn agroecosystems. *Journal of Economic Entomology* **78**, 1389–1392.

Büchs, W., Harenberg, A. & Zimmermann, J. (1997) The invertebrate ecology of farmland as a mirror of the intensity of the impact of man? – an approach to interpreting results of field experiments carried out in different crop management intensities of a sugar beet and an oilseed rape rotation including set-aside. *Biological Agriculture and Horticulture* **15**, 83–108.

Campbell, L. H., Avery, M. I., Donald, P., Evans, A. D., Green, R. E. & Wilson, J. D. (1997) *A review of the indirect effects of pesticides on birds.* JNCC Report, no. 227. Joint Nature Conservation Committee, Peterborough, UK. 148 pp.

Cárcamo, H. A. (1995) Effect of tillage on ground beetles (Coleoptera: Carabidae): a farm-scale study in central Alberta. *Canadian Entomologist* **127**, 631–639.

Chapman, J. & Sheail, J. (1994) Field margins – an historical perspective. In: *Field margins – integrating agriculture and conservation* (ed. N. D. Boatman), pp. 3–12. British Crop Protection Council, Farnham, UK.

Chiverton, P. A. & Sotherton, N. W. (1991) The effects on beneficial arthropods of the exclusion of herbicides from cereal crop edges. *Journal of Applied Ecology* **28**, 1027–1039.

Cilgi, T., Wratten, S. D., Robertson, J. L., Turner, D. E., Holland, J. M. & Frampton, G. K. (1996) Residual toxicities of three insecticides to four species (Coleoptera: Carabidae) of arthropod predators. *Canadian Entomologist* **128**, 1115–1124.

Clark, M. S., Luna, J. M., Stone, N. D. & Youngman, R. R. (1993) Habitat preferences of generalist predators in reduced-tillage corn. *Journal of Entomological Science* **28**, 404–416.

Clark, M. S., Gage, S. H. & Spence, J. R. (1997) Habitats and management associated with common ground beetles (Coleoptera: Carabidae) in a Michigan agricultural landscape. *Environmental Entomology* **26**, 519–527.

Coombes, D. S. & Sotherton, N. W. (1986) The dispersal and distribution of predatory Coleoptera in cereals. *Annals of Applied Biology* **108**, 461–474.

Cooper, D A. (1990) Development of an experimental programme to pursue the results of the Boxworth project. *Proceedings of the Brighton Crop Protection Conference, Pests and Diseases, 1998* **1**, 153–162.

Cowgill, S. E., Wratten, S. D. & Sotherton, N. W. (1993) The selective use of floral resources by the hoverfly *Episyrphus balteatus* (Diptera: Syrphidae) on farmland. *Annals of Applied Biology* **122**, 499–515.

Croxford, A. C. (1998) Buffer zones to protect the aquatic environment. *Proceedings of the Brighton Crop Protection Conference, Pests and Diseases, 1998* **3**, 923–930.

Cuthbertson, P. (1988) The pattern and level of pesticide drift into conservation and fully sprayed arable crop headlands. *Aspects of Applied Biology* **17**, 273–275.

Cuthbertson, P. S. & Jepson, P. C. (1988) Reducing pesticide drift into the hedgerow by the inclusion of an unsprayed field margin. *Proceeedings of the Brighton Crop Protection Conference, Pests and Diseases, 1988* **2**, 747–751.

Denholm, I., Birnie, L. C ., Kennedy, P. J., Shaw, K. E., Perry, J. N. & Powell, W. (1998) The complementary roles of laboratory and field testing in ecotoxicological risk assessment. *Proceedings of the Brighton Crop Protection Conference, Pests and Diseases, 1998* **1**, 585–588.

Desender, K. (1982) Ecological and faunal studies on Coleoptera in agricultural land. II. Hibernation of Carabidae in agro-ecosystems. *Pedobiologia* **23**, 295–303.

Desender, K., Maelfait, J.-P., D'hulster, M. & Vanhercke, L. (1981) Ecological and faunal studies on Coleoptera in agricultural land. I. Seasonal occurrence of Carabidae in the grassy edge of pasture. *Pedobiologia* **22**, 379–284.

De Snoo, G. R. (1994) Unsprayed field margins: implications for environment, biodiversity and agricultural practice. PhD thesis, Rijksuniversiteit Leiden, The Netherlands.

De Snoo, G. R. & De Leeuw, J. (1996) Non-target insects in unsprayed cereal edges and aphid dispersal to the adjacent crop. *Journal of Applied Entomology* **120**, 501–504.

De Snoo, G. R., Dobbelstein, R. T. J. M. & Koelewijm, S. (1994) Effects of unsprayed crop edges on farmland birds. In: *Field margins – integrating agriculture and conservation* (ed. N. D. Boatman), pp. 221–226. British Crop Protection Conference Monograph No. 58, Farnham, UK.

De Snoo, G. R., van der Poll, R. J. & De Leeuw, J. (1995) Carabids in sprayed and unsprayed crop edges of winter wheat, sugar beet and potatoes. *Acta Jutlandica* **70**, 199–211.

Dinter, A. & Poehling, H. M. (1995) Side-effects of insecticides on two erigonid species. *Entomologia Experimentalis et Applicata* **74**, 151–163.

Donald, P. F. & Evans, A. D. (1994) Habitat selection by corn buntings *Miliaria calandra* in winter. *Bird Study* **41**, 199–210.

Dover, J. W. (1996) Factors affecting the distribution of satyrid butterflies on arable farmland. *Journal of Applied Ecology* **33**, 723–734.

Dover, J. W., Sotherton, N. W. & Gobbett, K. (1990) Reduced pesticide inputs on cereal field margins: the effects on butterfly abundance. *Ecological Entomology* **15**, 17–24.

Dramstad, W. & Fry, G. (1995) Foraging activity of bumblebees (*Bombus*) in relation to flower resources on arable land. *Agriculture, Ecosystems and Environment* **53**, 123–135.

Duffield, S. J. & Aebischer, N. J. (1994) The effect of spatial scale of treatment with dimethoate on invertebrate population recovery in winter wheat. *Journal of Applied Ecology* **31**, 263–281.

Ekschmitt, K., Wolters, V. & Weber, M. (1997) Spiders, carabids, and staphylinids: The ecological potential of predatory macroarthropods. In: *Fauna in soil ecosystems: recycling processes, nutrient fluxes and agricultural production* (ed. G. Benckiser), pp. 307–361. Marcel Dekker, New York, USA.

El Titi, A. (1991). The Lautenbach project 1978–89: Integrated wheat production on a commercial arable farm, south-west Germany. In: *The ecology of temperate cereal fields* (eds. L. G. Firbank, N. Carter, J. F. Darbyshire & G. R. Potts), pp. 399–411. Blackwell Scientific Publications, Oxford, UK.

Evans, A .D. & Smith, K. W. (1994) Habitat selection of cirl buntings *Emberiza cirlus* wintering in Britain. *Bird Study* **41**, 81–87.

Ewald, J. A. & Aebischer, N. J. (1999) *Avian food, birds and pesticides*. JNCC Report no. 296. JNCC, Peterborough, UK. 103 pp.

Ewald, J. A. & Aebischer, N. J. (2000) Trends in pesticide use and efficacy during 26 years of changing agriculture in southern England. *Environmental Monitoring and Assessment* **64**, 493–529.

Fisher, N. M., Dyson, P. W., Winham, J. M., Davies, D. H. K. & Lee, K. (1992) A botanical survey of set-aside in Scotland. In: *Set-aside* (ed. J. Clarke), pp. 67–72. British Crop Proetection Council, Farnham, UK.

Fletcher, M. R., Jones, S. A., Greig-Smith, P. W., Hardy, A. R. & Hart, A. M. D. (1992) Population density and breeding success of birds. In: *Pesticides, cereal farming and the environment: the Boxworth Project* (eds. P. W. Greig-Smith, G. Frampton & A. R. Hardy), pp. 160–174. HMSO, London, UK.

Frampton, G. K. (2002) Long-term impacts of an organophosphate-based regime of pesticides on field and field-edge collembola communities. *Pest Management Science* **58**, 991–1001.

Frank, T. (1997) Species diversity of ground beetles (Carabidae) in sown weed strips and adjacent fields. *Biological Agriculture and Horticulture* **15**, 297–310.

Frank, T. (1998) Attractiveness of sown weed strips on hoverflies (Syrphidae, Diptera), butterflies (Rhopalocera, Lepidoptera), wild bees (Apoidea, Hymenoptera) and thread-waisted wasps (Sphecidae, Hymenoptera). *Mitteilungen der Schweizerischen Entomologishen Gesellschaft* **71**, 11–20.

Frank, T. & Nentwig, W. (1995) Ground-dwelling spiders (Araneae) in sown weed strips and adjacent fields. *Acta Oecologica* **16**, 179–193.

Gliessman, S. R. (ed.) (1990) *Agroecology*. Springer, New York, USA. 380 pp.

Green, R. E. (1984) The feeding ecology and survival of partridge chicks (*Alectoris rufa* and *Perdix perdix*) on arable farmland in East Anglia. *Journal of Applied Ecology* **21**, 817–830.

Greig-Smith, P. W. & Griffin, M. J. (1992) Summary and recommendations. In: *Pesticides, cereal farming and the environment: the Boxworth Project* (eds. P. W. Greig-Smith, G. Frampton & A. R. Hardy), pp. 200–215. HMSO, London, UK.

Hance, T. & Gregoire-Wibo, C. (1987) Effect of agricultural practices on carabid populations. *Acta Phytopathologica Entomologica Hungarica* **22**, 147–160.

Hance, T., Gregoire-Wibo, C. & Lebrun, Ph. (1990) Agriculture and ground-beetle populations. *Pedobiologia* **34**, 337–346.

Hart, A. D. M., Thompson, H. M., Fletcher, M. R., Greig-Smith, P. W., Hardy, A. R. & Langton, S. D. (1992) Effects of summer aphicides on tree sparrows. In: *Pesticides, cereal farming and the environment: the Boxworth Project* (eds. P. W. Greig-Smith, G. Frampton & A. R. Hardy), pp. 175–193. HMSO, London, UK.

Hickman, J. M. & Wratten, S. D. (1996) Use of *Phacelia tanacetifolia* strips to enhance biological control of aphids by hoverfly larvae in cereal fields. *Journal of Economic Entomology* **89**, 832–840.

Hislop, E. C. (1991) Review of air-assisted spraying. In: *Air-assisted spraying in crop protection* (eds. A. Lavers, P. Herrington & E. S. E. Southcombe), pp. 3–14. British Crop Protection Council, Bracknell, UK.

Hokkanen, H. & Holopainen, J. K. (1986) Carabid species and activity densities in biologically and conventionally managed cabbage fields. *Journal of Applied Entomology* **102**, 353–363.

Holland, J. M. & Luff, M. L. (2000) The effects of agricultural practices on Carabidae in temperate agroecosystems. *Integrated Pest Management Reviews* **5**, 109–129.

Holland, J. M. & Thomas, S. R. (1996) *Phacelia tanacetifolia* flower strips: their effect on beneficial invertebrates and game bird chick food in an integrated farming system. *Acta Jutlandica* **71**, 171–182.

Holland, J. M., Frampton, G. K., Cilgi, T. & Wratten, S. D. (1994a) Arable acronyms analysed – a review of integrated farming systems research in western Europe. *Annals of Applied Biology* **125**, 399–438.

Holland, J. M., Hewitt, M. V. & Drysdale, A. (1994b) Predator populations and the influence of crop type and preliminary impact of integrated farming systems. *Aspects of Applied Biology* **40**, 217–224.

Holland, J. M., Jepson, P. C., Jones, E. C. & Turner, C. (1997) A comparison of spinning disc atomisers and flat fan pressure nozzles in terms of pesticide deposition and biological efficacy within cereal crops. *Crop Protection* **16**, 179–185.

Holland, J. M., Cook, S. K., Drysdale, A., Hewitt, M. V., Spink, J. & Turley, D. (1998) The impact on non-target arthropods of integrated compared to conventional farming: results from the LINK Integrated Farming Systems project. *Proceedings of the Brighton Crop Protection Conference, Pests & Diseases, 1998* **2**, 625–630.

Holland, J. M., Perry, J. N. & Winder, L. (1999a) The within-field spatial and temporal distribution of arthropods within winter wheat. *Bulletin of Entomological Research* **89**, 499–513.

Holland, J. M., Winder, L. & Perry, J. N. (1999b) Arthropod prey of farmland birds: their spatial distribution within a sprayed field with and without buffer zones. *Aspects of Applied Biology* **54**, 53–60.

Holland, J. M., Winder, L. & Perry, J. N. (2000) The impact of an insecticide on the spatial distribution of beneficial arthropods and their reinvasion in winter wheat. *Annals of Applied Biology* **136**, 93–105.

Hopper, R. & Doberski, J. (1992) Set-aside fallow or grassland: reservoirs of beneficial invertebrates? In: *Set-aside* (ed. J. Clarke), pp. 159–164. British Crop Protection Council, Farnham, UK.

House, G. J. & Parmelee, R. W. (1985) Comparison of oil arthropods and earthworms from conventional and no-tillage agroecosystems. *Soil Tillage Research* **5**, 351–360.

Humphreys, I. C. & Mowat, D. J. (1994) Effect of some organic treatments on predators (Coleoptera: Carabidae) of cabbage root fly *Delia radicum* (L.) (Diptera: Anthomyiidae) and on alternative prey species. *Pedobiologia* **38**, 513–518.

Huusela-Vesitola, E. (1996) Effects of pesticide use and cultivation techniques on ground beetles (Col. Carabidae) in cereal fields. *Annales Zoologici Fennici* **33**, 197–205.

Idinger, J., Kromp, B. & Steinberger, K.-H. (1996) Ground photoeclector evaluation of the numbers of carabid beetles and spiders found in and around cereal fields treated with either inorganic or compost fertilisers. *Acta Jutlandica* **71**, 255–267.

Jensen, T. S., Dyring, L., Kristensen, B., Nielsen, B. O. & Rasmussen, E. R. (1989) Spring dispersal and summer habitat distribution of *Agonum dorsale* (Col. Carabidae). *Pedobiologia* **33**, 115–165.

Jmhasly, P. & Nentwig, W. (1995) Habitat management in winter wheat and evaluation of subsequent spider predation on insect pests. *Acta Oecologica* **16**, 389–403.

Joyce, K. A. (1998) The role of hedgerows in the ecology of invertebrates in arable landscapes. PhD thesis, University of Southampton, UK.

Joyce, K., Jepson, P., Doncaster, C. P. & Holland, J. M. (1997) Arthropod distribution patterns and dispersal processes within the hedgerow. In: *Species dispersal and land use processes* (eds. A. Cooper & J. Power), pp. 103–110. Proceedings of the 6th Annual Conference of the International Association for Landscape Ecology, University of Ulster, Coleraine, Northern Ireland, 9–12 September, 1997.

Kabacik-Wasylik, D. (1975) Research into the number, biomass and energy flow of Carabidae (Coleoptera) communities in rye and potato fields. *Polish Ecological Studies* **1**, 111–121.

Kendall, D. A., Chinn, N. E., Glen, D. M., Wiltshire, C. W., Winstone, L. & Tidboald, C. (1995) Effects of soil management on cereal pests and their natural enemies. In: *Ecology and integrated farming systems* (eds. D. M. Glen, M. P. Greaves & M. H. Anderson), pp. 83–102. Wiley, London, UK.

Kleijn, D. & van der Voort, L. A. C. (1997) Conservation headlands for rare arable weeds: the effects of fertiliser application and light penetration on plant growth. *Biological Conservation* **81**, 57–67.

Kromp, B. (1989) Carabid beetle communities (Carabidae, Coleoptera) in biologically and conventionally farmed agroecosytems. *Agriculture, Ecosystems and Environment* **27**, 241–251.

Kromp, B. & Steinberger, K.-H. (1992) Grassy field margins and arthropod diversity: a case study on ground beetles and spiders in eastern Austria (Coleoptera: Carabidae; Arachnida: Aranei, Opoliones). *Agriculture, Ecosystems and Environment* **40**, 71–93.

Lagerlöf, J. & Wallin, H. (1993) The abundance of arthropods along two field margins with different types of vegetation composition: an experimental study. *Agriculture, Ecosystems and Environment* **43**, 141–154.

Longley, M., Cilgi, T., Jepson, P. C. & Sotherton, N. W. (1997) Measurements of pesticide spray drift deposition into field boundaries and hedgerows. II. Summer applications. *Environmental Toxicology and Contamination* **16**, 165–172.

Lovei, G. L., Hickman, J. M., McDougall, D. & Wratten, S. D. (1993) Field penetration of beneficial insects from habitat islands: hoverfly dispersal from flowering crop strips. *Proceedings of the 46th New Zealand Plant Protection Conference, Christchurch, New Zealand, August 1993*, 325–328.

Luff, M. L. (1966) The abundance and diversity of the beetle fauna of grass tussocks. *Journal of Animal Ecology* **35**, 189–208.

Luff, M. L. (1987) Biology of polyphagous ground beetles in agriculture. *Agricultural Zoology Reviews* **2**, 237–278.

Luff, M. L. & Sanderson, R. A. (1992) Analysis of data on cereal invertebrates. MAFF Report. MAFF, London, UK. 39 pp.

Lys, J.-A. & Nentwig, W. (1992) Augmentation of beneficial arthropods by strip-management. 4. Surface activity, movements and activity density of abundant carabid beetles in a cereal field. *Oecologia* **92**, 373–382.

Lys, J.-A. & Nentwig, W. (1994) Improvement of the overwintering sites for Carabidae, Staphylinidae and Araneae by strip-management in a cereal field. *Pedobiologia* **38**, 238–242.

Lys, J.-A., Zimmermann, M. & Nentwig, W. (1994) Increase in activity density and species number of carabid beetles in cereals as a result of strip-management. *Entomologia Experimentalis et Applicata* **73**, 1–9.

Macleod, A. (1992) Alternative crops as floral resources for beneficial hoverflies (Diptera: Syrphidae). *Proceedings of the Brighton Crop Protection Conference, Pests and Diseases, 1992* **2**, 997–1002.

Mann, B. P., Wratten, S. D., Poehling, M. & Borgemeister, C. (1991) The economics of reduced-rate insecticide applications to control aphids in wheat. *Annals of Applied Biology* **119**, 451–464.

Maudsley, M., West, T., Rowcliffe, H. & Marshall, E. J. P. (1997) Spatial variability in plant and insect (Heteroptera) communities in hedgerows in Great Britain. In: *Species dispersal and land use processes* (eds. A. Cooper & J. Power), pp. 229–236. Proceedings of the Sixth Annual Conference of the International Association for Landscape Ecology, University of Ulster, Coleraine, Northern Ireland, 9–12 September, 1997.

McKinlay, R. G. & McCreath, M. (1995) Some biological alternatives to synthetic insecticides for sustainable agriculture. *Pesticide Outlook* **6**, 31–35.

Moreby, S. J. (1995) Heteroptera distribution and diversity within the cereal ecosystem. In: *Integrated crop protection: towards sustainability* (eds. R. G. McKinlay & D. Atkinson), pp. 151–158. British Crop Protection Council, Farnham, UK.

Moreby, S. J. (1997) The effects of herbicide use within cereal headlands on the availability of food for arable birds. *Proceeedings of the Brighton Crop Protection Conference, Weeds, 1997* **3**, 1197–1202.

Moreby, S. J. (2003) Birds of lowland arable farmland: the importance and identification of invertebrate diversity in the diet of chicks (*this volume*).

Moreby, S. J. & Aebischer, N .J. (1992) Invertebrate abundance on cereal fields and set-aside land: implications for wild game bird chicks. In: *Set-aside* (ed. J. Clarke), pp. 181–187. British Crop Protection Council, Farnham, UK.

Moreby, S. J. & Sotherton, N. W. (1995) The management of set-aside land as brood-rearing habitats for game birds. In: *Insects, plants and set-aside* (eds. A. Colston & F. Perring), pp. 41–43. Botanical Society of the British Isles, London, UK.

Moreby, S. J. & Sotherton, N. W. (1997) A comparison of some important chick-food insect groups found in organic and conventionally grown winter wheat fields in southern England. *Entomological Research in Organic Agriculture* **15**, 51–60.

Moreby, S. J., Aebischer, N. J., Southway, S. E. & Sotherton, N. W. (1994) A comparison of the flora and arthropod fauna of organically and conventionally grown winter wheat in southern England. *Annals of Applied Biology* **125**, 13–27.

Moreby, S. J., Sotherton, N. W. & Jepson, P. C. (1997) The effects of pesticides on species of non-target heteroptera inhabiting cereal fields in southern England. *Pesticide Science* **51**, 39–48.

Nentwig, W. (1989) Augmentation of beneficial arthropods by strip-management. 2. Successional strips in a winter wheat field. *Journal of Plant Diseases and Protection* **96**, 89–99.

Newton, I. (1998) *Population limitation in birds.* Academic Press, London, UK. 597 pp.

Nicolaus, L. K. & Lee, H. S. (1999) Low acute exposure to organophosphate produces long-term changes in bird feeding behaviour. *Ecological Applications* **9**, 1039–1049.

O'Connor, R. J. & Shrubb, M. (1986) *Farming and birds.* Cambridge University Press, Cambridge, UK. 290 pp.

Panek, M. (1997) The effect of agricultural landscape structure on food resources and survival of grey partridge *Perdix perdix* chicks in Poland. *Journal of Applied Ecology* **34**, 787–792.

Poehling, H.-M. (1989) Selective application strategies for insecticides in agricultural crops. In: *Pesticides and non-target invertebrates* (ed. P. C. Jepson), pp. 151–176. Intercept, Wimborne, UK.

Pollard, E. (1968) Hedges III. The effect of removal of the bottom flora of a hawthorn hedge on the Carabidae of the hedge bottom. *Journal of Applied Ecology* **5**, 125–139.

Pollard, E. & Yates, T. J. (1993) *Monitoring butterflies for ecology and conservation.* Chapman & Hall, London, UK. 274 pp.

Potts, G. R. (1977) Some effects of increasing the monoculture of cereals. In: *Origins of pest, parasite, disease and weed problems* (eds. J. M. Cherret & G. R. Sagar), pp. 183–202. Blackwell Scientific Publications, Oxford, UK.

Potts, G. R. (1986) *The partridge: pesticides, predation and conservation.* Collins, London, UK. 274 pp.

Potts, G. R. (1997) Cereal farming, pesticides and grey partridges. In: *Farming and birds in Europe* (eds. D. J. Pain & M. W. Pienkowski), pp. 151–177. Academic Press, London, UK.

Potts, G. R. & Vickerman, G. P. (1974) Studies on the cereal ecosystem. *Advances in Ecological Research* **8**, 107–197.

Poulsen, J. G., Sotherton, N. W. & Aebischer, N. J. (1998) Comparative nesting and feeding ecology of skylarks *Alauda arvensis* on arable farmland in southern England with special reference to set-aside. *Journal of Applied Ecology* **35**, 131–147.

Purvis, G. & Curry, J. P. (1984) The influence of weeds and farmyard manure on the activity of Carabidae and other ground-dwelling arthropods in a sugar beet crop. *Journal of Applied Ecology* **21**, 271–283.

Rands, M. R. W. (1985a) The benefits of pesticide-free headlands to wild game birds. In: *The impact of pesticides on the wild flora of agro-ecosystems* (eds. A. B. Hald & J. Kjolholt), pp. 19–26. Centre for Terrestrial Ecology, Copenhagen, Denmark.

Rands, M. R. W. (1985b) Pesticide use on cereals and the survival of grey partridge chicks: a field experiment. *Journal of Applied Ecology* **22**, 49–54.

Rands, M. R. W. & Sotherton, N. W. (1986) Pesticide use on cereal crops and changes in the abundance of butterflies on arable farmland in England. *Biological Conservation* **36**, 71–82.

Reddersen, J. (1997) The arthropod fauna of organic versus conventional cereal fields in Denmark. *Biological Agriculture and Horticulture* **15**, 61–72.

Ryszkowski, L., Karg, J., Margalit, G., Paolettie, M. G. & Zlotin, R. (1993) Above-ground insect biomass in agricultural landscapes of Europe. In: *Landscape ecology and agroecology* (eds. R. G. H. Bunce, L. Ryszkowski & M. G. Paolettie), pp. 71–82. CRC Press, Boca Raton, Florida, USA.

Salveter, R. (1998) The influence of sown herb strips and spontaneous weeds on the larval stages of aphidophagous hoverflies (Dipt., Syrphidae). *Journal of Applied Entomology* **122**, 103–114.

Schumacher, W. (1987) Measures taken to preserve arable weeds and their associated communities in central Europe. In: *Field margins* (eds. J. M. Way & P. W. Greig-Smith), pp. 11–22. British Crop Protection Council, Farnham, UK.

Sherratt, T. N. & Jepson, P. C. (1993) A metapopulation approach to modelling the long-term impact of pesticides on invertebrates. *Journal of Applied Ecology* **30**, 696–705.

Simpson, V. R. (1984) Chlorpyrifos and wildlife. *Veterinary Record* **114**, 101–102.

Skuhravý, V. (1958) Einfluss landwirtschlaftlicher Massnahmen auf die Phänologie der Feldcarabiden. *Folia Zoologica* **7**, 325–338.

Sotherton, N. W. (1982) Observations on the biology and ecology of the chrysomelid beetle *Gastrophysa polygoni* in cereals. *Ecological Entomology* **7**, 197–206.

Sotherton, N. W. (1984) The distribution and abundance of predatory arthropods overwintering on farmland. *Annals of Applied Biology* **105**, 423–429.

Sotherton, N. W. (1985) The distribution and abundance of predatory Coleoptera overwintering in field boundaries. *Annals of Applied Biology* **106**, 17–21.

Sotherton, N. W. (1990) The effects of six insecticides used in UK cereal fields on sawfly larvae (Hymenoptera: Tenthredinidae). *Proceedings of the Brighton Crop Protection Conference, Pests and Diseases, 1990* **3**, 999–1005.

Sotherton, N. W. (1991) Conservation Headlands: a practical combination of intensive cereal farming and conservation. In: *The ecology of temperate cereal fields* (eds. L. G. Firbank, N. Carter, J. F. Darbyshire & G.R. Potts), pp. 373–397. Blackwell Scientific Publications, Oxford, UK.

Sotherton, N. W. (1998a) Biodiversity in field margin strips: the ecological and agronomic perspectives. In: *Proceedings of the 23rd Congress of the International Union of Game Biologists* (eds. N. W. Sotherton, P. Granval, P. Havet & N. J. Aebischer), pp. 11–14. IUGB/Office National de la Chasse, Paris, France.

Sotherton, N. W. (1998b) Land use changes and the decline of farmland wildlife: An appraisal of the set-aside approach. *Biological Conservation* **83**, 259–268.

Sotherton, N. W. & Moreby, S. J. (1992) The importance of beneficial arthropods other than natural enemies in cereals. *Aspects of Applied Biology* **31**, 11–18.

Sotherton, N. W., Robertson, P. A. & Dowell, S. D. (1993) Manipulating pesticide use to increase the production of wild game birds in Britain. In: *Quail III: National Quail Symposium* (eds. K. E. Church & T. V. Dailey), pp. 92–101. Missouri Department of Conservation, Jefferson City, Missouri, USA.

Stinner, B. R. & House, G. J. (1990) Arthropods and other invertebrates in conservation-tillage agriculture. *Annual Review of Entomology* **35**, 299–318.

Theiling, K. M. & Croft, B. A. (1988) Pesticide side-effects on arthropod natural enemies: A database summary. *Agriculture, Ecosystems and Environment* **21**, 191–218.

Theunissen, J. (1997) Applications of intercropping in organic agriculture. *Biological Agriculture and Horticulture* **15**, 251–260.

Thiele, H.-U. (1977) *Carabid beetles in their environments.* Springer, Berlin, Germany. 369 pp.

Thomas, M. B., Wratten, S. D. & Sotherton, N. W. (1991) Creation of 'island' habitats in farmland to manipulate populations of beneficial arthropods: densities and emigration. *Journal of Applied Ecology* **28**, 906–917.

Tischler, W. (1955) Effect of agricultural practice on the soil fauna. In: *Soil zoology* (ed. D. K. M. Kevan), pp. 215–230. Butterworth, London, UK.

Trumper, E. V. & Holt, J. (1998) Modelling pest population resurgence due to recolonisation of fields following an insecticide application. *Journal of Applied Ecology* **35**, 273–285.

Tucker, G. M. & Heath, M. F. (1994) *Birds in Europe: their conservation status. Birdlife Conservation Series, No. 3.* Birdlife International, Cambridge, UK.

Tyler, B. M. J. & Ellis, C. R. (1979) Ground beetle in three tillage plots in Ontario and observations on their importance as predators of the northern corn rootworm *Diabrotica longicornis* (Coleoptera: Chrysomelidae). *Proceedings of the Entomological Society of Ontario* **110**, 65–73.

Vickerman, G. P. (1978) The arthropod fauna of undersown grass and cereal fields. *Scientific Proceedings, Royal Dublin Society, Series A* **6**, 273–283.

Vickerman, G. P. (1992) The effects of different pesticide regimes on the invertebrate fauna of winter wheat. In: *Pesticides, cereal farming and the environment: the Boxworth Project* (eds. P. W. Greig-Smith, G. Frampton & A. R. Hardy), pp. 82–108. HMSO, London, UK.

Weiss, M. J., Balsbaugh, E. U., French, E. W. & Hoag, B. K. (1990) Influence of tillage management and cropping systems on ground beetle (Coleoptera: Carabidae) fauna in the northern Great Plains. *Environmental Entomology* **19**, 1388–1391.

Wiggington, M. (1999) *British red data books: 1. Vascular plants (3rd Edition).* JNCC, Peterborough, UK. 468 pp.

Wiles, J. A. & Jepson, P. C. (1994) Sustrate-mediated toxicity of deltamethrin residues to beneficial invertebrates: estimation of toxicity factors to aid risk assessment. *Archives of Environmental Contamination and Toxicology* **27**, 384–391.

Wilson, P. J. (1992) The natural regeneration of vegetation under set-aside in Southern England. In: *Set-aside* (ed. J. Clarke), pp.73–79. British Crop Protection Council, Farnham, UK.

Wilson, P. J. (1994) Managing field margins for the conservation of the arable flora. In: *Field margins – integrating agriculture and conservation* (ed. N. D. Boatman), pp. 253–259. British Crop Protection Council, Farnham, UK.

Wilson, P. J. (1995) The potential for herbicide use in the conservation of Britain's arable flora. *Proceedings of the Crop Protection Conference, Weeds, 1995* **3**, 967–972.

Winder, L., Holland, J. M. & Perry, J. N. (1998) The within-field spatial and temporal distribution of the grain aphid (*Sitobion avenae*) in winter wheat. *Proceedings of the Brighton Crop Protection Conference, Pests and Diseases, 1998* **3**, 1089–1094.

Winstone, L., Iles, D. R. & Kendall, D. J. (1996) Effects of rotation and cultivation on polyphagous predators in conventional and integrated farming systems. *Aspects of Applied Biology* **47**, 111–117.

Zangger, A., Lys, J.-A. & Nentwig, W. (1994) Increasing the availability of food and the reproduction of *Poecilus cupreus* in a cereal field by strip management. *Entomologia Experimentalis et Applicata* **71**, 111–120.

Zimmermann, J. & Büchs, W. (1996) Management of arable crops and its effects on rove-beetles (Coleoptera: Staphylinidae), with special reference to the effects of insecticide treatments. *Acta Jutlandica* **71**, 183–194.

5
Productivity and Profitability: The Effects of Farming Practices on the Prey of Insectivorous Birds

DAVID I. McCRACKEN[1], ERIC M. BIGNAL[2], SHONA BLAKE[1] AND GARTH N. FOSTER[1]

[1]Research Division, Scottish Agricultural College, Auchincruive, Ayr, KA6 5HW, UK and [2]European Forum on Nature Conservation and Pastoralism, Kindrochaid, Gruinart, Bridgend, Isle of Islay, Argyll, PA44 7PT, UK

Introduction

The nature conservation value and importance of certain types of European farmland and farming systems have been brought to the attention of a wider audience in recent years. Reports arising from collaborative work between the Joint Nature Conservation Committee, Worldwide Fund for Nature, Institute for European Environmental Policy and Dutch Ministry for Agriculture, Nature Management and Fisheries (e.g. Beaufoy et al., 1994; Bignal & McCracken, 1996a), the various outputs from the meetings of the European Forum on Nature Conservation and Pastoralism (e.g. Curtis et al., 1991; Bignal & McCracken, 1992, 1996b; Bignal et al., 1994; McCracken et al., 1995a; Poole et al., 1998; Pienkowski & Jones, 1999), and the publications arising from work by others in this field (e.g. Goriup et al., 1991; Hötker et al., 1991; Dixon et al., 1993; Pain & Pienkowski, 1997) have all served to emphasise the fact that farm management practices have created, and are now essential to the continued maintenance of, many landscapes, habitats, and wildlife communities of value across Europe. However, simply having a broad appreciation of which farming systems are good for certain species or species assemblages is of little use without a detailed understanding of how each particular farming system functions and integrates with all the species reliant upon that system as a whole. Many species have intimate and complex interactions with the annual farming cycle, and their presence on any one piece of farmland is determined not only by the farm management occurring at that time, but also by the management practised over the previous weeks, months, and even years (Bignal & McCracken, 1996b, 2000; McCracken & Bignal, 1998).

Obtaining such detailed understanding of the relationships between farm management practices and farmland wildlife has been the focus of much of our own research work to date. We have a particular interest in the ecology and requirements of the chough (*Pyrrhocorax pyrrhocorax*) and in the role that surface-active and soil-dwelling invertebrates play as prey for this and other ground-feeding farmland birds (for example, waders such as snipe (*Gallinago gallinago*), lapwing (*Vanellus vanellus*), curlew (*Numenius arquata*), and redshank (*Tringa totanus*)), which exhibit intimate associations with grassland habitats. The need for such work has arisen because populations of many of these birds are declining across Europe (e.g. Hötker *et al.*, 1991; O'Brien & Smith, 1992; Tucker *et al.*, 1994; Pain *et al.*, 1997), and it has been recognised (not only by conservation science bodies such as the British Trust for Ornithology, but also by national agricultural agencies such as the UK Ministry of Agriculture, Fisheries and Food) that there are substantial gaps in knowledge about the effects of grassland management on such birds (e.g. Fuller *et al.*, 1995; Milsom *et al.*, 1998). Without such a detailed understanding of the role that farm management and wider environmental factors play in influencing the invertebrate prey items of these birds, it will be difficult (if not impossible) to be sure how to put effective measures in place on farmland for their conservation. These issues are discussed and illustrated below with reference to some of the subtle interactions with farm management that have been revealed through studies on the ecology and requirements of the chough, and studies on the effects of farm management practices on ground beetle adults and larvae (Coleoptera: Carabidae) and leatherjackets (the larvae of craneflies – Diptera: Tipulidae), which together can form important prey items for birds associated with grassland habitats.

The links between farming practices and chough on Islay

The chough is the rarest member of the crow family in the British Isles and, although once much more widespread, is now confined to the western seaboards of Ireland, Scotland, and Wales. The Hebridean island of Islay is the stronghold in Scotland, holding about 200 individuals consisting of 60–80 pairs and 50–60 sub-adult birds. Islay holds over 95% of the Scottish population, and maintaining suitable conditions on the island will therefore be crucial for the survival of this species in Scotland. The population on Islay has been subject to continuous study since the 1980s (e.g. Bignal *et al.*, 1987, 1989, 1996, 1997, 1998; Bignal & Curtis, 1989; McCracken *et al.*, 1992; McCracken & Foster, 1993, 1994), and during this time an intensive annual colour-ringing programme has ensured that many individual birds are readily identifiable. The information obtained from this long-term study has helped establish some of the detailed ecological relationships between farm management practices and chough requirements.

CHOUGH FEEDING ECOLOGY AND STRATEGIES

The chough's bill is a specialised tool for digging and probing in the soil, and for pecking immobile or relatively slow-moving food items off the ground. Choughs feed principally on soil-living, surface-active, and dung-associated invertebrates, but the species taken as food differ across Europe. In addition, many of these invertebrates exhibit strong seasonal peaks in abundance, and therefore the species consumed in any area also change markedly throughout the year (e.g. Garcia-Dory, 1983; Roberts, 1985; Soler &

Soler, 1993; Bignal *et al.*, 1996; Sanchez-Alonso *et al.*, 1996). Grain, bulbs, or fruit may also be taken when invertebrate prey is scarce (e.g. McCracken *et al.*, 1992; Blanco *et al.*, 1994; Rolando & Laiolo, 1997). The wide prey spectrum of the chough should not be interpreted as indicating a 'catholic' feeder, but rather a specialist that is associated with areas of ecological complexity and temporal variability.

Choughs select their food using visual clues, and therefore vegetation structure is vital in determining suitable foraging areas. Observations across Europe have shown that optimal foraging conditions can occur in a wide range of arid to temperate vegetation types, where bare ground and short or open swards allow the birds access to the soil surface. The majority of such areas are created and maintained by the extensive grazing of domestic livestock, or livestock rearing associated with the cultivation of crops (especially cereals) within an intricate mosaic of grasslands managed in different ways. On Islay these include sheep- and cattle-grazed pastures, silage and hay aftermaths, and cereal stubbles, as well as more natural, permanent herb-rich and heath pastures grazed by livestock. The chough's continued survival across Europe is intimately bound up with the maintenance of less intensive livestock farming systems and the annual cycles associated with these. The detailed studies on Islay have revealed a complicated relationship between farmland management, prey abundance and availability, and chough foraging success.

'BOTTLENECKS' TO RECRUITMENT AFFECTING THE ISLAY POPULATION

Previous studies on Islay have identified three major factors affecting the population throughout the year: nest site availability, suitable feeding areas, and two main 'bottlenecks' to recruitment. These 'bottlenecks' are high nestling mortality (with, for example, in some instances only two out of six nestlings surviving to fledging), and variable post-fledging survival (with, in some years, as many as 75% of the young that fledge failing to survive the first six months of life). Much of the chough research on the island to date has been targeted at understanding the high nestling mortality 'bottleneck', and it is on this issue that the remainder of this section is focused.

Many choughs on Islay nest in cracks and crevices in cliffs, and feed in nearby areas of short pasture grazed by cattle and/or sheep. However, many suitable feeding areas are located quite a distance from cliff-nesting sites, and in recent years the birds have managed to exploit these areas by nesting in old buildings. It has proved possible to help them by maintaining the roofs of such buildings and re-roofing others. However, even the provision of nest sites within areas of suitable feeding habitat provides no guarantee of breeding success. At many sites (both natural and artificial), the female lays eggs, the majority of the eggs hatch, the nestlings grow well for a few weeks before they are due to fledge, and then they die. Predation or disturbance cannot be blamed in the majority of cases; it appears that the adults are unable to provision their young adequately at this stage, and that the nestlings simply die of starvation. The research considered below was focused during this critical period in the breeding cycle, and answers were sought to the following questions:

- Where did the adults spend their time?
- How were they spending their time?
- On what prey were they feeding?
- Why were they feeding in some areas in preference to others?

Chough feeding ecology during the breeding season

The example presented here arose from an investigation of the feeding behaviour of three pairs during the 1992 breeding season and reported by Bignal *et al.* (1996). This was undertaken to explore in more detail the general patterns evident from the work on larger samples. The birds were kept under observation on most of the days between April and June, and their behaviour recorded at one minute intervals. The farmland utilised by the birds was mapped by farm management compartment, i.e. an area enclosed within livestock-proof boundaries. In many cases, a compartment was an individual field, but a compartment could also consist of a much larger moorland area, or a group of fields where the internal boundaries had broken down. Behavioural observations were recorded in nine activity categories, together with use of ten habitats. Each recording team spent approximately six hours in the field each day, and over the three months, a total of over 14,000 minute observations was collected.

Where and how did the adults spend their time?

The results showed that, although a total of 8, 17, and 25 compartments were visited by each of the three study pairs during this period, only a small number were used to any extent by each pair (with over half the observations being recorded in only 1, 2, or 4 compartments respectively). In addition, there was no relationship between time spent in a compartment and distance from the nest; so the birds were not simply using those compartments closest to their nest sites. A comparison of habitat availability and selection showed that the birds were actively selecting grassland as feeding areas, although the selection for old and new grassland varied between pairs. It was also evident that the birds were not simply feeding in any grassland compartment, but were only feeding in certain of the grassland areas available to them (indeed, one of the pairs travelled over 1 km to do so). *Figure 5.1* provides an indication of what the birds were doing when under observation. It is striking that over 90% of each pair's time was spent in food collection activity. This suggests that the birds knew exactly where to go to find food, and also that territorial boundaries with other chough had been set well *before* the nesting period began.

On what prey were they feeding?

All three pairs primarily collected food from the soil during this study period, and sampling the areas within compartments where they were seen to feed established that they were feeding exclusively on leatherjackets (larvae of Tipulidae) during this particular breeding season. Note that the birds would also have been expected to be feeding on surface-active and dung-dwelling insects at this time of year (e.g. McCracken *et al.*, 1992), but spring came very late to Islay in 1992 and regular sampling in the areas concerned indicated that the spring rise of such insects did not start until quite near the end of the chough breeding season.

Why were they feeding in some areas in preference to others?

Detailed information was therefore available on where the birds were, what they were doing, and what they were feeding on – but none of these factors on their own helped

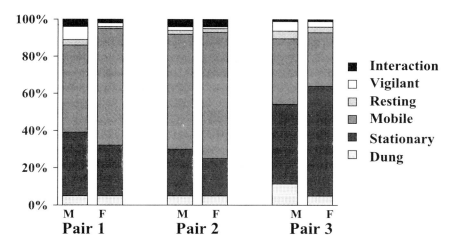

Figure 5.1.   Percentage of time spent engaged in 6 activities by the male (M) and female (F) belonging to three pairs of chough under intensive observation during the 1992 breeding season (adapted from Bignal *et al.*, 1996). Interaction: territorial interaction with other chough or other bird species; Vigilant: alert and on look-out for danger; Resting: resting or inactivity occurring outside of the nest-site; Mobile: feeding in soil or on soil surface and moving on quickly from place to place; Stationary: feeding in soil or on soil surface but staying in the one place; Dung: feeding in old or fresh dung pats.

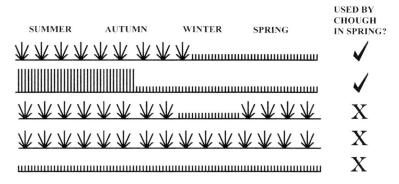

Figure 5.2.   The effects of past and current management history and vegetation structure on the utilisation of grasslands by chough during the breeding season (adapted from Bignal *et al.*, 1996). The relative height and structure of the vegetation during the nine months preceding the 1992 chough breeding season is illustrated for the five main categories of grassland recognised (on the basis of management history), together with an indication of whether these different types of grassland were utilised by chough during the 1992 breeding season.

explain why they used certain grassland compartments in preference to others, when most of the grassland compartments available appeared to be similar at that time of year. However, data were collected also on the farm management being practised in each compartment during the year previous to the observations. Once each compartment was considered in association with others undergoing similar management, then patterns started to become much clearer.

Figure 5.2 illustrates that the birds' choice of pasture not only reflected the low sward heights at the time, but was also influenced by the farm management during the preceding nine months. It was found that the pastures used preferentially by these three pairs of chough were those in which 1) the pasture management during the previous late summer/early autumn produced medium to high grass sward heights (either through a low stocking density of cattle or late-cut silage production), and 2) the pasture management during the winter and spring reduced the sward heights markedly (through relatively intensive grazing by sheep, cattle, or overwintering barnacle geese (*Branta leucopsis*)). Pastures from which livestock had been removed in late winter and those where sward height was tall in spring were used very little by the birds.

The interpretation (based on the findings from past work on leatherjacket ecology) for their choice of pasture during this period was that medium to high swards in the autumn (when adult crane flies are active and laying their eggs) tend to encourage high numbers of leatherjacket larvae in the soil during the following winter and spring, whereas low swards in the late winter and early spring allow the birds access to the leatherjackets in the soil not only during the breeding season (when rapid and easy access to food is essential) but also prior to this (when they are prospecting for good feeding sites). In some pastures, the management history dictated that, although leatherjackets may have been abundant in the soil in spring 1992, they were not available to the birds (because the sward was too high), while in others it dictated that, although the birds had easy access to the soil, there were few, or no, leatherjackets to collect (because sward conditions were unsuitable the previous autumn).

It is important to note that the above findings only provide an indication of chough requirements on Islay at this time of year and in that particular breeding season. This must be stressed, since we know that these three study pairs ranged more widely as soon as the young were fledged, and previous work (e.g. McCracken *et al.*, 1992) has also indicated that chough on Islay utilise different habitats and feeding strategies at different times of the year. For example, they would normally be expected to feed on a greater variety of surface-active insects during the summer, then switch largely to feeding on insects in dung in late summer/early autumn, and then move on to cereal grains during the winter (a time on Islay when there are few, if any, insects available). This therefore complicates matters even further, because it means that a whole combination of different farm management practices are required over a full year in order to meet all the needs of the birds.

Prey profitability and farmland birds

Both current and previous farm management can, therefore, have an important influence on the abundance and availability of insect prey items of importance to a wide range of farmland birds. But simply ensuring that many insect prey items are available to birds is not necessarily a guarantee of success. The size of these prey items is also of great importance, since both foraging and handling times are reduced and energy intake increased when birds feed on larger prey, as compared with a similar overall volume of smaller items. Biomass increases approximately as the cube of the length, and there is a negative correlation between body mass and the proportion of indigestible material on adult beetles – so longer beetles are of greater food value. For example, Beintema (1991) estimated that, in order to maintain energy intake requirements, a redshank chick near to fledging would need to collect

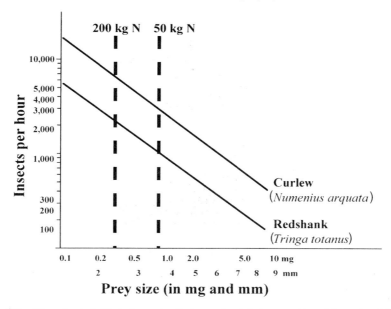

Figure 5.3.   The estimated effect of prey size (in terms of weight in mg and length in mm) and fertiliser application on the number of prey items per hour that would be required to be collected by curlew and redshank chicks close to fledging in order to maintain energy intake requirement. Adapted from Beintema *et al.*, 1990 and Beintema, 1991.

100 insect prey items per hour if each of these were 9 mm long, but would need over 700 items per hour if they were all only 4 mm in length (*Figure 5.3*).

THE EFFECT OF MANAGEMENT PRACTICES ON THE SIZE OF AVAILABLE PREY ITEMS

Beintema *et al.* (1990) also hypothesised that the intensity of grassland management could affect prey size (and hence profitability) – with higher inputs of fertiliser and intensive grazing or mowing being detrimental to larger insect species, and only smaller insect species being able to utilise these areas. Ground beetles have been shown to form important prey for the chicks of a number of wader species (e.g. Blake & Foster, 1998) and, in order to test the theory of Beintema *et al.* (1990), Blake *et al.* (1994, 1996) conducted a number of studies in southern Scotland into the effect of farm management practices on the ground beetle fauna of grasslands. For example, in addition to assessing the composition of the ground beetle fauna at 55 grassland sites in southern Scotland, Blake *et al.* (1994) also obtained details of the farm manage-ment practised at each site. The management information was used to allocate an overall Management Intensity Score of between 1 and 5 to each site (with 5 being the most intensive), and the ground beetle body size distribution (and hence biomass) on each site was expressed in terms of the Weight Median Length (WML) of all the individuals caught. This involved calculating the weight of each species (using information from the literature on known mean lengths, together with a formula describing the weight:length relationship for adult ground beetles). Each weight value was multiplied by the number of individuals of each species caught, the products summed, and the weights expressed as proportions of the total ground beetle

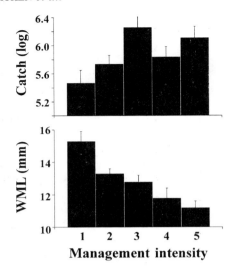

Figure 5.4. The effect of field management intensity on the mean number (log) and the Weight Median Length (mm) of ground beetle individuals obtained from intensive pitfall sampling of 55 grassland fields in southern Scotland (adapted from information provided in Blake *et al.*, 1994). Details of field management history were used to allocate an overall Management Intensity Score of between 1 and 5 (with 5 being the most intensive).

biomass at each site. The WML value for each site was the equivalent body length where the cumulative percentage biomass figure was equal to 50%.

Figure 5.4 indicates that beetle numbers were found to be higher on the more intensively managed swards, and therefore, at first sight, this might suggest that such swards would be ideal feeding grounds for wader chicks. However, beetle WML was found to be significantly related to intensity of farm management, with the largest WML occurring at the lowest level of management. This reduction in WML from 15 mm to 11 mm translates as a reduction in dry weight from 45 mg to 18 mg – hence, the average beetle in the least intensively managed sward was more than twice as heavy as its counterpart in the most intensive regime. Consequently, wader chicks would probably have found it more profitable to feed on the smaller number of larger and slower moving species (such as *Carabus violaceus, Pterostichus niger*, and *Pterostichus madidus*) associated with the unimproved sites, than on the large number of smaller and faster moving species (such as *Amara communis, Loricera pilicornis*, and *Bembidion aeneum*) found on the more improved and intensively managed sites.

The importance of additional factors

THE INFLUENCE OF WIDER ENVIRONMENTAL CHARACTERISTICS

Management can therefore have an important influence on the abundance, availability, and profitability of insect prey occurring on farmland. But management is not the only influence, and other factors (less amenable to human manipulation) can be just as important. For example, leatherjacket populations in grassland can also form important prey for a wide range of ground-feeding birds, and when investigating the feasibility of predicting the size of leatherjacket populations in grassland, McCracken *et al.* (1995b) found that population size was affected by aspects of past management history (with fields which had been used for silage and received applications of dung or slurry the previous year tending to have higher infestation levels). However, of even greater importance were the wider environmental characteristics of each site itself (with higher infestations being more likely in fields which had a tendency to waterlogging, on east- or south-facing slopes, and where sites experienced milder winter weather by being nearer the coast).

It is important to highlight this fact, that site and soil conditions can have a direct effect on leatherjacket numbers, because such factors are often not taken into account when, for example, farmers are being paid to manage fields on their farm specifically to benefit birds (e.g. under an Environmentally Sensitive Area or Countryside Stewardship Scheme agreement in the UK). In many instances, the choice of field to enter into the scheme is made by the farmer, and the management prescribed is generally confined to providing undisturbed areas to nest, with no consideration given to specifically enhancing insect prey items. Therefore, the potential to maximise wildlife benefit (in terms of bird breeding success) may not necessarily be fulfilled in all cases.

THE QUALITY OF SITES AND EXPERIENCE OF INDIVIDUALS

It is also worth emphasising that, even if the site and its management are both regarded as suitable for the birds and the invertebrates that they prey upon, it must be appreciated that there is still no guarantee that the relevant invertebrates will be present in sufficient numbers each year, or even that the birds will be able to locate and feed on them. *Figure 5.5* shows the main factors believed to affect chough breeding success (and hence productivity) on the island of Islay (from Bignal *et al.*, 1997). Imagine this diagram as a dartboard, where an individual dart has to be placed in each sector, and scores have to be high across all five sectors to ensure a 'good' breeding year for each pair of chough. Farm management plays a large role in influencing the score in the 'Home Range' sector, but that can count for nothing if other factors are not suitable – for example, the size of leatherjacket and other insect populations fluctuate naturally from year to year (influencing the score in the 'Macro' sector), and the age and experience of the birds can affect how good they are at finding certain prey items (and therefore 'Pair Quality' also has an important role to play).

The latter is of particular relevance with regard to the intensive study of three pairs of chough during the 1992 breeding season mentioned above. Two out of three of these study pairs did very well in 1992 (fledging 4 young each), whereas the third pair only fledged 2 young. This may seem rather surprising, especially when it was known that some of the fields in the vicinity of this nest site and within the birds' foraging territory had very high leatherjacket populations (e.g. two fields had over 1.6 million leatherjackets/ha, which would be doing considerable damage to the grass swards), and the management was such that these should have been readily accessible to the birds. However, the male at this site was only three years old, and this was his first breeding attempt. He was therefore inexperienced at having to provision young, and we think, from his actions, he was also inexperienced at having to find leatherjackets.

Young, inexperienced birds normally get a lot of their food requirements from

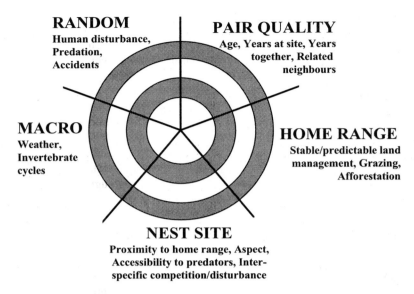

RANDOM
Human disturbance,
Predation,
Accidents

PAIR QUALITY
Age, Years at site, Years
together, Related
neighbours

MACRO
Weather,
Invertebrate
cycles

HOME RANGE
Stable/predictable land
management, Grazing,
Afforestation

NEST SITE
Proximity to home range, Aspect,
Accessibility to predators, Inter-
specific competition/disturbance

Figure 5.5.   Factors affecting chough breeding success on the island of Islay in the Scottish Hebrides (from Bignal *et al.*, 1997). For each pair of chough, successful breeding depends on 'scores' being high across all five categories (i.e. towards the centre of each segment). In most years, at least one category thwarts the breeding attempt of many pairs.

dung pats (large and easily located) and, under normal circumstances, there should have been insects available in dung at this time of year. As already mentioned, however, in 1992 there was an especially late spring and, consequently, there were no insects colonising fresh pats during the early part of the breeding season. Despite this, the male of this pair spent a large amount of time fruitlessly searching for food in dung pats, and it is therefore believed that this probably had a detrimental effect on the young through the reduction in food brought back to nestlings. In this regard, it is important to note that, even though the female was older and more experienced, chough behaviour means that the female will follow the male to feeding sites of his choosing – therefore, she may not have had as good an opportunity to collect leatherjackets as may have been expected.

Implications for conservation management

The fact to highlight here is that there is an extremely complex relationship between birds, their insect prey items, and farm management practices. The fact that choughs on Islay eat insects has been known for many years, although it is only recently that we have established a better indication of what exactly they eat and when. But knowing the broad basics that chough eat leatherjackets or dung beetles, or whatever, at any particular time of the year is only part of the process – it is also essential to know and understand the fine detail of how, and why, farm management affects utilisation of these prey items. It is, however, important to emphasise that, although much background information is available for many different types of wildlife species with regard to simple habitat associations or correlations between population densities and broadscale farm management practices, detailed information on the day-to-day management which produces the conditions wildlife species require is lacking. Consequently, we would argue that it is impossible, currently, to advise farmers on how best to manage the larger extent of their farms in order to benefit most individual species or species assemblages.

Much more in the way of detailed studies is therefore required for developing policy and advice, but this should not be taken to suggest that studies have to be instigated for every single wildlife species of concern. While some new studies would need to be implemented, it should also prove possible to take advantage of the findings from past or current studies of key species and draw relevant implications for those other species which 'function' in farmland habitats in a similar way. For example, the findings from the detailed research on chough on Islay will be of relevance to the conservation and enhancement of other ground-feeding birds associated with farmland (especially waders such as lapwing, snipe, curlew, and redshank).

We would also argue that it is essential that any such detailed studies take into account the requirements of the wildlife species under consideration throughout the whole year, and are not confined solely to one particular period or season. For example, it would be totally meaningless (and highly inappropriate) to suggest chough requirements on Islay for the whole year by simple extrapolation from the results of the breeding season study mentioned above, since we know from other research on the island that chough utilise different habitats, feeding strategies, and prey items at other times of the year. Since many wildlife species (but especially most birds and many invertebrates) are highly mobile, this would require a whole farm (or even region) approach in order to establish the associations (and hence implications) for the farm management practices and farming system throughout the year.

Only when such information is to hand will it really be possible to put together an appropriate conservation strategy for any one species, or suite of species, associated with farmland – a strategy that, by necessity, involves the integration of nature conservation needs into farming practices, and also takes on board the fact that the relevant management may need to be instigated some time prior to when the species may be expected to utilise an area.

Acknowledgements

The Scottish Agricultural College (SAC) receives financial support from the Scottish Executive Environment and Rural Affairs Department. The European Forum on Nature Conservation and Pastoralism exists to increase understanding of the high nature conservation and cultural value of certain farming systems, and to inform work on their maintenance; it arose out of initiatives taken by the Nature Conservancy Council and, subsequently, the Joint Nature Conservation Committee. This article is based on the results from a variety of research projects in which we have worked closely with a number of colleagues across these four organisations. We are particularly grateful to Sue Bignal, Gy Ovenden, Mike Pienkowski, Richard Stillman, and the late Colin Tubbs for their contributions to these research projects, and the development of many of the views presented here.

86   D. I. MCCRACKEN *et al.*

References

Beaufoy, G., Baldock, D. & Clark, J. (1994) *The nature of farming: low-intensity farming systems in nine European countries.* Institute for European Environmental Policy, London, UK. 66 pp.

Beintema, A. J. (1991) Insect fauna and grassland birds. In: *Birds and pastoral agriculture in Europe* (eds. D. J. Curtis, E. M. Bignal & A. M. Curtis), pp. 97–101. Scottish Chough Study Group, Argyll and Joint Nature Conservation Committee, Peterborough, UK.

Beintema, A. J., Thissen, J. B., Tensen, D. & Visser, G. H. (1990) Feeding ecology of charadriiform chicks in agricultural grassland. *Ardea* **79**, 31–44.

Bignal, E. M. & Curtis, D. J. (1989) *Choughs and land-use in Europe.* Scottish Chough Study Group, Argyll, UK. 112 pp.

Bignal, E. M. & McCracken, D. I. (1992) Nature conservation and pastoral farming in the British uplands. *British Wildlife* **4**, 367–376.

Bignal, E. M. & McCracken, D. I. (1996a) The ecological resources of European farmland. In: *The European environment and CAP reform: policies and proposals for conservation* (ed. M. Whitby), pp. 26–42. Centre for Agriculture and Biosciences International, Wallingford, UK.

Bignal, E. M. & McCracken, D. I. (1996b) Low-intensity farming systems in the conservation of the countryside. *Journal of Applied Ecology* **33**, 413–424.

Bignal, E. M. & McCracken, D. I. (2000) The nature conservation value of European traditional farming systems. *Environmental Reviews* **8**, 149–171.

Bignal, E. M., Monaghan, P., Benn, S., Bignal, S., Still, E. & Thompson, P. M. (1987) Breeding success and post-fledging survival in the chough *Pyrrhocorax pyrrhocorax. Bird Study* **34**, 39–42.

Bignal, E. M., Bignal, S. & Curtis, D. J. (1989) Functional unit systems and support ground for the choughs – the nature conservation requirements. In: *Choughs and land-use in Europe* (eds. E. M. Bignal & D. J. Curtis), pp. 102–109. Scottish Chough Study Group, Argyll, UK.

Bignal, E. M., McCracken, D. I. & Curtis, D. J. (eds.) (1994) *Nature conservation and pastoralism in Europe.* Joint Nature Conservation Committee, Peterborough, UK. 156 pp.

Bignal, E. M., McCracken, D. I., Stillman, R. A. & Ovenden, G. N. (1996) Feeding behaviour of nesting choughs in the Scottish Hebrides. *Journal of Field Ornithology* **67**, 25–43.

Bignal, E. M., Bignal, S. & McCracken, D. I. (1997) The social life of the chough. *British Wildlife* **8**, 373–383.

Bignal, E. M., Stillman, R. A., McCracken, D. I. & Ovenden, G. N. (1998) Clutch and egg size in the chough *Pyrrhocorax pyrrhocorax* on Islay, Scotland. *Bird Study* **45**, 122–126.

Blake, S. & Foster, G. N. (1998) The influence of grassland management on body size in ground beetles and its bearing on the conservation of wading birds. In: *European wet grasslands: biodiversity, management and restoration* (eds. C. B. Joyce & P. M. Wade), pp. 163–169. John Wiley, Chichester, UK.

Blake, S., Foster, G. N., Eyre, M. D. & Luff, M. L. (1994) Effects of habitat type and grassland management practices on the body size distribution of carabid beetles. *Pedobiologia* **38**, 502–512.

Blake, S., Foster, G. N., Fisher, G. E. J. & Ligertwood, G. E. L. (1996) Effects of management practices on the carabid faunas of newly-established wildflower meadows in southern Scotland. *Annales Zoologici Fennici* **33**, 139–147.

Blanco, G., Fargallo, J. A. & Cuevas, D. J. (1994) Consumption rate of olives by choughs *Pyrrhocorax pyrrhocorax* in central Spain: variations and importance. *Journal of Field Ornithology* **65**, 482–489.

Curtis, D. J., Bignal, E. M. & Curtis, M. A. (eds.) (1991) *Birds and pastoral agriculture in Europe.* Scottish Chough Study Group, Argyll and Joint Nature Conservation Committee, Peterborough, UK. 137 pp.

Dixon, J., Stanes, A. J. & Hepburn, I. (eds.) (1993) *A future for Europe's farmed countryside.* Royal Society for the Protection of Birds, Sandy, UK.

Fuller, R. J., Gregory, R. D., Gibbons, D. W., Marchant, J. H., Wilson, J. D., Baillie, S. R. & Carter, N. (1995) Population declines and range contractions among lowland farmland birds in Britain. *Conservation Biology* **9**, 1425–1441.

Garcia-Dory, M. A. (1983) Datos sobre la ecologia del género *Pyrrhocorax* (*P. pyrrhocorax* i *P. graculus*) en el Parque nacional de la Montaña de Covadonga. *Alytes* **1**, 411–448.

Goriup, P. D., Batten, L. A. & Norton, J. A. (eds.) (1991) *The conservation of lowland dry grassland birds in Europe*. Joint Nature Conservation Committee, Peterborough, UK. 136 pp.

Hötker, H., Davidson, N. C. & Fleet, D. M. (eds.) (1991) Waders breeding on wet grasslands. *Wader Study Group Bulletin* **61** (Suppl.), 1–107.

McCracken, D. I. & Bignal, E. M. (1998) Applying the results of ecological studies to land-use policies and practices. *Journal of Applied Ecology* **35**, 961–967.

McCracken, D. I. & Foster, G. N. (1993) Surface-active invertebrate communities and the availability of potential food for the chough *Pyrrhocorax pyrrhocorax* on pastures in north-west Islay. *Pedobiologia* **37**, 141–158.

McCracken, D. I. & Foster, G. N. (1994) Invertebrates, cow dung and the availability of potential food for the chough *Pyrrhocorax pyrrhocorax* L. on pastures in north-west Islay. *Environmental Conservation* **21**, 262–266.

McCracken, D. I., Foster, G. N., Bignal, E. M. & Bignal, S. (1992) An assessment of chough *Pyrrhocorax pyrrhocorax* diet using multivariate analysis methods. *Avocetta* **16**, 19–29.

McCracken, D. I., Bignal, E. M. & Wenlock, S. E. (eds.) (1995a) *Farming on the edge: the nature of traditional farmland in Europe*. Joint Nature Conservation Committee, Peterborough, UK. 216 pp.

McCracken, D. I., Foster, G. N. & Kelly, A. (1995b) Factors affecting the size of leatherjacket (Diptera: Tipulidae) populations in pastures in the west of Scotland. *Applied Soil Ecology* **2**, 203–213.

Milsom, T. P., Ennis, D. C., Haskell, D. J., Langton, S. D. & McKay, H. V. (1998) Design of grassland feeding areas for waders during winter: the relative importance of sward, landscape factors and human disturbance. *Biological Conservation* **84**, 119–129.

O'Brien, M. & Smith, K. W. (1992) Changes in the status of waders breeding on wet lowland grasslands in England and Wales between 1982 and 1989. *Bird Study* **39**, 165–176.

Pain, D. J. & Pienkowski, M. W. (eds.) (1997) *Farming and birds in Europe: the common agricultural policy and its implications for bird conservation*. Academic Press, London, UK. 436 pp.

Pain, D. J., Hill, D. A. & McCracken, D. I. (1997) Impact of agricultural intensification of pastoral systems on bird distributions in Britain 1970–1990. *Agriculture, Ecosystems and Environment* **64**, 19–32.

Pienkowski, M. W. & Jones, D. G. L. (eds.) (1999) *Managing high-nature-conservation farmland: policies and practices*. European Forum on Nature Conservation and Pastoralism, Isle of Islay, Argyll, UK. 176 pp.

Poole, A., Petretti, F., Pienkowski, M. W., McCracken, D. I., Petretti, F., Bredy, C. & Deffeyes, C. (eds.) (1998) *Mountain livestock farming and EU policy development*. European Forum on Nature Conservation and Pastoralism, Isle of Islay, Argyll, UK. 199 pp.

Roberts, P. J. (1985) The choughs of Bardsey. *British Birds* **78**, 217–232.

Rolando, A. & Laiolo, P. (1997) A comparative analysis of the diet of the chough *Pyrrhocorax pyrrhocorax* and the Alpine chough *P. graculus* co-existing in the Alps. *Ibis* **139**, 388–395.

Sanchez-Alonso, C., Ruiz, X., Blanco, G. & Torre, I. (1996) An analysis of the diet of red-billed chough *Pyrrhocorax pyrrhocorax* nestlings in NE Spain using neck ligatures. *Ornis Fennica* **73**, 179–185.

Soler, J. J. & Soler, M. (1993) Diet of the red-billed chough *Pyrrhocorax pyrrhocorax* in south-east Spain. *Bird Study* **40**, 216–222.

Tucker, G. M., Heath, M. F., Tomialocjc, L. & Grimmett, R. F. A. (1994) *Birds in Europe: their conservation status*. BirdLife International, Cambridge, UK. 596 pp.

6
Birds as Predators of Lepidopterous Larvae

DAVID M. GLEN

Cardiff University, School of Biosciences (Present address: Styloma Research & Consulting, Phoebe, The Lippiatt, Cheddar, Somerset, BS27 3QP, UK)

Introduction

Insectivorous birds feed on a wide range of foods, but studies of their diets clearly demonstrate the importance of lepidopterous larvae as prey. For example, Royama (1970) concluded that Lepidoptera, particularly larvae, were the major food of nestling great tits (*Parus major*) in his study in mixed woodland near Oxford, UK, as well as in other published studies in oak, Scots pine, Corsican pine, and larch woodland in various countries. Also, an analysis of the diet of 40 bird species living in forest and hedgerow habitats in Slovakia (Kristin & Patocka, 1997) revealed that larvae of the moth families Noctuidae, Geometridae and Tortricidae were the predominant insect prey in the diets of these birds. In particular, hairless, or relatively hairless (= glabrous), larvae more than 5 mm in length were preferred as food to smaller larvae, or those covered in hairs. This preference for glabrous larvae was borne out in an experimental study where three species of North American warblers and the American redstart (*Setophaga ruticilla*) were offered a choice between two hairy lymantriid larvae, gypsy moth (*Lymantria dispar*) and tussock moth (*Orgyia leucostigma* – Lymantriidae) and two glabrous larvae, cabbage looper (*Trichoplusia ni* – Noctuidae) and mealworms (*Tenebrio* sp. – Coleoptera, Tenebrionidae) (Whelan *et al.*, 1989).

The importance of bird predation in the evolution of Lepidoptera has been clearly demonstrated by studies on industrial melanism in adult moths (Kettlewell, 1956). Other adaptations for avoiding predation, such as choice of substrate on which to hide, feeding schedules, and timing of life cycles, are also thought to result from the evolutionary pressure exerted by insectivorous birds (Holmes *et al.*, 1979; Holmes, 1990). The important impact of insectivorous birds on numbers of lepidopterous larvae and pupae was emphasised by Otvos (1979) and Holmes (1990), who reviewed studies of the effects of insectivorous birds in forest ecosystems. Both authors concluded that insectivorous birds exert their greatest influence when insect populations are at low levels, by suppressing populations and delaying build-up to epidemic numbers. Kirk *et al.* (1996) arrived at similar conclusions in their comprehensive review of birds as predators of insect pests in temperate agriculture. Otvos

Insect and Bird Interactions
© Intercept Ltd., PO Box 716, Andover, Hampshire, SP10 1YG, UK.

(1979) considered that birds might increase the time period between pest outbreaks and accelerate the decline of an outbreak. They might also have a significant impact on isolated pest outbreaks as a result of large numbers of birds moving into the outbreak area from surrounding habitats, but he believed that birds have little impact on insect pest outbreaks covering large areas because bird functional and numerical responses are inadequate in such circumstances.

Otvos (1979) concluded that, if insects are exposed to bird predation during a relatively short period, at the time when insects are growing fast, any region of stability in the dynamics of the prey population induced by bird predation must be very limited, if it exists at all. In contrast, if birds prey on overwintering larvae or pupae, then because the prey are available for longer, this predation is more likely to stabilise the prey population over a wider range of densities, and birds may exert a considerable controlling influence. Holmes (1990) arrived at the rather different conclusion that the impact of birds on their prey is likely to be strongest during the breeding season, when the birds' food demands are greatest. Otvos (1979) considered that birds may complement, rather than compete with, other natural enemies of lepidopterous larvae.

The population dynamics of many species of Lepidoptera have been studied because of their importance as pests, or as species of conservation interest. Dempster (1983) reviewed published studies that had lasted for a minimum of six generations, and concluded that predators (mainly birds) acted as key factors determining the size of adult populations in five out of 14 species where key factors had been identified. However, he emphasised that this probably underestimates the importance of predators, since key mortalities in five other species may well have been due to their action. Predators tended to have their greatest impact during the older larval and pupal stages, but Dempster (1983) emphasised that bird predators, like other key factors, may themselves be greatly affected by weather. He illustrated this with the example of the white admiral butterfly (*Ladoga camilla*), where Pollard (1979) showed that loss of older larvae, probably due to bird predation, was the key factor influencing numbers of adult butterflies. However, the operation of this key mortality was influenced by summer temperature, because in colder weather larvae developed more slowly and were available to birds for longer. Dempster (1983) concluded that, for such reasons, the abundance of many species of Lepidoptera may be closely correlated with variations in weather.

Despite the importance of birds as key factors in population dynamics, Dempster (1983) concluded that the impact of natural enemies could possibly be density dependent in only three out of 24 life table studies of Lepidoptera. He rejected the concept of regulation about an equilibrium density in the Lepidoptera in these studies. However, the methods used by Dempster (1983) and others for analysing life tables often fail to detect the regulatory role of natural enemies (Hassell, 1985; Dempster & Pollard, 1986). In the case of generalist predators such as birds, Hassell (1985) suggested that this failure to detect regulation can result from the underlying density-dependent nature of the relationship between predation and prey density being obscured by stochastic elements in the interaction.

This chapter aims to review evidence on bird predation on lepidopterous larvae not considered, or considered only briefly, by Otvos (1979) and Dempster (1983), and to assess the role of insectivorous birds as predators of lepidopterous larvae in the light of this additional evidence. Following on from the conclusions of Otvos (1979), studies of bird predation on overwintering larvae will be considered first, followed by studies of predation on feeding larvae and, finally, the implications of predation by birds on other natural enemies of lepidopterous larvae will be addressed. Most studies discussed here are of species belonging to the three families (Noctuidae, Geometridae, and Tortricidae) shown to be the preferred food of insectivorous birds, and on which we might expect bird predation to have greatest impact. Some references will be made to bird predation on eggs and pupae.

Bird predation on overwintering stages

CODLING MOTH

Because the codling moth (*Cydia pomonella* – Tortricidae) is a major pest of apple orchards, its population dynamics have been widely studied, even in apple orchards where its densities are atypically low. Adult moths lay eggs on developing fruits and nearby leaves, the larvae tunnel into the fruit, eventually reaching the core, where they feed on the developing seeds. At this stage, the larvae are protected from natural enemies, and mortality is relatively low. Mature larvae leave the fruit to search for cocooning sites, and substantial numbers fail to build cocoons (MacLellan, 1958; Geier, 1964; Wearing, 1975, 1979; Glen & Milsom, 1978), but the causes of losses at this stage are often uncertain. Under relatively cool weather conditions, these larvae enter diapause and hibernate, before developing to pupae and adults the following year. However, under warm conditions, many larvae develop immediately, and give rise to a second or third generation of adults.

Larvae in cocoons beneath bark appear to the human eye to be well hidden. However, they are not safe from bird predation, and studies have shown that they are vulnerable to woodpeckers (*Dendrocopus* spp.) in Canada (MacLellan, 1958, 1959, 1970), titmice (*Parus* spp.) in England (Solomon *et al.*, 1976; Glen & Milsom, 1978; Solomon & Glen, 1979; Glen *et al.*, 1981) and silvereyes (*Zosterops lateralis*) in New Zealand (Wearing & McCarthy, 1992). Overwintering larvae are especially vulnerable because they are available as prey to birds for several months of the year (from August/September of one year to May/June the following year in England).

In Nova Scotia, Canada, MacLellan (1958) found that, over a seven-year period, woodpeckers killed 52% of overwintering larvae, and this mortality was important in reducing the numbers of larvae attacking fruit in unsprayed orchards to low levels. Clark *et al.* (1967) attributed an increase in the numbers of larvae attacking fruit, following the use of broad-spectrum insecticides, to a substantial improvement in survival of mature larvae. They pointed out that, if a reduction in bird predation were responsible for increased survival of mature larvae, it would be necessary to propose that woodpeckers did not search sprayed orchards as effectively as unsprayed orchards. This seems plausible, given that the general level of insect abundance could have been greatly reduced in sprayed orchards, which might therefore make the orchard an unproductive habitat for woodpeckers to search for food. In Australia, where bird predators were absent, numbers of overwintering codling moth larvae were determined largely by competition for cocooning sites (Geier, 1964).

Bird predation on codling moth larvae in England has been investigated in a series

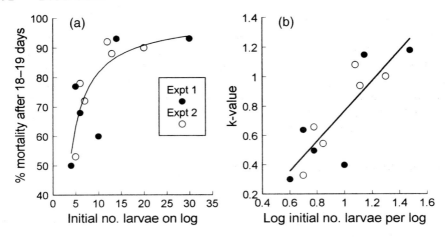

Figure 6.1. (a) Per cent mortality of codling moth larvae beneath bark on logs in relation to initial larval number, after logs were placed in an apple orchard (Bristol, UK) in January, 1972 (Expt 1) and December, 1972 (Expt 2) (Solomon & Glen, 1979). The curve shows 1.8 larvae surviving per log. (b) Mortality expressed as *k*-values in relation to log initial larval number. The line shows the fitted regression with slope = 1.02 ($P < 0.001$).

of experiments and in life table studies at Long Ashton, near Bristol. In three initial experiments (Solomon *et al.*, 1976; Solomon & Glen, 1979), logs were cut from apple trees and mature codling moth larvae were introduced and watched until they built cocoons beneath the bark. The position of each larval cocoon was recorded on a map of each log, with one to 30 larvae being allowed to settle per log. The logs were then tied into natural-looking positions on apple trees in winter. A total of 44 logs were exposed to bird predation in this way, whilst 13 were enclosed in bird-proof netting. Logs protected from birds were examined the following spring, and only a few larvae were found to have died. Logs exposed to birds were examined at regular intervals, and larvae were found to disappear rapidly, normally with characteristic signs of bird predation. Blue tits (*Parus caeruleus*) and great tits were considered to be the main predators. The numbers of larvae on logs with four or more per log initially were reduced to an average of just under two larvae surviving per log after 18–19 days, as shown in *Figure 6.1a*, where the curve represents 1.8 larvae surviving per log at this time. When this mortality is expressed as a *k*-value (Varley *et al.*, 1973) ($\log_{10}$ initial number of larvae – $\log_{10}$ final number of larvae) and plotted against the logarithm of initial larval density, the relationship has a slope of 1 (*Figure 6.1b*). This confirms that bird predation was a fully compensating, spatially density-dependent mortality (Varley *et al.*, 1973). It seems reasonable to suggest that, once there were just under two larvae per log, the time required to find the remaining larvae did not, at that time, justify further time spent on the log. However, despite this, numbers of surviving larvae continued to decline thereafter, indicating that birds searched the logs throughout the winter and spring.

Further experiments in 1975–1976 (Glen & Milsom, 1978) showed that, in the same orchard used for the two winter experiments just described, 56% of mature codling moth larvae leaving the fruit survived to build cocoons beneath the bark. Of those, 86% survived to emerge as adults on trees where the bark was protected from bird predation, but only 14% survived on trees exposed to birds. When cocoons

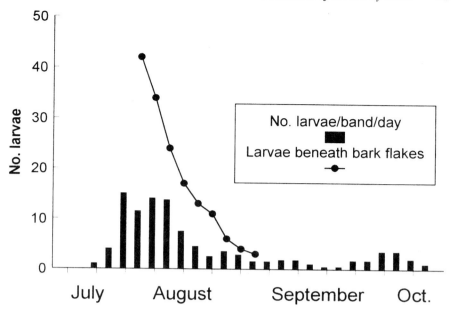

Figure 6.2.   Survival of codling moth larvae beneath flakes of bark glued to trunks and branches of trees in an apple orchard (Bristol, UK) in 1975 in relation to the numbers of mature larvae emerging from fruit, as shown by numbers forming cocoons in corrugated bands around tree trunks (Glen & Milsom, 1978).

beneath flakes of bark were glued to the trunks and branches of trees in the orchard in early August 1975 and 1976, birds soon started to take larvae from the cocoons, and almost all larvae had disappeared by late August (1975) (*Figure 6.2*) or late September (1976). This covered the main period when mature larvae were emerging from fruit in the orchard. Although birds undoubtedly would have found the larvae beneath glued flakes of bark more readily than naturally cocooned larvae, the fact that the former larvae were taken so rapidly indicated that tits were already searching the bark from early August onwards. Thus, it is possible that they had some impact on larvae that developed without diapause to give a second generation of moths. Nevertheless, this impact would inevitably have been much less than on diapausing larvae, because of the great difference in duration of exposure to predation. The rapid reduction in larval numbers in all the above experiments suggests that codling moth larvae are a preferred food, readily taken through late summer, autumn, winter, and early spring.

Spatially density-dependent mortality by polyphagous predators such as birds does not necessarily lead to regulation of the prey, for which density-dependent responses over time are required (Dempster & Pollard, 1986; Latto & Hassell, 1988). Codling moth populations in the experimental orchard in England were studied over six years from 1972 to 1977. Numbers of adults per tree emerging from larvae overwintering beneath the bark were estimated using emergence traps each year (Glen, 1976; Glen & Milsom, 1978), and the numbers of larvae in fruit per tree were also estimated each year during this period. Mortality from the time when mature larvae left fruit until adult moths emerged the following year was expressed as k-values and was shown to be of considerable importance in influencing the number of larva attacking fruit (Glen, 1982). There was no evidence of density dependence over this period of time

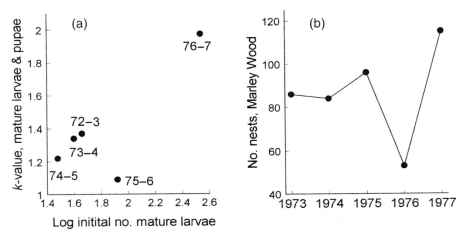

Figure 6.3.   (a) Overwinter mortality of mature larvae and pupae (expressed as *k*-values) over five successive years from 1972 to 1977 in relation to numbers of larvae maturing in fruit (Bristol, UK). (b) Total numbers of blue tits and great tits nesting each year in Marley Wood, Oxford. Bird data from Minot & Perrins, 1986.

(*Figure 6.3a*), but this was largely because the *k*-value for one year (1975–1976) was well below that expected from the trend in the other four. Interestingly, the lower than expected larval mortality in 1975–1976 corresponded with a decline in the numbers of blue and great tits nesting in Marley Wood, near Oxford in 1976 (Perrins, 1979; Minot & Perrins, 1986). Since tit populations are thought to fluctuate synchronously over wide areas (Lack, 1966), it seems possible that a decline in the numbers of blue and great tits may have accounted for the lower than expected mortality of codling moth larvae at Long Ashton in 1975–1976. Fluctuations in breeding numbers of tits are primarily due to variations in post-fledgling survival of young from the previous year, but causes of fluctuations are unknown (Perrins, 1980)). Thus, spatial density dependence did not result in temporal density dependence and regulation, possibly because of a reduction in numbers of tits visiting the orchard in one winter out of five. We can only speculate what might happen over the longer term. Given that codling moth larvae appear to be a preferred food, then it is possible that normal fluctuations in the availability of other food have little impact on the tendency for individual birds to search for and prey on codling moth larvae. However, availability of other foods may well influence overwintering survival and, thus, the numbers of birds searching orchards and the intensity of predation on codling moth larvae. The conclusion from these spatial and temporal studies is that bird predation on overwintering larvae resulted in a relatively sparse population of adult moths emerging in the orchard the following year, with a relatively uniform spatial distribution pattern on trees throughout the orchard, irrespective of the differences in numbers of larvae that had matured on particular trees. This stable distribution in space was not, however, translated into stable numbers of adult moths over time, because disruption to density-dependent predation activities in one year out of five resulted in a considerable increase in adult moth numbers. We can only speculate about how such disruption to density-dependent predation might be expected to occur over time, perhaps in parallel with fluctuations in numbers of blue and great tits over wide geographical areas.

Rather curiously, over the six years of study, a biased sex ratio of 1.8:1 male:female codling moths were found to emerge from bark (Glen *et al.*, 1981). The only exception to this was on trees protected from bird predation in 1975–1976, where equal numbers of both sexes emerged. Moreover, approximately equal numbers of each sex were found amongst larvae reared from eggs laid by wild moths (codling moth larvae can be readily sexed because the male gonads are visible through the dorsal wall of the abdomen). When blue tits were introduced to an aviary containing logs with equal numbers of either male or female larvae, the birds took more female than male larvae (Glen *et al.*, 1981). Thus, the impact of bird predation on the reproductive potential of the codling moth was rather greater than indicated by the figures for percentage predation alone. Because female larvae are, on average, heavier than males, it was initially thought that larger size might account for the greater susceptibility of female larvae to predation. However, when tits were exposed to logs with male and female larvae of similar weights, significantly more female larvae were still taken, and there was no difference to the situation where tits were exposed to heavy females and light males. Head-capsule width appears to be the critical difference that makes the females more susceptible than males. Males have narrower head capsules than females and so are able to squeeze through smaller holes. Thus, it is likely that female larvae tend to build their cocoons in more exposed positions than males. Presumably, in evolutionary terms, the larger size of the female results from a balance between the advantages of larger size, such as greater egg-laying capacity, and the disadvantage of greater susceptibility to predation.

In New Zealand, Wearing (1979) found that silvereyes destroyed, on average, 53% of overwintering codling moth larvae over a seven-year period, with only 5% larval mortality inside bird exclusion cages, which were put in place on certain trees in four years. Wearing & McCarthy (1992) studied predation by silvereyes in one year (1979) on larvae on logs in cages and in the field. Silvereyes readily fed on larvae on logs in cages and the number of larvae removed per bird-hour was linearly related to the initial number of larvae per log. As in studies of predation by tits in England, silvereyes rapidly removed larvae from cocoons beneath bark on logs tied onto trees in an apple orchard in winter, and this predation was spatially density dependent. Predation on logs was in excess of 95%, substantially higher than under natural conditions (60–70%) in Nova Scotia (MacLellan, 1970) or New Zealand (Wearing, 1975). However, in the UK, Glen & Milsom (1978) estimated 83% mortality of larvae in cocoons under natural conditions in 1975–1976, and it is likely, for reasons discussed above, that this was lower than would have been expected from the trend in other years. Wearing & McCarthy (1992) suggest that, where there was lower intensity of predation under natural conditions, this probably reflects the percentage of larvae building their cocoons in natural sites inaccessible to birds on trees, compared to logs.

PINE CONE MOTH

Lack (1966) gives a lucid account of detailed studies by Gibb (1958, 1960) of predation by blue tits and coal tits (*Parus ater*) on larvae of the pine cone moth (*Cydia conicolana* – Tortricidae), which both Betts (1958) and Gibb (1958) called *Ernarmonia conicolana* (Eucosmidae). However, Bradley *et al.* (1979) list this species as

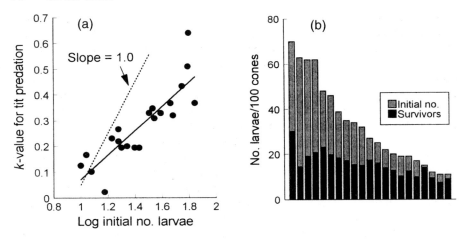

Figure 6.4.   (a) Predation by titmice (expressed as *k*-values) on overwintering larvae of the pine cone moth in relation to initial larval numbers per 100 cones, 1955–1957, Breckland. The line shows the fitted regression with slope = 0.47 (*P* <0.001). (b) Initial number of larvae per 100 cones and number surviving predation by tits. After Lack, 1966; data from Gibb, 1958.

C. conicolana; thus, it belongs to the same genus as the codling moth, and its larvae have similar feeding habits in that they penetrate pine cones and feed on the developing seed. In autumn, the fully-grown larva tunnels to the apical part of the scale, where it hollows out a chamber and overwinters, before pupating the following spring. Predation by tits is the main cause of larval mortality, once the larvae have entered their overwintering chamber. The birds find the hidden larvae mainly by tapping the pine cones. By carefully examining fallen pine cones, Gibb (1958) obtained an accurate record of the numbers of larvae killed by birds, larval parasitoids, and squirrels, as well as the numbers of moths that successfully emerged. The *k*-value for predation on larvae by tits was positively related to the number of larvae per hundred pine cones, as shown in *Figure 6.4a*. This evened out the large differences in initial numbers of larvae per hundred cones (*Figure 6.4b*). However, with this species, the slope of the relationship between the *k*-value for bird predation and log initial density was significantly less than one, indicating that bird predation did not compensate fully for greater larval numbers, and that final numbers increased slightly as initial numbers increased. In one year (1952–1953), when other foods were unusually abundant, birds took only 23% of larvae, less than half that in the three subsequent years (Gibb, 1960).

GYPSY MOTH

Gypsy moth overwinters as egg masses, mainly on tree trunks and branches. Birds are thought to be important predators of these egg masses in Europe, Japan, and North America. Brown & Cameron (1982) reported 35–90% predation by birds in a review of five studies and, in a more recent study in North America, 53% of eggs were taken (Cooper & Smith, 1995). Higashiura (1989) found that birds (mainly nuthatches (*Sitta europaea*)) took 4–71% of eggs over a nine-year period (mean 39%) in Japan. There was evidence of a significant positive relationship between per cent predation and egg mass density over the last five years of study, but not over the full nine year period. The data show percentage predation increasing at low densities, then levelling off. Higashiura (1989) considered that the egg masses were unpalatable to birds, and were therefore consumed mainly during winters with heavy snowfalls, as a last resort in order to avoid starvation. Egg masses under the snow-line survived predation.

Bird predation on foliar-feeding larvae

Larvae of the winter moth (*Operophtera brumata* – Geometridae), feeding on oak and other trees and shrubs, are particularly important prey for titmice in broad-leaved woodland when rearing young. Because the larvae are abundant only for a short period each spring, there is a significant correlation between the timing of egg laying by the great tit and the timing of caterpillar availability, as measured by median pupation date of the winter moth (Lack, 1966; van Noordwijk *et al.*, 1995). Also, clutch size in the great tit is directly related to the number of winter moth larvae (Perrins, 1991). Despite the importance of winter moth caterpillars as prey, Betts (1955) found that blue tits, great tits, and coal tits took less than 3% of feeding larvae in Wytham Wood, near Oxford. The relatively small impact of bird predation on feeding larvae was confirmed by a long-term study of winter moth population dynamics on oak trees in Wytham Wood, where there was no obvious change in numbers between the young and mature larval stages on the foliage (Varley *et al.*, 1973). In an experimental study in Sweden, where birds were excluded from the canopies of oak and birch trees, bird predation had no influence on the abundance of lepidopterous larvae on oak, but significantly reduced numbers on birch (Gunnarsson & Hake, 1999).

In a major study, Royama (1970) investigated the food of nestling great tits in Wytham Wood, which contains a number of tree and shrub species in addition to oak. He found that the composition of different species of Lepidoptera in the diet was not directly related to either seasonal or local abundance of these insects, and he suggested that complex factors governed prey selection. In early season, winter moth larvae predominated, together with *Oporinia dilutata* (Geometridae), in the food of nestlings. Various noctuid larvae were then taken in substantial numbers until, in early June, pupae, and to some extent larvae, of tortricoid species were taken, the pupae being mainly the green oak tortrix (*Tortrix viridana* – Tortricidae). Royama (1970) noted that the most cryptic larvae, which the human eye finds most difficult to locate, appeared to be the most susceptible to predation by great tits. In contrast, only a few feeding larvae of tortricoid species were fed to nestlings, even though larvae were abundant and obvious to the human eye. Royama (1970) suggested that changes in behaviour of green oak tortrix caused by moulting or pupation could have made them more vulnerable or that, probably, declines in availability of other, larger, more profitable prey may have been more important. The major prey species fed to early broods were all tree-feeding larvae, which did not appear in the diet of late broods (after mid-June), which were fed fewer species. Moreover, the total biomass fed to each chick was less than for early broods, confirming the suggestions of earlier authors that the food supply was scarcer for late than early broods. Given that the birds appeared to have difficulty in finding sufficient food for nestlings in later broods, it seems reasonable to suggest that the birds may well have removed a substantially

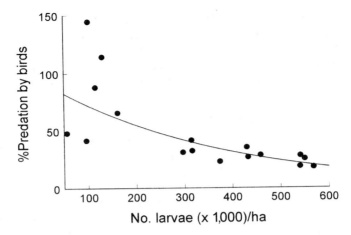

Figure 6.5. Per cent predation by birds on spruce budworm larvae in relation to density of fourth instar larvae in forests in New Hampshire and Maine, USA (Crawford & Jennings, 1989).

greater percentage of these later-appearing larvae than the larvae of species fed to earlier broods.

Spruce budworm larvae (*Choristoneura fumiferana* – Tortricidae) are important prey for North American warblers and other insectivorous birds, which often feed their young largely on this species. Spruce budworm frequently reaches very high and damaging numbers and, at the peak of an outbreak, with 8 million larvae per hectare, birds took only 1% of larvae (Morris *et al.*, 1958). Despite this lack of impact at high budworm density, several workers have suggested that birds may restrict the rate of increase of low-density populations of the spruce budworm (Crawford & Jennings, 1989). Royama (1984) showed that spruce budworm in New Brunswick underwent population cycles with an average periodicity of 35 years. He concluded that predation was not the primary cause of such cycles, but suggested that predation may influence the mean magnitude of oscillations. Crawford & Jennings (1989) showed that 84% of larvae and pupae were taken by birds when budworm densities were low, but the percentage taken by birds declined as prey density increased (*Figure 6.5*). Crawford & Jennings (1989) concluded that the effects of birds in the most suitable habitats were even greater than this average picture, and they suggested that birds can have a powerful influence on spruce budworm populations, but only in woodland habitats capable of supporting substantial numbers of birds when budworm numbers are low. Crawford & Titterington (1979) found that dense homogeneous fir stands supported few birds, but the greater the mixture of spruce and hardwoods in fir stands, the better the habitat was for birds that are important predators of spruce budworm. Crawford & Jennings (1989) suggested that differences in the quality of different areas of forest as bird habitats could explain the patchwork-quilt effect of amplitude in oscillations in budworm numbers reported by Royama (1984). They further suggested that outbreaks could result from budworm populations being released from the control exerted by insectivorous birds; a release due, for example, to a large influx of budworms from other areas with fewer birds, late spring storms causing death of substantial numbers of migrating or nesting birds, or high bird mortality on overwintering grounds.

Experiments where populations of the forest tent caterpillar (*Malacosoma disstria* – Lasiocampidae) were artificially increased in trembling aspen forest in North America (Parry *et al.*, 1997) support the hypothesis that birds can suppress isolated outbreaks of defoliators. In this study, natural populations of the caterpillar were at low density, and birds, especially the northern oriole (*Icterus galbula*), virtually eliminated the experimental populations.

In addition to the above evidence, several studies in North America, Europe, and Japan support the hypothesis that birds often exert considerable influence on leaf-feeding lepidopterous larvae on trees, forest understorey, and field layers. This evidence is summarised briefly below.

Torgersen & Campbell (1982) found that predators, thought to be several species of birds, killed 66–74% of larvae of the western spruce budworm (*Choristoneura occidentalis* – Tortricidae) at two sites in Washington State, USA. No reduction was recorded at a third site, where initial larval densities were twice as great. Torgersen *et al.* (1983) found that 2–49% of artificially placed pupae of the Douglas fir tussock moth (*Orgyia pseudotsugata* – Lymantriidae) were taken by predators, mainly foliar-feeding birds. Over a six-year period (11 generations), this mortality was inversely related to natural pupal densities.

Holmes *et al.* (1979) conducted enclosure experiments in mixed forest in New Hampshire, USA, which demonstrated clearly the impact of insectivorous birds on lepidopterous larvae (mainly Geometridae, Noctuidae, and Tortricidae) feeding on the foliage of striped maple, an understorey shrub. Significantly fewer lepidopterous larvae were recorded on striped maple on which birds were able to feed, than inside enclosures of bird-proof netting. Several species of birds were responsible for this predation and, on average, larval numbers were reduced by 37%. Similarly, white oak saplings enclosed in netting to exclude birds had more chewing insects (mainly lepidopterous larvae) than uncaged or insecticide-treated trees (Marquis & Whelan, 1994). In 1989 and 1990, the greatest leaf area lost was from caged plants, compared to moderate losses for uncaged plants and least damage for insecticide-treated plants (*Figure 6.6a* shows results for 1990). As a result, caged plants produced one-third less above-ground biomass than insecticide-treated plants, with uncaged, unsprayed trees intermediate between these extremes (*Figure 6.6b*). Marquis & Whelan (1994) suggested that observed declines in populations of North American insectivorous birds may result in a long-term reduction in forest productivity, because of the negative effect of larger numbers of chewing insects on tree growth. Similar results have recently been reported on unfertilised willow (*Salix phylicifolia*), which has weak chemical defences against insect herbivores. Bird predation considerably reduced densities of chewing insects and leaf damage, and increased willow growth (Sipura, 1999). However, bird predation had no such influences on fertilised *S. phylicifolia*, nor on a willow species with strong chemical defences against insect attack (*S. myrsinifolia*).

Conner *et al.* (1999) reported that predation by the Carolina chickadee (*Poecile carolinensis*) on the leaf-mining larvae of *Cameraria hamadryadella* – Gracillariidae was inversely density dependent at three spatial scales (leaves, trees, and woodlands). They suggested that this inversely density-dependent predation complements the strongly density-dependent mortality caused by interspecific competition when leaf-miners are at outbreak levels, and therefore predation contributes more to the suppression of the moth than if predation were density dependent.

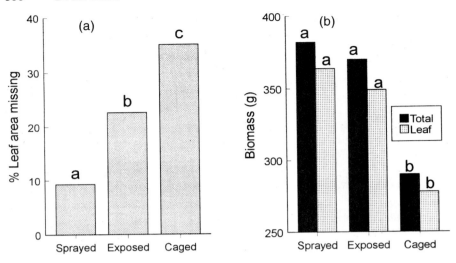

Figure 6.6. Effects of bird predation on chewing insects (mainly lepidopterous larvae) on (a) insect damage to leaves of white oak seedlings, and (b) tree biomass, Missouri, USA, 1990 (Marquis & Whelan, 1994). (Different letters above bars indicate significant differences ($p = 0.05$)).

When patches of bilberry in the field layer of forest at five sites in Sweden were either enclosed in bird-proof netting or left exposed, birds (mainly hazel hen (*Tetrastes bonasia*) chicks, great tits, and pied flycatchers (*Ficedula hypoleuca*) reduced the total density of lepidopterous and sawfly (Hymenoptera: Symphyta) larvae by 63% (Atlegrim, 1989). In particular, geometrid and sawfly larvae, whose feeding behaviour exposed them to birds, were found in lower numbers in unprotected plots than in caged plots. In contrast, tortricid larvae were not significantly reduced in numbers, which Atlegrim attributed to their concealed feeding habit (many tortricid larvae are known as leaf-rollers due to their habit of living within an enclosed space produced by the larva spinning leaves, etc., together). However, it is possible that their low density, and perhaps smaller size, in comparison to geometrids and sawflies could also have contributed, as suggested by Royama (1970) for green oak tortrix (see earlier). The defensive habit of leaf-roller larvae in wriggling vigorously backwards and dropping downwards when disturbed would also make them difficult to catch. Atlegrim (1989) found that larval damage to bilberry plants was much greater on caged than exposed plots. In a study in Japan (Murakami, 1999), birds visited Japanese lilac plants less frequently when leaf rolls produced by tortricid larvae were removed, indicating that the birds used the rolled leaves as visual cues. When birds were excluded, survival of one tortricid species (*Homonopsis foedenratana*) increased, but two other species were unaffected.

Tremblay *et al.* (2001) observed several species of birds visiting maize fields in Quebec Province, Canada and found that cutworm (*Agrotis* spp.) and European corn borer (*Ostrinia nubilalis*) larvae (both Noctuidae) were less abundant in open plots in maize fields than in plots where birds were excluded. This was especially notable in plots near the field edge, which birds were more likely to visit than those in the middle of the field. Interestingly, larvae of the northern corn rootworm (*Diabrotica longicornis* – Coleoptera: Chrysomelidae) were not reduced in numbers by birds.

Despite the importance of lepidopterous larvae as food at nesting time, experiments where selective insecticides, harmful to lepidopterous larvae but not birds, have been used in forests, have shown that the reduced numbers of lepidopterous larvae have not resulted in reduced local breeding densities of songbirds (Newton, 1994). Moreover, the use of selective insecticides has had very little effect on the breeding success of North American warblers (Nagy & Smith, 1997; Holmes, 1998). Nestling diet was not affected in the study by Holmes (1998), although females spent more time foraging. Nevertheless, shifts in diet from lepidopterous larvae to other, less nutritious prey have been recorded by Sample *et al.* (1993), together with a reduction in fat reserves (Whitmore *et al.*, 1993), presumably due to a reduction in overall food availability, increased costs of obtaining food, and a reduction in food quality in these cases.

Effects of birds on predatory and parasitic arthropods

In considering the impact of insectivorous birds on the population dynamics of their prey, Lack (1966) raised the important question that there may also be unknown and unrecognised effects of birds in preying on arthropod natural enemies of moth larvae. For example, how does predation by tits on overwintering larvae of the pine cone moth influence an ichneumon parasitoid and its impact on the moth population? Lack noted that, in studies of predation by woodpeckers on codling larvae in Canada, only 3% of larvae taken by woodpeckers were parasitised by *Ascogaster quadridentata* (Hymenoptera: Braconidae), whereas 14% of the larvae not taken by woodpeckers were parasitised (MacLellan, 1958). In this case, the lower susceptibility of parasitised larvae to attack presumably relates to their being better hidden due to their small size (and head capsule width) in comparison to unparasitised larvae. In addition, Otvos (1979) cites studies where birds were less likely to prey on parasitised larvae of gypsy moth and jackpine budworm (*Choristoneura pinus* – Tortricidae). These are two examples where bird predation should increase natural enemy:pest ratios to the benefit of biological control. Otvos (1979) cites instances where birds have been shown to assist in the spread of insect virus diseases, by eating infected insects and passing live virus particles in their faeces.

An intriguing example of complementary interaction, though in different directions, between bird predation and parasitoids is found in the African butterfly *Danaus chrysippus*, whose larvae feed on various species of milkweed that differ in cardenolide contents. As a consequence, the larvae from various plants differ in their toxicity to birds. Birds selectively prey on larvae with low concentrations of cardenolides, using the shape of the plant where the larvae are feeding as the basis for selection. In contrast, parasitoids selectively attack larvae feeding on plants that are toxic to vertebrate predators, where their chances of survival are therefore greatest. It has been proposed that this contrasting susceptibility of larvae to predation and parasitism is the main mechanism for maintaining polymorphism for palatability in this species (Gibson & Mani, 1984).

In contrast to these examples, Rosenheim (1998) and Sih *et al.* (1998) suggest that higher order predators may often disrupt effective regulation of herbivore populations by lower order natural enemies, such as arthropod predators and parasitoids. For codling moth larvae attacked by birds in England, rates of parasitism were low (unpublished data), but parasitoids *Pimpla turionellae* (Ichneumonidae) attacked pupae beneath flakes of bark glued to trees in spring (Glen & Curtis, 1978). The techniques used in studies of codling moth population dynamics would almost certainly have greatly underestimated the impact of these pupal parasites, and it is quite possible that their impact would have been considerably greater in the absence of bird predation on larvae. Similarly, in a study of *Synaxis cervinaria* (Geometridae) in California, larval parasitoids had little impact because parasitised larvae were more likely to be eaten by predators than unparasitised larvae. However, predators and pathogens reduced numbers of released eggs, larvae, and pupae by more than 99%, and maintained this moth at low densities (Valenti *et al.*, 1998). In spruce forests in south west Sweden, birds (mainly willow tit (*Parus montanus*), crested tit (*Parus cristatus*), coal tit, goldcrest (*Regulus regulus*), and treecreeper (*Certhia familiaris*)) substantially reduced the numbers of spiders over the winter period, particularly those longer than 2.5 mm (Askenmo *et al.*, 1977; Jansson & von Brömssen, 1981). Jansson & von Brömssen (1981) found that bird density in autumn was correlated with the fresh biomass of spiders >2.5 mm in length, and that the mean number of spiders was reduced by 66% over winter. However, numbers of other major prey were reduced to an even greater extent: lepidopterous larvae (91%), aphids (99%), adult and larval Diptera (73%), and Coleoptera (97%).

It is tempting to conclude that, where the proportional impact of insectivorous birds on arthropod natural enemies does not exceed the proportional impact on the herbivorous prey species, then birds are likely to have a neutral effect on interactions between arthropod natural enemies and their prey. Also, as pointed out by Otvos (1979), they may have a benign effect. However, Sih *et al.* (1998) have emphasised that predators feeding on each other can have 'emergent' impacts on prey, i.e. effects that cannot be predicted by simply summing the effects of individual predators.

Conclusions

Studies of insectivorous birds feeding on two species of tortricid moth larvae and gypsy moth eggs over winter reinforce the suggestion by Lack (1966) that birds, at times, destroy a substantial percentage of their prey outside the breeding season. The results are also consistent with his suggestion that insectivorous bird populations may be limited by food shortage at this time of year. In this respect, it is notable that, in experiments where additional winter food has been provided, breeding numbers of titmice increased, compared to control areas, in 8 out of 13 cases (Newton, 1994). Predation by birds (titmice, woodpeckers, or silvereyes) on overwintering larvae was important in the population dynamics of codling moth, because it substantially reduced the number of adult moths (especially females), eggs, and larvae attacking fruit. Although bird predation on codling moth, pine cone moth and, to a lesser extent, gypsy moth was spatially density dependent, there was no evidence of temporal density dependence and, therefore, there was no evidence that it regulated populations. Failure to detect regulation could have resulted from the relatively short periods of study in each case (Hassell, 1985). However, it is likely that lack of regulation resulted from variability in titmice numbers (the codling moth example), and variability in other food supplies (the pine cone moth and gypsy moth examples) from year to year, influencing the relationship between predation and prey density.

Whilst insectivorous birds appear to have a negligible impact at nesting time on feeding larvae of species such as winter moth and spruce budworm when they are abundant, a number of studies have shown that, as suggested by Otvos (1979) and others, bird predation can have substantial impacts on feeding larvae of several moth families when at lower densities. However, leaf-rolling tortricid larvae appear to be less affected by bird predation than noctuid and geometrid larvae. If there is any spatial relationship at this feeding stage between predation and larval density, it is inversely density dependent. Predation on leaf-feeding larvae at low to moderate densities can result in a direct reduction in leaf damage and increased tree growth (Marquis & Whelan, 1994; Sipura, 1999), which will improve economic returns from forestry. The study on willows by Sipura (1999) suggests that bird predation may have direct practical benefits for the use of fast-growing willows as biomass crops for renewable energy. Increased wood production in both forests and willow plantations could contribute to amelioration of global warming, through reductions in atmospheric CO_2. The quality of woodland and farmland as habitats for insectivorous birds undoubtedly has a considerable influence on their ability to suppress larval numbers, and thus stimulate tree growth. As noted by Kirk *et al.* (1996), birds that inhabit farmland make important contributions to pest suppression, and thus birds should be protected for more than aesthetic reasons, or their contribution to local biodiversity.

Interactions between birds and other natural enemies of lepidopterous larvae have received only limited attention, and warrant further study to elucidate the consequences of bird predation, including 'emergent' effects.

References

Askenmo, C., von Brömssen, A., Ekman, J. & Jansson, C. (1977) Impact of some wintering birds on spider abundance in spruce. *Oikos* **28**, 90–94.

Atlegrim, O. (1989) Exclusion of birds from bilberry stands: impact on insect larval density and damage to the bilberry. *Oecologia* **79**, 136–139.

Betts, M. M. (1955) The food of titmice in oak woodland. *Journal of Animal Ecology* **24**, 282–323.

Betts, M. M. (1958) Notes on the life history of *Ernarmonia conicolana* (Heyl.) (Lep., Eucosmidae). *Entomologists Monthly Magazine* **94**, 134–137.

Bradley, J. D., Tremewan, W. G. & Smith, A. (1979) *British tortricoid moths, Tortricidae: Olethreutinae.* Ray Society, c/o British Museum (Natural History), London, UK. 336 pp.

Brown, M. W. & Cameron, E. A. (1982). Natural enemies of *Lymantria dispar* eggs in central Pennsylvania, USA, and a review of the world literature on natural enemies of *L. dispar* eggs. *Entomophaga* **27**, 311–321.

Clark, L. R., Geier, P. W., Hughes, R. D. & Morris, R. F. (1967) *The ecology of insect populations in theory and practice.* Methuen, London, UK. 232 pp.

Conner, E. F., Yoder, J. M. & May, J. A. (1999) Density-related predation by the Carolina chickadee *Poecile carolinensis* on the leaf-mining moth *Cameraria hamadryadella* at three spatial scales. *Oikos* **87**, 105–112.

Cooper, R. J. & Smith, H. R. (1995) Predation on gypsy moth (Lepidoptera: Lymantriidae) egg masses by birds. *Environmental Entomology* **24**, 571–575.

Crawford, H. S. & Jennings, D. T. (1989) Predation by birds on spruce budworm *Choristoneura fumiferana*: functional, numerical, and total responses. *Ecology* **70**, 152–163.

Crawford, H. S. & Titterington, R. W. (1979) Effects of silvicultural practices on bird communities in upland spruce-fir stands. In: *Proceedings management of north central and northeastern forests for nongame birds* (eds. R. M. DeGraaf & K. E. Evans), pp. 110–119. General Technical Report NC-51, United States Forest Service, Washington, DC, USA.

104   D. M. GLEN

Dempster, J. P. (1983) The natural control of populations of butterflies and moths. *Biological Reviews* **58**, 461–481.

Dempster, J. P. & Pollard, E. (1986) Spatial heterogeneity, stochasticity and the detection of density dependence in animal populations. *Oikos* **46**, 413–416.

Geier, P. W. (1964) Population dynamics of codling moth *Cydia pomonella* (L.) (Tortricidae) in the Australian Capital Territory. *Australian Journal of Zoology* **12**, 381–416.

Gibb, J. (1958) Predation by tits and squirrels on the eucosmid *Ernarmonia conicolana* (Heyl.). *Journal of Animal Ecology* **27**, 375–396.

Gibb, J. (1960) Populations of tits and goldcrests and their food supply in pine plantations. *Ibis* **102**, 163–208.

Gibson, D. O. & Mani, G. S. (1984) An experimental investigation of the effects of selective predation by birds and parasitoid attack on the butterfly *Danaus chrysippus* (L.). *Proceedings of the Royal Society of London, B* **221**, 31–51.

Glen, D. M. (1976) An emergence trap for bark-dwelling insects, its efficiency and effects on temperature. *Ecological Entomology* **1**, 91–94.

Glen, D. M. (1982) Effects of natural enemies on a population of codling moth *Cydia pomonella*. In: *Natural enemies and insect pest dynamics* (ed. S. D. Wratten). *Annals of Applied Biology* **101**, 199–201.

Glen, D. M. & Curtis, D. E. (1978) Pupal parasites of codling moth. In: *The use of integrated control and the sterile insect technique for control of the codling moth* (ed. E. Dickler). *Mitteilungen aus der Biologischen Bundesanstalt fur Land- und Forstwirtschaft, Berlin-Dahlem* **180**, 95–96.

Glen, D. M. & Milsom, N. F. (1978) Survival of mature larvae of codling moth (*Cydia pomonella*) on apple trees and ground. *Annals of Applied Biology* **90**, 133–146.

Glen, D. M., Milsom, N. F. & Wiltshire, C. W. (1981) The effect of predation by blue-tits (*Parus caeruleus*) on the sex-ratio of codling moth (*Cydia pomonella*). *Journal of Applied Ecology* **18**, 133–140.

Gunnarsson, B. & Hake, M. (1999) Bird predation affects canopy-living arthropods in city parks. *Canadian Journal of Zoology* **77**, 1419–1428.

Hassell, M. P. (1985) Insect natural enemies as regulating factors. *Journal of Animal Ecology* **54**, 323–334.

Higashiura, Y. (1989) Survival of eggs in the gypsy moth *Lymantria dispar*. I. Predation by birds. *Journal of Animal Ecology* **58**, 403–412.

Holmes, R. T. (1990) Ecological and evolutionary impacts of bird predation on forest insects: an overview. *Studies in Avian Biology* **13**, 6–13.

Holmes, R. T., Schultz, J. C. & Nothnagle, P. (1979) Bird predation on forest insects: an exclosure experiment. *Science, New York* **206**, 462–463.

Holmes, S. B. (1998) Reproduction and nest behaviour of Tennessee warblers *Vermivora peregrina* in forests treated with Lepidoptera-specific insecticides. *Journal of Applied Ecology* **35**, 185–194.

Jansson, C. & von Brömssen, A. (1981) Winter decline of spiders and insects in spruce *Picea abies* and its relation to predation by birds. *Holarctic Ecology* **4**, 82–93.

Kettlewell, H. B. D. (1956) Further selection experiments on industrial melanism in the Lepidoptera. *Heredity* **10**, 287–301.

Kirk, D. A., Evenden, M. D. & Mineau, P. (1996) Past and current attempts to evaluate the role of birds as predators of insect pests in temperate agriculture. In: *Current ornithology, Vol. 13* (eds. V. Nolan & E. D. Ketterson), pp. 175–269. Plenum, New York, USA.

Kristin, A. & Patocka, J. (1997) Birds as predators of Lepidoptera: selected examples. *Biologia* **52**, 319–326.

Lack, D. (1966) *Population studies of birds*. Oxford University Press, London, UK. 341 pp.

Latto, J. & Hassell, M. P. (1988) Generalist predators and the importance of spatial density dependence. *Oecologia* **77**, 375–377.

MacLellan, C. R. (1958) Role of woodpeckers in control of the codling moth in Nova Scotia. *Canadian Entomologist* **90**, 18–22.

MacLellan, C. R. (1959) Woodpeckers as predators of the codling moth in Nova Scotia. *Canadian Entomologist* **91**, 673–680.

MacLellan, C. R. (1970) Woodpecker ecology in the apple orchard environment. *Proceedings of the Tall Timbers Conference on Ecological Animal Control by Habitat Management* **2**, 273–284.

Marquis, R. J. & Whelan, C. J. (1994) Insectivorous birds increase growth of white oak through consumption of leaf-chewing insects. *Ecology* **75**, 2007–2014.

Minot, E. O. & Perrins, C. M. (1986) Interspecific interference competition – nest sites for blue and great tits. *Journal of Animal Ecology* **55**, 331–350.

Morris, R. F., Cheshire, W. F., Miller, C. A. & Mott, D. G. (1958) The numerical response of avian and mammalian predators during a gradation of the spruce budworm. *Ecology* **39**, 487–494.

Murakami, M. (1999) Effect of avian predation on survival of leaf-rolling lepidopterous larvae. *Researches on Population Ecology* **41**, 135–138.

Nagy, L. R. & Smith, K. G. (1997) Effects of insecticide-reduced reduction in lepidopteran larvae on reproductive success of hooded warblers. *Auk* **144**, 619–627.

Newton, I. (1994) Experiments on the limitation of bird breeding densities: a review. *Ibis* **136**, 397–411.

Otvos, I. S. (1979) The effects of insectivorous bird activities in forest ecosystems: an evaluation. In: *The role of insectivorous birds in forest ecosystems* (eds. J. G. Dickson, R. N. Connor, R. R. Fleet, J. A. Jackson & J. C. Krou), pp. 341–374. Academic Press, New York, USA.

Parry, D., Spence, J. R. & Volney, W. J. A. (1997) Responses of natural enemies to experimentally increased populations of the forest tent caterpillar *Malacosoma disstria*. *Ecological Entomology* **22**, 97–108.

Perrins, C. M. (1979) *British tits*. Collins, London, UK. 304 pp.

Perrins, C. M. (1980) Survival of young great tits *Parus major*. *Proceedings of the 17th International Ornithology Congress, Berlin, 1978*, 159–174.

Perrins, C. M. (1991) Tits and their caterpillar food-supply. *Ibis* **133** (supplement), 49–54.

Pollard, E. (1979) Population ecology and change in range of the white admiral butterfly *Ladoga camilla* L. in England. *Ecological Entomology* **4**, 61–74.

Rosenheim, J. A. (1998) Higher-order predators and the regulation of insect herbivore populations. *Annual Review of Entomology* **43**, 421–447.

Royama, T. (1970) Factors governing the hunting behaviour and selection of food by the great tit (*Parus major* L.). *Journal of Animal Ecology* **39**, 619–668.

Royama, T. (1984) Population dynamics of the spruce budworm *Choristoneura fumiferana*. *Ecological Monographs* **54**, 429–462.

Sample, B. E., Cooper, R. J. & Whitmore, R. C. (1993) Dietary shifts among songbirds from a diflubenzuron-treated forest. *The Condor* **95**, 616–624.

Sih, A., Englund, G. & Wooster, D. (1998) Emergent impacts of multiple predators on prey. *Trends in Ecology and Evolution* **13**, 350–355.

Sipura, M. (1999) Tritrophic interactions: willows, herbivorous insects and insectivorous birds. *Oecologia* **121**, 537–545.

Solomon, M. E. & Glen, D. M. (1979) Prey density and rates of predation by tits (*Parus* spp.) on larvae of codling moth (*Cydia pomonella*) under bark. *Journal of Applied Ecology* **16**, 49–59.

Solomon, M. E., Glen, D. M., Kendall, D. A. & Milsom, N. F. (1976) Predation of overwintering larvae of codling moth (*Cydia pomonella* (L.)) by birds. *Journal of Applied Ecology* **13**, 341–352.

Torgersen, T. R. & Campbell, R.W. (1982) Some effects of avian predators on the western spruce budworm in North Central Washington. *Environmental Entomology* **11**, 429–431.

Torgersen, T. R., Mason, R. R. & Paul, H. G. (1983) Predation on pupae of Douglas-fir tussock moth *Orgyia pseudotsugata* (McDunnough) (Lepidoptera: Lymantriidae). *Environmental Entomology* **12**, 1678–1682.

Tremblay, A., Mineau, P. & Stewart, R. K. (2001) Effects of bird predation on some pest insect populations in corn. *Agriculture, Ecosystems and Environment* **83**, 143–152.

Valenti, M. A., Berryman, A. A. & Ferrell, G. T. (1998) Natural enemy effects on the survival of *Synaxis cervinaria* (Lepidoptera: Geometridae). *Environmental Entomology* **27**, 305–311.

van Noordwijk, A. J., McCleery, R. H. & Perrins, C. M. (1995) Selection for the timing of great tit breeding in relation to caterpillar growth and temperature. *Journal of Animal Ecology* **64**, 451–458.

Varley, G. C., Gradwell, G. R. & Hassell, M. P. (1973) *Insect population ecology: an analytical approach*. Blackwells, Oxford, UK. 212 pp.

Wearing, C. H. (1975) Integrated control of apple pests in New Zealand. 2. Field estimation of fifth-instar larval and pupal mortalities of codling moth by tagging with cobalt-58. *New Zealand Journal of Zoology* **2**, 135–149.

Wearing, C. H. (1979) Integrated control of apple pests in New Zealand. 10. Population dynamics of codling moth in Nelson. *New Zealand Journal of Zoology* **6**, 165–199.

Wearing, C. H. & McCarthy, K. (1992) Predation of codling moth *Cydia pomonella* L. by the silvereye *Zosterops lateralis* (Latham). *Biocontrol Science and Technology* **2**, 285–295.

Whelan, C. J., Holmes, R. T. & Smith, H. R. (1989) Bird predation on gypsy moth (Lepidoptera: Lymantriidae) larvae: an aviary study. *Environmental Entomology* **18**, 43–45.

Whitmore, R. C., Cooper, R. J. & Sample, B. E. (1993) Bird fat reductions in forests treated with Dimilin. *Environmental Toxicology and Chemistry* **12**, 2059–2064.

PART 2

Effects of Insecticides on Bird Populations

7
Can Pesticides Cause Reductions in Bird Populations?

MARK I. AVERY, ANDY D. EVANS AND LENNOX H. CAMPBELL

Royal Society for the Protection of Birds, The Lodge, Sandy, Bedfordshire, SG19 2DL, UK

Introduction

In the 1960s and early 1970s, a major battle was fought and won by conservationists in the UK. Widespread agricultural usage of organochlorine products for the purposes of pest control caused mass mortality of game birds, pigeons and seed-eating passerines (Cadbury *et al.*, 1988). The most dramatic effect, however, was the devastation of populations of birds of prey as a result of direct toxicity and the impairment of breeding performance (see also, Walker, 2003). Happily, the severity of the situation was rapidly acknowledged and restrictions placed on the use of these compounds from the late 1960s onwards. Although DDT was not finally withdrawn from all agricultural uses until 1985, bird recovery was well under way, and today UK populations of peregrine falcon (*Falco peregrinus*) and sparrowhawk (*Accipiter nisus*) are at an all-time high.

The same cannot be said for all our bird populations; indeed, many species associated with farmland are in steep decline (*Table 7.1*) and have been since the mid-1970s. Why? When organochlorine compounds were phased out, the need for pest control did not go away. Farmers did not stop using pesticides. The UK government was determined to see the country self-sufficient in terms of food production, and this meant encouraging farmers to maximise crop yields. Furthermore, as a result of the perceived need to protect farmers from fluctuating world prices and American competition, the EU implemented a guaranteed price for grain; this, too, spurred farmers to generate as much produce as physically possible. In order to maximise yields, it is necessary to eliminate both competition for nutrients from non-crop plants and damage by 'pest' insects. The infant agrochemical industry grew quickly, spawning multi-national companies which developed a new generation of pesticides that (if used correctly) were going to poison neither tractor drivers nor wild birds. They were designed, of course, to poison non-crop flora and pest insects, and often were not specific in terms of target species. In fifteen years, cereal fields moved from being ecosystems to virtual monocultures. What effect did all this have on populations

Insect and Bird Interactions
© Intercept Ltd., PO Box 716, Andover, Hampshire, SP10 1YG, UK.

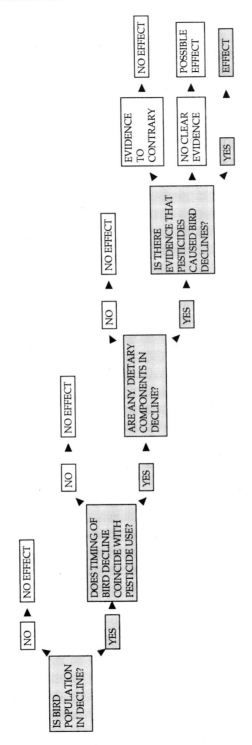

Figure 7.1. Logical screen applied to data in order to assess the degree of likelihood that indirect effects of pesticides were depressing bird populations. Adapted from Campbell *et al.*, 1977.

of farmland birds, species whose diet consists of weed seed in winter and which feed their young on insects that live in arable crops? We might be reassured by official monitoring of the toxicity of new products that farmland birds are no longer being directly poisoned, but have pesticides had a more insidious, yet nonetheless lethal, indirect effect? Have birds simply been starved to death? The work described here was undertaken by the Royal Society for the Protection of Birds (RSPB) as a lead partner of a consortium contracted to assess the degree of the indirect effects of pesticides on birds, and was funded by the then Department of the Environment and English Nature on behalf of the Joint Nature Conservation Committee. Demonstrating the existence of the indirect effects of pesticides is extremely difficult; it requires detailed measures of food availability, demographic parameters, and, ideally, experimental manipulation. We adopted a two-part approach to the review; first, a likelihood assessment procedure whereby a logical 'screen' was imposed on existing data sets to see whether individual species might have been subject to indirect effects, and second, detailed examination and review of existing ecological case studies.

Likelihood assessment procedure

For the purposes of this study, we defined indirect effects of pesticides on birds as: 'a reduction in the size of a population brought about by impaired breeding success and/or full-grown survival as a result of a reduction in food availability caused by pesticide applications, whether or not the food items were the target of those applications'. From this definition, we were able to construct a logical screen to assess whether or not an individual species might have been affected indirectly by pesticide use (*Figure 7.1*). If the answers to all four questions posed in *Figure 7.1* are 'yes' for a given species, then it is possible that it has suffered from the indirect effects of pesticides. This approach is likely to be conservative because, although the indirect effects of pesticides reduce the potential size of the population in subsequent years, they need not necessarily depress population size. For instance, impaired productivity as a result of pesticide use might just slow the rate of increase of an expanding population. There are several ways in which pesticides might indirectly affect bird populations (Campbell *et al.*, 1997); here, we consider just two:

(1) insecticides reduce the abundance of invertebrates which act as a food source for adults and young;
(2) herbicides reduce the abundance of weeds which act as host plants for immature stages of insects, and hence abundance of insects which act as a food source for young birds.

Population trends of farmland birds

We are fortunate in the UK that the populations of most of our breeding birds have been monitored for the past 40 years. We looked at changes in population size (as estimated by the Common Birds Census [CBC] of the British Trust for Ornithology [BTO]) and range (Gibbons *et al.*, 1993) of 40 species which breed on farmland habitats (*Table 7.1*). Of these, 24 have declined and 16 are stable or increasing. Some of the declines have been very severe; numbers of tree sparrow (*Passer montanus*), grey partridge (*Perdix perdix*), turtle dove (*Streptopelia turtur*), bullfinch (*Pyrrhula*

Table 7.1. Population trends of farmland birds in the UK between 1968 and 1999 (data from the BTO Common Birds Census) and range change 1968–1992 (from Gibbons *et al.*, 1993). Species on the Amber List of Birds of Conservation Concern: 2002–2007 are underlined, those on the Red List are in bold (Gregory *et al.*, 2002). id indicates insufficient data available to calculate a trend. Adapted from Campbell *et al.*, 1977.

Species	% UK population change 1968–1999			% UK range change 1968–1992
	Farmland	(All habitats)	Woodland	
Hobby (*Falco subbuteo*)	id		id	+141
Red-legged partridge (*Alectoris rufa*)	−33		id	+32
Grey partridge (*Perdix perdix*)	**−84**		**id**	**−19**
Pheasant (*Phasianus colchicus*)	+36		+24	+1
Stone curlew (*Burhinus oedicnemus*)	**id**		**id**	**−42**
Lapwing (*Vanellus vanellus*)	−40		n/a	−9
Stock dove (*Columba oenas*)	+199		+221	−7
Wood pigeon (*Columba palumbus*)	+66		+228	−2
Collared dove (*Streptopelia decaocto*)	+1581		+761	+7
Turtle dove (*Streptopelia turtur*)	**−80**		**−73**	**−25**
Barn owl (*Tyto alba*)	id		id	−48
Little owl (*Athene noctua*)		(−31)		−11
Skylark (*Alauda arvensis*)	**−58**		**n/a**	**−2**
Swallow (*Hirundo rustica*)	+34		n/a	+1
House martin (*Delichon urbica*)		(−42)		−1
Sand martin (*Riparia riparia*)	id		id	−24
Meadow pipit (*Anthus pratensis*)		(−30)		−3
Pied wagtail (*Motacilla alba yarrellii*)	+77		n/a	0
Yellow wagtail (*Motacilla flava flavissima*)		(−48)		−9
Blackbird (*Turdus merula*)	−38		−9	−2
Song thrush (*Turdus philomelus*)	**−69**		**−46**	**−2**
Mistle thrush (*Turdus viscivorus*)	−54		−15	−2
Robin (*Erithacus rubecula*)	+22		+39	+1
Dunnock (*Prunella modularis*)	−38		−56	−3
Wren (*Troglodytes troglodytes*)	+82		+33	0
Spotted flycatcher (*Muscicapa striata*)	**−80**		**−81**	**−25**
Red-backed shrike (*Lanius collurio*) Effectively extinct in UK: occasional breeder only				
Starling (*Sturnus vulgaris*)	**−64**		**−90**	**−4**
House sparrow (*Passer domesticus*)		**(−34)**		**−5**
Tree sparrow (*Passer montanus*)	**−93**		**id**	**−20**
Chaffinch (*Fringilla coelebs*)	+94		+85	+1
Linnet (*Carduelis cannabina*)	**−47**		**−87**	**−5**
Greenfinch (*Carduelis chloris*)	+19		+6	−3
Goldfinch (*Carduelis carduelis*)	+32		id	−5
Bullfinch (*Pyrrhula pyrrhula*)	**−65**		**−39**	**−7**
Yellowhammer (*Emberiza citrinella*)	**−39**		**−76**	**−7**
Cirl bunting (*Emberiza cirlus*)	**id**		**id**	**−83**
Reed bunting (*Emberiza schoeniclus*)	**−41**		**id**	**−12**
Corn bunting (*Miliaria calandra*)	**−88**		**id**	**−32**

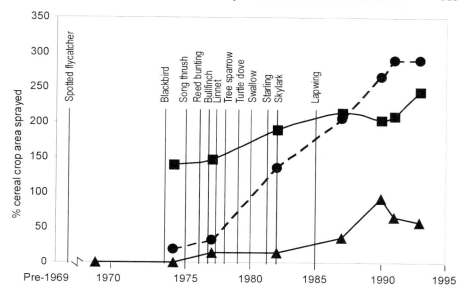

Figure 7.2.   Percentage of cereal crop area sprayed in the UK with herbicides (●), fungicides (■) and insecticides (▲). The years of the estimated start of the decline of twelve species of bird are also shown. Data from Pesticide Usage Survey Group. Adapted from Campbell *et al.*, 1977.

pyrrhula), song thrush (*Turdus philomelus*), and spotted flycatcher (*Muscicapa striata*) had fallen by more than 70%, and of skylark (*Alauda arvensis*), linnet (*Carduelis cannabina*), and reed bunting (*Emberiza schoeniclus*) by more than 50%.

Timing of declines

Data published by the Pesticide Usage Survey Group were used to determine overall trends in herbicides, insecticides, fungicides, and molluscicides (*Figure 7.2*). The years in which declines of bird species began were estimated in a blind experiment involving visual inspection of CBC trends by 12 independent volunteers (Campbell *et al.*, 1997). This procedure allowed the start of declines to be identified for 12 out of the 18 declining species. Temporal associations between the start of bird declines and pesticide usage were ascribed for all four product categories according to the following conditions (*Table 7.2*):

> Decline started when less than 10% of cereal area was sprayed = no association
> Decline started when 10%–50% of cereal area was sprayed = probable association
> Decline started when more than 50% of cereal area was sprayed = association

Trends in dietary components

The principal dietary components of 37 of the 40 species considered (excluding the three predatory species, hobby (*Falco subbuteo*), barn owl (*Tyto alba*), and little owl (*Athene noctua*)) were identified through a comprehensive literature review (Wilson *et al.*, 1996). The invertebrate components of these species' diets are summarised in *Table 7.3*. There were three invertebrate groups which appeared proportionately

Table 7.2.   Assessment of temporal associations between declines of farmland bird species and the extent of the use of the four main groups of pesticides. Adapted from Campbell *et al.*, 1977.

	Herbicides	Fungicides	Insecticides	Molluscicides
Spotted flycatcher	?	X	X	X
Blackbird	Y	P	X	X
Song thrush	Y	P	X	X
Reed bunting	Y	P	P	X
Bullfinch	Y	P	P	X
Linnet	Y	P	P	X
Tree sparrow	Y	Y	P	X
Turtle dove	Y	Y	P	X
Swallow	Y	Y	P	X
Starling	Y	Y	P	P
Skylark	Y	Y	P	P
Lapwing	Y	Y	P	P

X, No association – less than 10% of cereals treated at the beginning of population decline.
P, Possible association – between 10% and 50% of cereals treated at the beginning of population decline.
Y, Association – more than 50% of cereals treated at the beginning of population decline.
?, Comparable usage data not available at start of decline, but may have been considerable.

Table 7.3.   Presence of invertebrate taxa and vertebrates in the diet of farmland birds. Bird species are arranged in order of magnitude of population change with the species in greatest decline on the left. Unshaded, not known to be taken as food; grey, present, but not an important dietary component; black, an important component. Adapted from Campbell *et al.*, 1977.

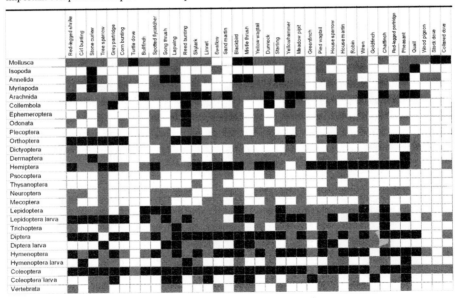

more often in the diets of declining than increasing species; Coleoptera, Orthoptera, and Lepidoptera (predominantly larvae).

Trends in the populations of farmland invertebrates

Unfortunately, there are few data pertaining to widespread and long-term changes in

Table 7.4.   Summary of long-term trends in invertebrate abundance. Adapted from Donald, 1998.

Category	Trend	Source
All invertebrates	--	GCT[1]
Springtails (Collembola)	++	GCT
Aphids (Hemiptera)	Overall =, 1 sp ++, 1 sp +++	RIS[2]
Aphids (Hemiptera)	---	GCT
Aphid specific predators[3]	=	GCT
Polyphagous predators[4]	--	GCT
Parasitoid wasps	--	GCT
Carabids	Overall =, *Bembidion lampros* ++	GCT
Chrysomelids	Overall -, *Gastrophysa polygoni* --	GCT
Staphylinids	--	GCT
Moths: woodland	=	RIS
Moths: farmland	--	RIS
Butterflies: widespread	Recent increases in some species	BMS
Butterflies: localised	Decrease[5]	BMS
Sawflies	--	GCT
Predatory flies	=	GCT
Spiders/harvestmen	--	GCT

GCT = Game Conservancy Trust Sussex data (1972–1989) (Aebischer, 1990, 1991; Aebischer & Potts, 1990).
RIS = Rothamsted Insect Survey (1962–1990) (Woiwod, 1991; Woiwod & Harrington, 1994).
BMS = Butterfly Monitoring Scheme (1976–1992) (Pollard & Yates, 1993).
- = decline of <25%, -- = decline of 25%–75%, --- = decline of >75%.
+ = increase of <25%, ++ = increase of 25%–75%, +++ = increase >75%.
= no detectable trend.
Notes: 1 Excluding springtails (Collembola) and mites (Acari).
2 *Metopolophium festucae* and *Sitobion avenae* increased significantly.
3 Ladybirds, lacewings, soldier beetles, and hoverflies.
4 Spiders, harvestmen, ground beetles, rove beetles, earwigs and predatory flies.
5 24 of 32 farmland species have declined.

farmland invertebrate populations in the UK. The findings of a literature review (Donald, 1998) are summarised in *Table 7.4*. Overall, the picture (although far from complete) is one of stability or decreasing abundance. Many more data are available on short-term effects of pesticides on invertebrate abundance. Although there are clearly differences in the magnitude of effect between different chemicals, the timing of applications, and the invertebrate species considered, the overall picture is consistently one of significant reductions in the abundance of a wide range of potential bird food species (Campbell *et al.*, 1997). Moreover, although recovery of such invertebrate species can occur in a matter of weeks (under experimental plot conditions at least), even a few weeks are long enough to have a significant negative effect on bird reproductive success if spraying occurs in the breeding season.

Case studies

GREY PARTRIDGE

The grey partridge has been the subject of a study by the Game Conservancy Trust (GCT) on the South Downs since 1961. As a result, more research has been done on its population ecology than for any other farmland bird. Despite being prized as a quarry species, grey partridge have undergone a severe decline on UK farmland (*Table 7.1*). Several possible reasons for this decline were investigated, including the recent dramatic changes in farming practice as a result of the drive for ever increasing crop yields, and an increase in predator populations resulting from a decline in gamekeeping. As part of the review (Campbell *et al.*, 1997), R. E. Green assessed the evidence from the GCT project for the indirect effects of pesticides having played a significant part in the decline of the grey partridge. In particular, he addressed two questions:

(1) how much of the decline in partridge population is attributable to the effects of pesticides on the food supply of partridge chicks?
(2) does reduction in pesticide application result in increased partridge populations via improved food supply, and hence chick survival?

The diet of the chicks has been well established as a result of the examination of faecal samples and crop contents. They eat a wide range of arthropods, including Arachnida, Collembola, aphids, heteropteran bugs, larvae of Lepidoptera and Symphyta (sawflies), adult Diptera, adults and larva of a range of beetles including carabids, chrysomelids, and staphylinids. Several studies have demonstrated that chick survival rate (as determined by August brood counts) is positively correlated with the abundance of a range of insect taxa (*Table 7.5*). It has also been established by several studies on a variety of scales that the abundance of partridge chick food in crops can be diminished by pesticide applications (Potts, 1986; Sotherton *et al.*, 1987, 1993). Perhaps the most convincing aspect of this study is, however, that if pesticide applications are reduced experimentally, partridge chick survival increases (Rands, 1985, 1986; Sotherton *et al.*, 1993). Moreover, in one study, the spring density of partridges doubled after reduction of pesticide use, but did not change in nearby areas where management remained unaltered (Sotherton *et al.*, 1993).

The overall conclusion is that this study provides good evidence for the operation of indirect effects of pesticides on birds. It should also be noted, however, that other aspects of agricultural change will also have served to reduce the availability of chick-

Table 7.5. Insect taxa whose abundance was correlated significantly with the survival rate of grey partridge chicks in three studies (Green, 1984; Potts & Aebischer, 1991). The table summarises the results of multiple regression models with chick survival rate (from brood counts) as the dependent variable and median (Potts) or mean (Aebischer & Green) densities of prey taxa from vacuum net and sweep net samples in cereals as the independent variables. Taxa are ranked in descending order of their unstandardised partial regression coefficients. Non-significant and negative coefficients are excluded.

Source	Rank	Taxon
Potts (1977) (Sussex)	1	preferred beetles: weevils, diurnal carabids and chrysomelids
	2	sawfly larvae
	3	large Hemiptera: mainly mirid bugs and planthoppers
	4	small Hemiptera: mainly aphids
Potts & Aebischer (1991) (Sussex)	1	carabid and elaterid beetles
	2	sawfly, lepidopteran and lacewing larvae
	3	weevils and chrysomelid beetles
	4	large Hemiptera: mainly heteropteran bugs and planthoppers
	5	aphids
Green (1984) (East Anglia)	1	sawfly and lepidopteran larvae
	2	chrysomelid larvae
	3	acalypterate diptera
	4	aphids
	5	other Hemiptera: mainly mirid bugs and leafhoppers

food insects. A good example is the loss of undersowing cereal crops with grass to provide a ley the following season. This will have reduced the availability of Symphyta larva to grey partridge chicks (Aebischer, 1990).

CORN BUNTING

Corn buntings are closely associated with arable farmland, and the UK population has also undergone a recent severe decline (*Table 7.1*). A recent study carried out in partnership between the GCT, RSPB, English Nature (EN) and Sussex University investigated the ecology of corn bunting on farmland. The study site was that used by the GCT in their long-term investigation of grey partridge ecology on the South Downs. The aims were to determine the primary reasons for the decline, and to devise practical management solutions.

Chick diet was assessed by faecal analysis and found to be similar to that of the grey partridge.

Arachnida, Orthoptera, larvae of Lepidoptera and Symphyta, and Coleoptera made up 95% of over 700 invertebrates counted (Brickle & Harper, 1999). Moreover, in cereal fields that were sprayed with pesticides, the abundance of chick-food insects was inversely related to the number of applications. Adults searching for food for their nestlings tended to forage where abundance of key invertebrate groups was high, and hence in those areas which received fewer pesticide applications. Where chick-food invertebrate abundance was low near the nest, adults tended to fly further and forage for longer than when it was higher. Moreover, when chick food abundance was low, the chicks themselves were in poorer body condition and their probability of fledging was lower than when chick food abundance was high. Although the scale of this study was such that an experimental manipulation was impractical, the results do offer compelling evidence that pesticides do indirectly affect breeding success of corn buntings.

Overall conclusions

The indirect effects of pesticides on grey partridge have been proven, and there is compelling evidence that breeding success of corn buntings has been reduced indirectly by pesticides. When the logical screen is applied to the evidence presented above, the most striking result is the size of the knowledge gaps due to data deficiency, particularly on long-term trends of invertebrates. There are, however, a number of species for which the indirect effects of pesticides cannot be ruled out (*Table 7.6*). For these species, there is no evidence to refute the hypothesis that pesticides cause population declines by reducing food availability, and thereby adversely affecting demographic parameters. On the other hand, there is much circumstantial evidence to support this hypothesis. It should also be remembered that we believe this analysis to be conservative, because pesticides could have indirect effects on the demography of a bird species without causing a population reduction. As pointed out earlier, a reduction in breeding success might, for example, simply slow the growth rate of an expanding population.

Conservation action

We believe that the circumstantial evidence for the indirect effects of pesticides on

Table 7.6. Conclusions: have farmland bird species declined because of the indirect effects of pesticides? Adapted from Campbell *et al.*, 1977.

Indirect effects of pesticides	Bird species[1]
• **are a major factor in population declines,** demonstrated by detailed field and experimental research.	Grey partridge
• **cannot be ruled out as a major factor in population declines,** because of close temporal associations between the decline of the species and the extent of pesticide use, and current knowledge of species' diet and ecology.	Lapwing Turtle dove Skylark Swallow Blackbird Song thrush Starling Tree sparrow Linnet Bullfinch Reed bunting
• **cannot be ruled out on the basis of current knowledge of diet and ecology,** though no close temporal association between the decline of the species and the extent of pesticide usage has been shown.	Sand martin[4] Yellow wagtail Mistle thrush Dunnock Spotted flycatcher[3] Red-backed shrike[2] Yellowhammer Corn bunting
• **can be ruled out as unlikely,** on the basis of current knowledge of the species' diet and ecology.	Stone curlew Barn owl Little owl Cirl bunting[4]
• **are either absent or of insufficient magnitude to cause population declines; and the species are not currently declining.**	Hobby Red-legged partridge[5] Pheasant[5] Stock dove Wood pigeon Collared dove House martin Meadow pipit Pied wagtail Robin Wren House sparrow Chaffinch Greenfinch Goldfinch Quail

[1] Each group of species is listed in order of change in population size or range, as in *Table 7.1*.
[2] Insufficient data available to assess patterns of trends.
[3] Decline started at time when, although data are not available, herbicide usage was expanding and a temporal association cannot be totally ruled out.
[4] Indirect effects may have contributed at start of decline.
[5] Real trends may be masked by widespread release of captive birds.

declining farmland birds is so strong that the government (who are committed through the Rio Convention to halting and reversing the declines) should adopt the precautionary principle. By this we mean that they should take action to reduce pesticide use without waiting for scientific proof of these damaging effects. Action might include provision of management prescriptions within agri-environment schemes to compensate farmers for profit foregone through reducing pesticide inputs on crops, especially crop margins. We are delighted to see such provision within some ESAs and the Countryside Stewardship Scheme, but would like to see the prescriptions more widely available through the Broad and Shallow Scheme. Another way of ameliorating these effects would be to encourage a more cautious and targeted use of pesticides. We believe this could be achieved by imposing a differential tax on pesticides to favour specific products which tend to be more expensive than broad-spectrum chemicals.

The recent growth in demand for organic produce is encouraging, and we are pleased to see the government's commitment to increasing the resources available to farmers who wish to convert from conventional farming. However, organic farming is far from perfect, and we feel that organic husbandry could be improved to benefit biodiversity. For instance, mechanical weeding is arguably worse for skylarks (it destroys their nests) than is spraying with a herbicide.

New technologies

The advent of precision farming appears to offer opportunities to reduce any indirect effects of pesticides. Precise mapping of pest accumulations and yield variation within fields could well lead to far more targeted pesticide applications. Other new technologies appear to pose serious threats to biodiversity. We are particularly concerned about the development of genetically modified herbicide-tolerant crops. These would appear to encourage wider use of broad-spectrum products; a move in the opposite direction to that which we would advocate. Insect resistant crops are also being produced by this technology; the biotechnology industry is claiming that these crops will lead to a reduction in pesticide use, which is bound to have environmental benefits. This position ignores the possibilities of

(1) potentially disastrous gene transfer to wild plants;
(2) the possibility of secondary toxicity through a build-up of toxins in the animals which prey on the target insects if the latter are not killed with sufficient rapidity.

Although there are well defined procedures for deciding whether a particular product, be it a pesticide or a GM crop, is suitable for release on a commercial scale, these do not as yet consider the indirect effects of the product on the environment. We believe that indirect effects should be taken into consideration in these procedures.

As well as adopting a precautionary approach, we believe it is essential to plug some of the knowledge gaps evident in this review. We are delighted that the Department of the Environment, Food and Rural Affairs (DEFRA) and the Pesticides Safety Directorate have commissioned further research on the subject of the indirect effects of pesticides on birds, but we remain concerned that a large-scale experiment, involving replicates and reversals, is the only way to demonstrate the scale of these effects in the wider countryside.

References

Aebischer, N. J. (1990) Assessing pesticide effects on non-target invertebrates using long-term monitoring and time-series modelling. *Functional Ecology* **4**, 369–373.

Aebischer, N. J. (1991) Twenty years of monitoring invertebrates and weeds in cereal fields in Sussex. In: *The ecology of temperate cereal fields* (eds. L. G. Firbank, N. Carter, J. F. Darbyshire & G. R. Potts), pp. 305–331. Blackwell Scientific Publications, Oxford, UK.

Aebischer, N. J. & Potts, G. R. (1990) Long-term changes in the number of invertebrates assessed by monitoring. *Proceedings of the Brighton Crop Protection Conference – Pests and Diseases – 1990* **3B–4**, 163–172.

Brickle, N. W. & Harper, D. C. G. (1999) Diet of nestling corn buntings *Miliaria calandra* in southern England examined by compositional analysis of faeces. *Bird Study* **46**, 319–329.

Cadbury, C. J., Elliot, G. & Harbard, P. (1988) Birds of prey conservation in the UK. *RSPB Conservation Review* **2**, 9–16.

Campbell, L. H., Avery, M. I., Donald, P., Evans, A. D., Green, R. E. & Wilson, J. D. (1997) *A review of the indirect effects of pesticides on birds*. JNCC Report, no. 227. Joint Nature Conservation Committee, Peterborough, UK. 147 pp.

Donald, P. J. (1998) Changes in the abundance of invertebrates and plants on British farmland. *British Wildlife* **5**, 279–289.

Gibbons, D. W., Reid, J. B. & Chapman, R. A. (1993) *The new atlas of breeding birds in Britain and Ireland: 1988–1991*. Poyser, London, UK. 520 pp.

Gregory, R. D., Wilkinson, N. I., Noble, D. G., Robinson, J. A., Brown, A. F., Hughes, J., Procter, D. A., Gibbons, D. W. & Galbraith, C. A. (2002) The population status of birds in the United Kingdom, Channel Islands and the Isle of Man: an analysis of conservation concern 2002–2007. *British Birds* **95**, 410–450.

Green, R. E. (1984) The feeding ecology and survival of partridge chicks *Alectoris rufa* and *Perdix perdix* on arable farmland in the UK. *Journal of Applied Ecology* **21**, 817–830.

Pollard, E. & Yates, T. J. (1993) *Monitoring butterflies for ecology and conservation*. Chapman & Hall, London, UK. 274 pp.

Potts, G. R. (1977) Some effects of increasing the monoculture of cereals. In: *Origins of pest, parasite, disease and weed problems* (eds. J. M. Cherret & G. R. Sagar), pp. 183–202. Blackwell Scientific Publications, Oxford, UK.

Potts, G. R. (1986) *The partridge: pesticides, predation and conservation*. Collins, London, UK. 274 pp.

Potts, G. R. & Aebischer, N. J. (1991) Modelling the population dynamics of the grey partridge: conservation and management. In: *Bird population studies: their relevance to conservation and management* (eds. C. M. Perrins, J.-D. Lebreton & G. J. M. Hirons), pp. 373–390. Oxford University Press, UK.

Rands, M. R. W. (1985) Pesticide use on cereals and survival of grey partridge chicks: a field experiment. *Journal of Applied Ecology* **22**, 49–54.

Rands, M. R. W. (1986) The survival of game bird chicks in relation to pesticide use on cereals. *Ibis* **128**, 57–64.

Sotherton, N. W., Moreby, S. J. & Langley, M. G. (1987) The effects of the foliar fungicide pyrazophos on beneficial arthropods in barley fields. *Annals of Applied Biology* **111**, 75–87.

Sotherton, N. W., Robertson, P. A. & Dowell, S. D. (1993) Manipulating pesticide use to increase the production of wild game birds in Britain. In: *Quail III: National Quail Symposium* (eds. K. E. Church & K. V. Dailey), pp. 92–101. Missouri Department of Conservation, Jefferson City, USA.

Walker, C. H. (2003) Organochlorine insecticides and raptors in Britain (*this volume*).

Wilson, J. D., Arroyo, B. E. & Clark, S. C. (1996) *The diet of bird species of lowland farmland: a literature review*. Report to Department of the Environment and English Nature, Department of Zoology, Oxford University, Oxford, UK. 86 pp.

Woiwod, I. P. (1991) The ecological importance of long-term synoptic monitoring. In: *The ecology of temperate cereal fields* (eds. L. G. Firbank, N. Carter, J. F. Darbyshire & G. R. Potts), pp. 275–304. Blackwell Scientific Publications, Oxford, UK.

Woiwod, I. P. & Harrington, R. (1994) Flying in the face of change: the Rothampsted Insect Survey. In: *Long-term experiments in agricultural and ecological sciences* (eds. R. A. Leigh & A. E. Johnston), pp. 321–342. CAB International, UK.

8

Interactions Between an Endangered Bird Species, Non-endemic Insect Pests, and Insecticides: The Deployment of Insect Growth Regulators in the Conservation of the Seychelles Magpie-Robin (*Copsychus sechellarum*)

JOHN P. EDWARDS

Central Science Laboratory, Department for Environment, Food and Rural Affairs, Sand Hutton, York, YO41 1LZ, UK

Introduction

The Seychelles magpie-robin (*Copsychus sechellarum*) (*Figure 8.1*) is one of the world's rarest birds (Mountfort, 1988). In 1991, the population was estimated to be only about 23 individuals, all inhabiting a single island (Frégate) in the Seychelles. The island of Frégate occupies about 210 hectares and comprises an interior granite hill (125 m) and three flat coastal areas. Human habitation on the island consisted of one 12-bedroom hotel and approximately 12 small dwelling houses. In 1991, the human population on the island was about 30 permanent residents (islanders), and about 20 tourists augmented this number each week during the peak holiday season.

Formerly, the magpie-robin was found on several other islands in the Seychelles group, including Mahé, Praslin, Marianne, Félicité, and La Digue (Penny, 1974). However, following the introduction of cats and rats some time after the first colonisation of the islands in the mid-eighteenth century, the last magpie-robin was seen on Mahé in 1867 (Penny, 1974) and, by the mid-1950s, the bird had become extinct on all islands in the Seychelles group, except Frégate. Earlier, a small population of magpie-robins had been introduced to the island of Alphonse in the Amirantes, where they apparently thrived until the 1940s, but had disappeared by 1965 (Penny, 1974).

It is likely that predation by rats and feral cats, coupled with the loss of suitable habitats, has been largely responsible for this decline in the majority of original habitats. Since 1940, the only remaining indigenous colony of the species has been confined to Frégate Island (which had never been colonised by rats), and in 1977/78,

Insect and Bird Interactions
© Intercept Ltd., PO Box 716, Andover, Hampshire, SP10 1YG, UK.

Figure 8.1.   The Seychelles magpie-robin (*Copsychus sechellarum*). Photo: J. P. Edwards.

about 40 birds occupied some 12 breeding territories. This despite the introduction of feral cats to the island in 1960 – although many of these were subsequently killed in a government campaign (Mountfort, 1988). In 1978, an attempt was made to introduce some of the Frégate birds to the island of Aride, but within a few years, only a single male remained. In 1981, there was a further outbreak of feral cats on the island, and the population on Frégate was reduced to about 20 birds (Mountfort, 1988). A cat eradication scheme was immediately introduced on the island. Despite the success of this scheme (cats were finally eliminated in 1983), the magpie-robin population has remained critically low, and in 1989 (Rands & Komdeur, 1990), there were about 26 birds occupying eight territories. Despite intensive conservation efforts by the International Council for Bird Preservation (ICBP, now BirdLife International) to revive the population, the magpie-robin continues to be a critically endangered species.

Factors contributing to the decline of the magpie-robin

The results of research studies undertaken by the International Council for Bird Preservation/BirdLife International prior to 1991 had identified a number of factors that may have contributed to the decline of this species on Frégate Island. Since neither cats nor rats had been present on the island for several years, reduction of population due to predation by these animals was not considered to be a factor.

Natural (endemic) predators include the Wright's skink and endemic snakes. Both are nest predators, capable of removing eggs and very small chicks, but are unlikely to take fledged young or adult birds. The relative impact of predation by these endemic predators (and by the introduced Indian mynah bird (*Acridotheres tristis*)) is not known, although it is thought that virtually all egg and small chick disappearances are attributable to one of these main predators. Apart from possible problems associated with the use of pesticides, and reduction of suitable habitat (see below), the recovery of the magpie-robin was not thought likely to be adversely influenced by humans. Islanders do not deliberately kill magpie-robins for food – in fact, the bird has become part of the folklore of Frégate (as an omen of good luck) and is encouraged by the islanders, who often feed the birds in the vicinity of their houses. Similarly, many of the tourists that visit the island do so specifically to see this rare bird, and their donations actively contribute to the conservation of the species. Moreover, the fact that the magpie-robin has become endeared to the islanders may have been a very important factor in ensuring its survival on Frégate. This active encouragement of a bird by islanders is by no means always the case, as there is evidence that another Seychelles species (e.g. the Seychelles kestrel (*Falco araea*), which is believed by some Seychellois to be unlucky and an omen of death) is killed if it approaches human habitation on Mahé (Penny, 1974).

The quality and quantity of both feeding and breeding habitats has declined on the island. This is mainly due to changes in vegetation associated with the commercial production of agricultural produce (e.g. banana and coconut). As a result, there are fewer ideal active feeding habitats (open ground beneath the tree canopy with abundant leaf litter), and fewer nesting sites (holes in older, rotted trees). Currently, the conservation agencies are continuing a management plan to increase and improve nesting and feeding sites. Although breeding takes place all year round, the clutch size is only one. Thus, if the population is to increase, breeding success must be maintained at a very high level. At present, this is not the case, for a variety of reasons. First, nest predation and lack of suitable nesting sites undoubtedly reduce breeding success. Second, the lack of abundant food supplies (primarily arthropods) close to breeding sites limits the time spent on the nest, and thus the time available for incubation and defence against predators. Third, disturbance and competition (possibly by introduced bird species, e.g. the Indian mynah) reduces successful breeding. Fourth, it has been shown that food supply is critical, so much so that supplementary feeding (with large cockroaches) markedly improves breeding performance. Finally, recent investigations have implicated pesticides used to control cockroaches and ants in dwelling houses and the hotel as a possible contributory factor in reducing breeding success.

The interaction between magpie-robins and insects on Frégate

It is clear that the magpie-robin is primarily an insectivorous bird. Native paintings on islanders' houses often depict the birds perched on the back of giant tortoises. Observations in the field suggest that the birds frequently associate with these large reptiles, and feed on insects exposed in leaf litter and decaying wood disturbed by the movement of these animals through the dense woodland habitat occurring over much of the island.

Little is known about the native insect fauna of Frégate or, indeed, of the Seychelles in general. Perhaps the most well known insect on Frégate is the indigenous giant tenebrionid beetle (*Pulposipes herculeanus*), which exists only on this island. In 1991, these strange beetles were still relatively common on the island, and the large larvae and pupae could be found in decaying wood amongst the leaf litter in forested areas, where they are sometimes exposed by the movement of giant tortoises. It is tempting to speculate that these insects may, originally, have formed a substantial part of the magpie-robin's natural diet. Observation on the island in 1991 suggested that, in addition to this remarkable coleopteran, a representative number of typical insect types (mosquitoes and other Diptera, Lepidoptera, and Hymenoptera) were also present. Significantly, reports from conservation workers on the island (subsequently confirmed by the author) indicated the presence of substantial populations of several non-indigenous insect species (primarily, ants and cockroaches) on the island (see below).

Insecticide use and the decline of the magpie-robin

Household insecticides began to be used regularly against indoor insect pest populations on Frégate during the 1980s. Two insect species (one cockroach and one ant) cause problems in the hotel and houses on Frégate. Examination of specimens sent to CSL from field workers on the island identified the cockroach as the American cockroach *Periplaneta americana*, and the ant as *Monomorium destructor*. The cockroach is an introduced, cosmopolitan pest species that can survive both inside and outside buildings on Frégate. The ant is also an introduced species, and may also be able to live outside buildings in such tropical climates. Both insect pests are potential disease carriers (*Salmonella*, etc.), are unpleasant in domestic dwellings, and cause fouling (cockroaches) and physical damage to electrical cables and installations (ants). Since tourism is a major source of income for the island, the use of insecticides to control insect populations in the hotel and in domestic dwellings was necessary to maintain adequate standards of public health. Thus, by 1991, conventional insecticides were frequently used against these pests in many domestic dwellings. These insecticides are potentially harmful to the magpie-robins for several reasons. First, unlike many rare species, the magpie-robin is often found feeding close to human habitation – indeed, there are confirmed reports of birds actually feeding inside the houses (huts) of agricultural workers and hotel staff. With the observed propensity of these birds to feed on cockroaches (including dead specimens), it is quite possible that they could consume insecticide-treated cockroaches, and thus receive insecticide more or less continuously in their diet. Eventually, this could result in the accumulation of a lethal dose. The insecticides known to have been used on the island include propoxur, dichlorvos, tetramethrin, d-phenothrin, fenitrothion, and bendiocarb. Of these, neither tetramethrin nor d-phenothrin are particularly toxic to birds (acute oral LD_{50} – bobwhite quail (*Colinus virginianus*) = >1,000 mg/kg and >2,500 mg/kg respectively). By contrast, propoxur is highly toxic to birds (LD_{50} – red-winged blackbird (*Agelaius phoeniceus*) = 2–6 mg/kg), fenitrothion has high avian toxicity (LD_{50} – bobwhite quail = 28 mg/kg), and dichlorvos is highly toxic to both mammals and birds (LD_{50} – chicken (*Gallus gallus domesticus*) = 15 mg/kg). Likewise, the avian toxicity of bendiocarb (LD_{50} – bobwhite quail = 19 mg/kg;

mallard duck (*Anas platyrrynchos*) = 3.1 mg/kg) is also relatively high. Thus, it became clear that a number of these compounds could pose a substantial toxic hazard to the magpie-robins. Moreover, as well as the possibility that the birds might accumulate a toxic dose of these pesticides, there may, in addition, be sub-acute effects on the birds' behaviour. Furthermore, it seemed likely that use of these insecticides inside buildings did not always conform to the manufacturers' recommended application rates; over-enthusiastic use of pesticide is not uncommon in such circumstances. For these reasons, it was considered that the widespread use of these pest control chemicals may have played a significant part in the continuing endangered status of the magpie-robin. In particular, concern was expressed over the use of the product containing dichlorvos (10 g a.i./kg), and evidence of its widespread use (the presence of aerosol cans) could be found in houses and hotel rooms throughout the island. Although there was no incontrovertible evidence that this, or any other insecticide, was directly implicated in the plight of the magpie-robin, the critically low population level of this rare bird was such that conservationists were determined to eliminate all possible threats to the birds' recovery. Moreover, some indirect evidence implicating the possible adverse effects of insecticide use on the birds was derived from the fact that, over a two-year period during which accurate records were kept, the mortality of adult birds was much higher (58%) in those populations occupying territories near to human habitation than that recorded (10%) in adults occupying territories away from human habitation (death in adults is relatively infrequent in this rather long-lived species). Thus, the conservation agency felt it essential to remove any possible hazard associated with insecticide use on the island. However, since the management authorities on the island did not feel it appropriate to remove the islanders' right to control cockroach and ant infestation in their homes and, since control of cockroaches and ants is essential in the hotel for reasons of public health and for the maintenance of appropriate standards for the tourist trade, alternative insect control measures were sought to combat the problems of ants and cockroaches, whilst eliminating the risk of acute or sub-acute pesticide poisoning of these rare birds. Furthermore, no large-scale eradication programme aimed at eliminating the outdoor population of *P. americana* (even if this was considered to be technically possible) was deemed advisable since conservation workers had shown that these insects form a substantial and important part of the birds' diet. Thus, there was an urgent need to implement alternative methods for the control of indoor populations of cockroaches and ants, whilst ensuring that these measures did not have an adverse effect on the recovery of the magpie-robin population.

Environmentally-sensitive alternatives to conventional insecticides

Since the early 1970s, scientists at the Central Science Laboratory have been investigating novel, environmentally-compatible insect control techniques, based on insect-specific, physiologically-active molecules. These studies have resulted in the commercial development of the insect juvenile hormone analogue, methoprene, for the eradication of Pharaoh's ant (*Monomorium pharaonis*) infestations in hospitals and other domestic premises (Edwards, 1975; Edwards & Clark, 1978). *Monomorium destructor* is very closely related to *M. pharaonis*, and it was thought highly likely that baits containing methoprene would prove effective in controlling infestations of the former species in buildings on Frégate. More recently, research at CSL has concentrated on the development of another novel molecule (hydroprene) for use against cockroach infestations in similar situations (Edwards & Short, 1988; Short & Edwards, 1992). Our studies have demonstrated that (S)-hydroprene will eradicate infestations of the Oriental cockroach *Blatta orientalis* in simulated domestic environments (Edwards & Short, 1993). The American cockroach *P. americana* is related to *B. orientalis*, and we were confident that the former species would be controlled by similar treatment. Both methoprene and hydroprene are essentially non-toxic to vertebrates (acute oral LD_{50} (methoprene and hydroprene) – rat = >34,000 mg/kg; acute oral LD_{50} (methoprene) mallard duck = >2,000 mg/kg – highest dose tested). In addition, these compounds appear to have no mutagenic, teratogenic, or endocrine effects in non-arthropods. Both compounds work by preventing reproduction and/or development in insect populations, and can be delivered to ants in the form of attractive (food-based) baits, and to cockroaches from simple point-sources (rather like 'Vapona®' strips). In addition, although both compounds are reasonably stable indoors, they are unstable in ultraviolet light and in non-sterile (natural) water (Quistad *et al.*, 1975; Schooley & Quistad, 1979), and degrade rapidly in the outdoor environment to form relatively harmless products like acetic acid and carbon dioxide. Thus, methoprene and hydroprene are ideal candidates for use against insect pest populations in situations where negligible environmental and toxicological impact are essential, and are particularly suitable for use in circumstances where endangered vertebrate species are potentially at risk from conventional insecticides. Consequently, we advised ICBP/BirdLife International that the use of these novel molecules would provide effective, long-term control of indoor cockroach and ant populations, and remove completely any hazard to the magpie-robin population posed by the existing use of conventional toxic pesticides.

Materials and methods

PROGRAMME FOR IMPLEMENTATION OF NOVEL INSECT CONTROL METHODS ON
FRÉGATE ISLAND

In July 1991, the author visited Frégate Island to establish a programme of ant and cockroach control in the houses and hotel based on the technologies described above. The programme had the following main objectives:

(1) to assess the level of cockroach and ant infestation in houses and in the hotel;
(2) to identify prime 'target sites' for implementation of cockroach and ant control measures;
(3) to implement cockroach and ant control measures in houses and hotel;
(4) to instruct appropriate persons in the continued use of prescribed measures;
(5) to discourage the use of conventional insecticides (especially those containing dichlorvos or propoxur) for ant and cockroach control;
(6) to develop a large-scale breeding unit for *P. americana* to aid the supplementary feeding programme;
(7) to contribute, where appropriate, to the existing conservation plan being undertaken on the island.

Assessment of cockroach and ant infestation levels

The presence and identity of ants and cockroaches in buildings on Frégate were determined by a combination of test baiting, trapping, visual inspection, and interrogation of local inhabitants. To test for the presence of ants, an attractive, food-based bait (a mixture of honey and peanut butter, 1:1 by weight) was placed in small plastic tubes which were taped in likely situations in all the inhabited buildings (including stores, hotel rooms, and the two shops) on the island. The tubes were capped with a plastic top through which a small (4 mm diameter) hole had been bored to permit the entry of ants, but which restricted the entry of larger insects and other creatures. Baits were not laid at any predetermined intensity, but a total of 190 test baits were laid in the buildings throughout the island. Baits were left in place for approximately 24 hours, whereupon they were inspected for the presence of ants. In addition, a subjective estimate of the numbers of ants in each tube was made to give an indication of the intensity of the infestation, and a record was kept of the infesting species. Initially, it was intended to use sticky traps (AgriSense BCS Ltd., Mid Glamorgan, Wales, UK) to identify the presence of cockroach infestations in buildings on the island. However, initial attempts to use these traps in the vicinity of the conservation project house resulted in the accidental trapping of three small reptiles (two endemic skinks and a gecko). Subsequently, the skinks were released alive, although the gecko died on the trap. For this reason, we felt it inappropriate to continue to use sticky traps on a large scale, and no further trapping of cockroaches took place. In the absence of sticky traps, cockroach infestation was assessed by visual inspection and by questioning residents and hotel staff. Although this method of assessing cockroach infestation is rather subjective and unreliable, it became clear that the island residents felt that ants were a more serious problem than cockroaches. Nevertheless, large numbers of cockroaches (always *P. americana*) were found in several locations, and it was decided to proceed as if all buildings harboured some cockroach infestation.

Treatment of ant and cockroach infestations and reduction of conventional insecticide usage in domestic premises

Ant infestations were treated with food-based baits (honey/peanut butter 1:1 w/w) containing 'Pharorid®' mixed together to give a final concentration of methoprene [(S)-isopropyl 11-methoxy-3,7,11-trimethyl-2(E),4(E)-tridecadienoate] of 0.25% a.i. Approximately 5 g of this mixture was placed in small plastic tubes with a perforated cap, and these tubes were taped in appropriate positions in infested buildings. These treatment baits were placed at an approximate intensity of 1 bait/3 sq metres floor area. In all, 220 treated baits were laid in buildings throughout the island.

Cockroach infestations were treated with hydroprene [(S)-ethyl-3,7,11-trimethyl-2(E),4(E)-tridecadienoate], delivered from prototype point-source releasers supplied by the manufacturers (Zoecon (Sandoz) Crop Protection, Dallas, Texas and Palo Alto, California, USA). Details of these devices remain confidential, but consist essentially of a hydroprene-impregnated circular release pad enclosed in a perforated plastic case. These devices were taped in accordance with the manufacturer's instructions in suitable positions in all buildings on the island. During the treatment, islanders were discouraged from using conventional insecticide sprays, and asked not to touch or otherwise interfere with baits and point-source releasers. The islanders were also encouraged to improve standards of hygiene and the structural integrity of buildings. In addition, towards the end of the treatment period, the total existing supply of aerosols containing dichlorvos was purchased by ICBP staff and removed from the island shop, and an undertaking was obtained from the island management that these stocks would not be replaced.

Provision of cockroaches for supplementary feeding

The provision of cockroaches for the supplementary feeding programme was tackled in two ways. Initially, because of the urgent need for cockroaches in order to feed a nest containing a newly hatched chick, cockroaches were trapped in the grounds around the project house. The traps used were simple pitfall traps constructed from empty glass food jars, the sides of which were coated with a thin film of a mixture of paraffin wax and paraffin oil (approx.1:1 v/v) to prevent trapped cockroaches from escaping. Subsequently, because it was found that trap catches were being depleted by endemic skinks feeding on the trapped cockroaches, these traps were equipped with metal lids in which a slit was made to permit entry of cockroaches, but to exclude predatory lizards. These traps were baited with sliced banana and honey, and placed in likely positions each evening. Each morning the traps were inspected and some adult cockroaches removed for the large-scale breeding colonies (see below). The remaining cockroaches were frozen in the freezing compartment of a domestic refrigerator, before being taken out into the field and fed to magpie-robins. Magpie-robins were fed from a simple, wooden bird table, which prevented skinks from eating the cockroaches before the birds. The birds adapted quickly to using the tables, and usually appeared within a few seconds of the arrival of each batch of cockroaches. When active nests were being fed, feeding was carried out twice a day (cockroaches permitting). Because local trapping of cockroaches produced small and variable numbers of insects for supplementary feeding, it was decided to set up large confined breeding colonies of *P. americana*. These colonies were housed in two large plastic dustbins, the sides of which were smeared with the paraffin wax/oil mixture to prevent escape. The captive cockroaches were fed on commercially available poultry feed, and water was provided in glass jars with cotton wool wicks. Initially, these colonies were stocked with adult cockroaches (predominantly females) trapped in the project house grounds as described above. Lids were placed on these dustbins to prevent other creatures entering the colonies, and the bins were placed in a shaded position to minimise fluctuations in temperature.

Results and discussion

ASSESSMENT OF ANT AND COCKROACH INFESTATION LEVELS

About 20% of the 190 baits laid to assess ant populations were negative (i.e. did not attract ants), and about 5% were empty (i.e. the bait had been completely removed, but no ants were present). With the exception of a few baits that were missing, the remaining baits all contained ants. The most frequently occurring species was *M. destructor* (present in 23.6% of bait tubes), although several other species were also detected. These were, in decreasing order of abundance, *Tapinoma melanocephalum* (20.6%), *Pheidole megacephala* (12.1%), *Paratrechina longicornis* (7.9%), *Tapinoma eraticum* (6.8%), and *Tetramorium guiniense* (1.6%). These are all common 'tramp' species that have been distributed worldwide, largely as a result of trade between nations. Interestingly, we found no evidence of any indigenous ant species, although it should be remembered that we did not bait extensively in outdoor areas. Importantly, we found no evidence of *Anoplolepis longipes* on the island. This latter species is known to be present on the main island (Mahé) where, because of its extremely aggressive nature, it poses serious threat to indigenous fauna, including ant species (Lewis *et al.*, 1976). More recently (C. Feare, pers. commun.), *A. longipes* has been found on another island (Bird Island), where its presence and aggressive nature is regarded as a potential threat to ground-nesting Sooty terns (*Sterna fuscata*).

One or more of the above pest ant species were detected in the majority of buildings on the island, and it was decided to bait all buildings with methoprene. During a period of about three weeks, baits were laid, as described above, in all buildings on the island. Visual inspection of baits approximately 24 hours after placement indicated that they were, indeed, attractive to foraging workers of *M. destructor*, *P. megacephala*, and *T. guiniense*. However, the baits appeared much less attractive to *Paratrechina longicornis* and *Tapinoma melanocephalum*.

For reasons described above, no sticky trap surveys were carried out to determine cockroach infestation levels. However, *P. americana* was present in several buildings at levels that were detectable on visual inspection. Most islanders that were interviewed said that they had cockroaches in their dwellings, but that these were not as serious a problem as ants. No other cockroach species were observed in buildings, although specimens of *Pycnoscelus surinamensis* and an unidentified member of the family Ectobiidae were occasionally trapped in jar traps placed outside the project house.

EFFECTIVENESS OF TREATMENTS AGAINST ANTS AND COCKROACHES

Since the objective of the project was not primarily to test the effectiveness of the molecules used (rather to remove the potential hazard posed by conventional pesticides to the magpie-robin), no formal post-treatment monitoring of insect populations was undertaken. In previous studies, both methoprene and hydroprene have been shown to be effective at controlling *M. pharaonis* and *P. americana*. Moreover, subsequent observations on the island by conservation workers indicated that *M. destructor* infestations had been substantially reduced six months after the initial baiting with methoprene. In addition, cockroach infestations appear to have been kept under control by the hydroprene treatments (A. Gretton, pers. commun.). Most importantly, use of conventional neurotoxic insecticides in buildings has been substantially reduced, and neither dichlorvos nor propoxur are being sold on the island.

PROVISION OF COCKROACHES FOR SUPPLEMENTARY FEEDING

Despite the establishment of large-scale breeding colonies of *P. americana* in plastic dustbins, population increase in these colonies was not as rapid as was initially expected. In part, this may have been due to problems with excessive condensation inside the dustbins, leading to substantial mould growth on the food material. Subsequently, it was suggested that the poultry food being used to feed these colonies may have been contaminated with insecticide and/or fungicide incorporated by the manufacturers. Although we were unable to substantiate this, as a precaution, an alternative supply of pesticide-free poultry food was obtained for subsequent use. Simultaneously, large numbers of cockroaches were being caught during evening visits to the pigsty, where they were simply swept from the walls into a large container (A. Gretton, unpublished). The combination of breeding colonies and various cock-roach trapping techniques provided sufficient material for supplementary feeding of the birds (particularly of active nesting pairs), and this undoubtedly improved breeding success in a number of cases.

Conclusions and current status of the Seychelles magpie-robin

This project has illustrated how novel, environmentally-friendly insecticides are not only less damaging to the environment, but may also be positively used in situations where other (non-target) organisms are endangered. The introduction of the insect juvenile hormone analogues, methoprene and hydroprene, on Frégate as replace-ments for conventional toxic pesticides has given good control of ants and cockroaches in domestic buildings, whilst posing negligible hazard to the endangered population of magpie-robins on the island. Such molecules may have an increasing role to play in other situations where insect pest management and wildlife conservation must be undertaken in the same habitat. The use of these 'environmentally-sensitive' pesti-cides, and the reduction in the use of more toxic materials, coupled with traditional conservation measures undertaken by BirdLife International, did result in an increase in the numbers of magpie-robins on Frégate, and by 1997 the population had increased to 37 individuals (Lucking *et al.*, 1997). Furthermore, at this time, 16 birds had been relocated to the island of Cousin, 7 to Cousine, and one to Aride, making a total population of 61 birds. Unfortunately, recent redevelopment of Frégate Island (including the construction of a new hotel) has led to the introduction of rats. These pressures on the Frégate population have accelerated the continued deliberate reloca-tion of the magpie-robins to other suitable islands in the group, where it is hoped that the bird will become re-established, and eventually thrive.

Thus, the future of this endangered species remains precarious, and the population remains potentially vulnerable to the adverse impact of declining habitat, predation, and the unconsidered or inappropriate use of conventional pesticides.

Acknowledgements

The author wishes to thank the following companies, organisations, and individuals for financial and/or technical support for this project. AgriSense-BCS Ltd., Treforest, Mid Glamorgan, UK; the British Agrochemicals Association, Peterborough, Lincs., UK; Central Science Laboratory, Sand Hutton, York, UK; the government of the Republic of the Seychelles; International Council for Bird Preservation (now BirdLife International), Cambridge, UK; Killgerm Chemicals Ltd., Ossett, West Yorkshire, UK; Royal Society for the Protection of Birds, Sandy, Beds., UK; Zoecon Corporation

(Sandoz), Dallas, Texas, and Palo Alto, California, USA. In addition, I wish to thank Dr. M. Rands (ICBP) and Drs. P. Stanley and A. Hardy (CSL) for their support and encouragement in the setting up of this project. Also, a special vote of thanks to Adam Gretton (ICBP, Frégate Island) for his friendship and enthusiastic help with this part of the magpie-robin recovery programme, and to Mrs. M. Rainey for help with the project.

References

Edwards, J. P. (1975) The effects of a juvenile hormone analogue on laboratory colonies of the pharaoh's ant *Monomorium pharaonis* (L) Hymenoptera, Formicidae. *Bulletin of Entomological Research* **65**, 75–80.

Edwards, J. P. & Clark, B. (1978) Eradication of Pharaoh's ants with baits containing the insect juvenile hormone analogue, methoprene. *International Pest Control* **20**, 5–10.

Edwards, J. P. & Short, J. E. (1988) Prospects for controlling cockroaches using insect juvenile hormones. *Proceedings of the 8th British Pest Control Conference, Stratford-upon-Avon, 1988, Paper No. 12.* British Pest Control Association, London, UK. 6 pp.

Edwards, J. P. & Short, J. E. (1993) Elimination of a population of the Oriental cockroach (Dictyoptera, Blattidae) in a simulated domestic environment with the insect juvenile hormone analogue (S)-hydroprene. *Journal of Economic Entomology* **86**, 436–443.

Lewis, T., Cherrett, J. M., Haines, I., Haines, J. B. & Mathias, P. L. (1976) The crazy ant (*Anoplolepis longipes* Jerd.) (Hymenoptera, Formicidae) in Seychelles, and its chemical control. *Bulletin of Entomological Research* **66**, 97–111.

Lucking, R., Lucking, V. & Sivell, D. (1997) *The Seychelles magpie-robin recovery plan: progress report, no. 27.* BirdLife International, Cambridge, UK. 14 pp.

Mountfort, G. (1988) *Rare birds of the world.* Collins, London, UK. 256 pp.

Penny, M. (1974) *The birds of Seychelles and the outlying islands.* Collins, London, UK. 160 pp.

Quistad, G. B., Staiger, L. E. & Schooley, D. A. (1975) Environmental degradation of the insect growth regulator methoprene isopropyl (2E,4E)-11-methoxy-3,7,11-trimethyl-2,4,-dodecadienoate). III. Photodecomposition. *Journal of Agricultural and Food Chemistry* **23**, 299–303.

Rands, M. & Komdeur, J. (1990) *Saving the Seychelles magpie-robin* Copsychus sechellarum*: a recovery plan for one of the world's most threatened birds.* International Council for Bird Preservation Report. ICBP, Cambridge, UK. 13 pp.

Schooley, D. A. & Quistad, G. B. (1979) Metabolism of insect growth regulators in aquatic organisms. In: *Pesticide and xenobiotic metabolism in aquatic organisms* (eds. M. A. Khan, J. J. Lech & J. J. Menn), pp. 161–176. ACS Symposium series, no. 99. American Chemical Society, Washington, DC, USA.

Short, J. E. & Edwards, J. P. (1992) Effects of hydroprene on development and reproduction in the Oriental cockroach *Blatta orientalis.* *Medical and Veterinary Entomology* **6**, 244–250.

9
Organochlorine Insecticides and Raptors in Britain

COLIN H. WALKER

Cissbury, Hillhead, Colyton, Devon, EX24 6NJ, UK

Introduction

DDT was the first of a number of different organochlorine insecticides introduced into agriculture following the end of the Second World War. These compounds soon came to be very widely used in western Europe and North America, for spraying crops, dressing seed, controlling soil pests, and dipping sheep. They were also used for vector control, rodent control, wood treatment, and other purposes. It soon became apparent that some of them were both strongly persistent and highly toxic to many vertebrate species. Given their wide pattern of use and their marked biological persistence, it was hardly surprising that they were found in many different species, and that some of them tended to undergo biomagnification with passage along food chains.

Because of this tendency to undergo biomagnification, the highest residues of the most persistent compounds were found in predators at the top of terrestrial and aquatic food chains. Thus, there was early concern about the possible harmful effects upon populations of predatory vertebrates. Amongst the earliest investigations of harmful effects of organochlorine (OC) insecticides upon predatory vertebrates in the field were those upon raptorial birds in Britain. These studies were originally undertaken by the Nature Conservancy at Monks Wood Experimental Station, near Huntingdon, UK. This was an appropriate place to undertake these wide-ranging investigations at this time. First, the laboratories, which officially opened in 1963, had an interdisciplinary Toxic Chemicals and Wildlife Section which was able to carry out all major aspects of the work – population studies, residue studies, and related work on metabolism and toxicological studies with birds. Further, Britain is a relatively small island, and there were, and still are, a considerable number of active amateur ornithologists. Thus, there were already substantial records of the population numbers and the breeding success of certain predatory birds before and during the period of intensive use of OCs. Also, there were many knowledgeable amateur ornithologists willing to offer their unpaid services to collect dead birds for analysis, and to participate in population studies on birds.

Insect and Bird Interactions
© Intercept Ltd., PO Box 716, Andover, Hampshire, SP10 1YG, UK.

The following account summarises the main findings from two studies of the kind described – one on the sparrowhawk (*Accipiter nisus*), the other on the peregrine (*Falco peregrinus*). In a concluding section, these findings will be related to results from other studies in other parts of the world, before considering the hazards of persistent lipophilic pollutants to predators in a more general way.

The organochlorine insecticides

Two main types of organochlorine insecticides (OCs) have caused environmental problems due to biological persistence and consequent biomagnification in food chains. These are (1) DDT and a few related compounds (principally DDD formulated as rhothane), and (2) the cyclodiene insecticides, principally aldrin, dieldrin, and heptachlor. Both groups act as neurotoxins, although they differ in their sites of action within the nervous system. DDT acts on sodium channels, thus prolonging action potentials, and thereby disturbing the transmission of nerve impulses. Poisoning is characterised by unco-ordinated tremors and twitches before more extreme neurotoxic symptoms are shown. The cyclodienes act upon GABA receptors which function as inhibitory receptors. The action of cyclodienes on GABA receptors can lead to convulsions (Eldefrawi & Eldefrawi, 1990).

The marked persistence associated with OCs is sometimes due to the parent insecticides, sometimes to their stable metabolites, and sometimes to both. The main insecticidal ingredient of DDT is the pp' isomer, which is strongly persistent. Its principal metabolite, pp' DDE, is even more persistent. The compound pp' DDD, which is produced by the reductive metabolism of pp' DDT, has a similar persistence to the parent compound. Aldrin is rapidly metabolised to dieldrin, which is highly persistent and neurotoxic. Similarly, heptachlor is rapidly metabolised to the persistent and neurotoxic compound, heptachlor epoxide. These highly persistent molecules have two basic properties in common; they are strongly lipophilic and very resistant to biotransformation, only undergoing very slow conversion to water soluble and readily excretable metabolites.

The relationship between these pesticides and their stable metabolites is shown in *Table 9.1*. It is evident that there were considerable differences between the different original insecticides with regard to their uses in agriculture, and therefore in their main routes of entry into ecosystems. DDT was used for a wide variety of purposes, whereas the more toxic cyclodienes were generally more restricted, their use as seed

Table 9.1.   Some organochlorine insecticides yielding persistent residues.

Insecticides	Principal residues	Main uses
DDT	pp'DDE, pp'DDT, pp'DDD	Crop sprays Vector control Veterinary (ectoparasites)
Aldrin (HHDN)	Dieldrin (HEOD)	Soil insecticide
Dieldrin	Dieldrin	Seed dressing Crop sprays Vector control Veterinary
Heptachlor	Heptachlor epoxide	Seed dressing

dressings probably representing their main route of entry into agricultural ecosystems in Britain. The concentrations of the cyclodiene compounds on newly dressed seed were typically of the order of 1,000 ppm, thus presenting a serious toxic hazard to seed-eating vertebrates. The introduction of DDT into Britain began shortly after the end of World War Two, whereas the cyclodienes did not appear until the mid 1950s.

Residues and their interpretation

The analysis of bird tissues and eggs for OC residues at Monks Wood commenced in 1963. Before that time, only a limited number of analyses had been carried out, mainly to investigate incidents reported on agricultural land as part of the Wildlife Incident Investigation Scheme run by the Ministry of Agriculture, Fisheries and Food (MAFF). Thus, continuous records of residues in sparrowhawks and peregrines go back to 1963.

In the following account, residue levels of pp' DDT, pp' DDE, dieldrin, and heptachlor epoxide in these two species will be presented, and their toxicological interpretation discussed. In the following section, the long-term effects on populations of the two raptors will be considered in the light of the OC residues found.

RESIDUES IN SPARROWHAWKS

Unfortunately, regular analysis of tissues of raptors found dead in the field did not commence until many years after the first reports that implicated the OCs in the poisoning of wild birds and mammals on agricultural land. By 1963, the use of OCs had been seriously restricted. Consequently, there are few residue data for the period when these compounds were in greatest use, and were evidently having their most serious effects (Newton & Haas, 1984; Newton, 1986). However, some use of OCs continued between 1963 and 1985, a period for which there are good residue data for the sparrowhawk. In particular, there was a substantial increase in the sowing of winter wheat, which was still dressed with dieldrin up to and including 1975.

The levels of dieldrin in the livers of sparrowhawks found dead in the field were related to the areas in which they were collected (*Figures 9.1* and *9.2*). Thus, in two areas where the insecticide was being widely used, the geometric mean for 1963 and 1964 was between 5 and 8 ppm, whereas in the two other areas, the corresponding values fell below 1 ppm (Newton, 1988). Following the restrictions placed on OC use at the beginning of the 1960s, dieldrin levels fell to below 1 ppm (geometric mean) in one of these areas by 1975. However, in the other area, which covered a large part of eastern England, the levels remained close to 5 ppm until 1976, when dieldrin was banned as a seed dressing for winter-sown cereals. After 1976, the dieldrin levels in this area fell to levels similar to those in other areas of Britain.

The dieldrin levels in sparrowhawk livers, representing different areas and periods, are plotted as distribution diagrams in *Figure 9.3* (Walker & Newton, 1999). In a few cases, these values are for dieldrin plus heptachlor epoxide, a compound with the same mode of action and of similar toxicity and persistence to dieldrin. For the period 1963–1986, values for birds collected in the eastern zone referred to above (Zone 4 in *Figure 9.2*, henceforward 'eastern area') are plotted seperately from those representing the two more western zones least affected by dieldrin (Zones 1 and 2 in *Figure 9.2*,

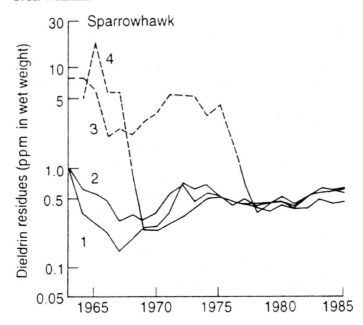

Figure 9.1.   HEOD levels in the livers of sparrowhawks found dead in the four zones shown in *Figure 9.2.*
HEOD is the active principal of dieldrin. Broken lines show periods when populations were depleted or
decreasing; solid lines show periods when populations were normal or increasing. Population increase
occurred when liver levels were less than about 1.0 ppm wet weight. From Newton (1988); reproduced from
Walker *et al.* (2001) with permission.

henceforward 'western area'), see *Figure 9.3*. Residues of 9 ppm or more were taken
as evidence of direct acute lethal poisoning by cyclodienes. Considering the period
1963–1975, in the eastern area of high cyclodiene use some 29% of individuals fell
into this category, in comparison with only 6% in the western area. Residues in the
range 3–9 ppm may represent individuals which experienced sublethal effects
(Walker & Newton, 1999), a point that will be returned to shortly. Considering again
the period 1963–1975, 35% of individuals fell into this category in the eastern area,
compared with 15% in the western area. Taking these two categories together, some
65% of individuals found dead in the eastern area during this period had cyclodiene
residues of 3 ppm or more, compared with only 21% in the western area.

   Two independent lines of evidence suggest that residues in the range 3–9 ppm may
be indicative of sublethal effects. First, dieldrin and related compounds which act
upon GABA receptors can cause a variety of sublethal neurotoxic effects, which have
been experienced by humans or observed in experimental animals. These effects,
which include loss of co-ordination, disorientation, and dizziness, occur at levels
considerably below those which cause severe convulsions and death (Walker &
Newton, 1999; Walker, 2001). The sparrowhawk depends upon sophisticated hunting skills to catch its prey, skills that would be severely impaired by sublethal effects
such as these, which could lead to death due to starvation. The other line of evidence
comes from field studies. Newton & Wyllie (1992) observed that sparrowhawk and
kestrel (*Falco tinnunculus*) populations in Britain that had been adversely affected by

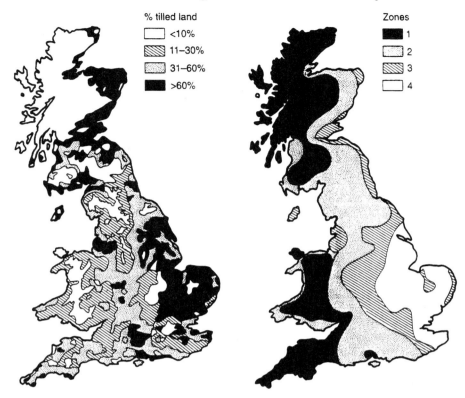

Figure 9.2.   Changes in the status of sparrowhawks in relation to agricultural land use and organochlorine use. The agricultural map (left) indicates the proportion of tilled land, where almost all pesticide is used. The sparrowhawk map (right) shows the status of the species in different regions and time periods:
Zone 1 – Sparrowhawks survived in greatest numbers through the height of the 'organochlorine era' around 1960; population decline judged at less than 50% and recovery effectively complete before 1970.
Zone 2 – Population decline more marked than in Zone 1, but recovered to more than 50% by 1970.
Zone 3 – Population decline more marked than in Zone 2, but recovered to more than 50% by 1980.
Zone 4 – Population almost extinct around 1960, and little or no recovery evident by 1980.
In general, population decline was most marked, and recovery latest, in areas with the greatest proportion of tilled land (based on agricultural statistics for 1966). From Newton & Haas (1984); reproduced from Walker *et al.* (2001) with permission.

dieldrin did not recover until the geometric mean of the liver dieldrin concentration in the population fell below 1 ppm. At this mean concentration, very few individuals (not more than 5%) could have had liver dieldrin concentrations of 9 ppm or above, a threshold proposed earlier for acute lethal toxicity. It seems doubtful that such a low contribution to mortality could hold population numbers down, and again the question is raised whether individuals with lower levels than this (s17 3–9 ppm) could have died as a consequence of sublethal effects, thus increasing the proportion of the population affected by cyclodienes.

Regarding DDT residues in sparrowhawks during the period under consideration, very few individuals had high enough residues to suggest direct lethal toxicity. DDT is considerably less toxic to birds than the cyclodiene insecticides aldrin, dieldrin, and heptachlor. However, relatively high levels of the stable metabolite pp' DDE were

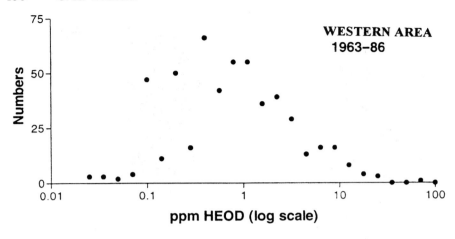

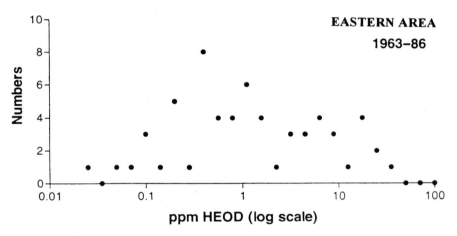

Figure 9.3.   Distribution of dieldrin (HEOD) residues (expressed as ppm on a log. scale) in the livers of sparrowhawks from two different areas of Britain. The numbers of individuals with dieldrin residues falling within the ranges of the concentrations represented by 0.15 log. units are given on the vertical axis. The concentrations plotted represent the mid-points of each log. range. Eastern area, highest cyclodiene use; western area, lowest cyclodiene use. Reproduced from Walker & Newton (1999) with permission.

found in dead sparrowhawks and their eggs. This compound has a much longer biological half-life in vertebrates than does pp' DDT, and was the principal DDT residue found in the longer term in terrestrial food chains (Walker *et al.*, 2001). Ratcliffe (1967) showed that the eggshells of British wild sparrowhawks became thinner shortly after World War Two, and suggested that this was caused by DDT, which was first used on a large scale at this time. Subsequent work confirmed this theory, when it was shown that (1) levels of DDE commonly found in the field could cause eggshell thinning in captive American kestrels (*Falco sparverius*), as well as in certain other birds (Wiemeyer & Porter, 1970; Peakall *et al.*, 1973; Peakall, 1993), and (2) there was a close relationship between pp' DDE concentrations and shell thickness in the eggs of sparrowhawks collected in the field in Britain (Newton,

1986). Of critical importance in the interpretation of pp' DDE residues in sparrowhawks is the relationship between the thickness of the eggshell and its stability. A small degree of thinning seems to have little significance, but at some critical point eggshells become so thin that they easily break. Peakall (1993) suggested that when peregrine eggshells thin by about 18%, breakage readily occurs and populations may decline (see later). During the period before the use of DDT was severely curtailed, DDE levels were high enough to cause eggshell thinning in British sparrowhawks, and in certain individuals the effect was great enough to lead to egg breakage.

RESIDUES IN PEREGRINES

The peregrine is far less numerous than the sparrowhawk in Britain, with only some 1,300 pairs in the entire country, even in recent years after the substantial recovery that occurred following bans and restrictions on OC insecticides (Ratcliffe, 1993). For this reason, few carcasses were found in the field and sent for analysis during the critical period between 1963 and 1975. Ratcliffe (1980) gives the OC residue data for the livers of 20 birds that were collected during this period. *Table 9.2* summarises these data in a way that permits comparison with the much more extensive sparrowhawk data that were discussed earlier. In a few earlier samples, dieldrin and heptachlor epoxide residues have been combined, as with the analysis of sparrowhawk data (Walker & Newton, 1998). Thus, *Table 9.2* shows the numbers of individuals containing cyclodiene residues which fall broadly within the ranges represented by the three major peaks identified in the sparrowhawk distribution diagrams.

With so few individuals, caution is needed in interpretation. If the same reasoning is applied as was to the sparrowhawk, it may be suggested that 65% of the sample would have experienced lethal or sublethal effects due to dieldrin. At first glance, only 15% had residues of 10 ppm or above, the levels normally associated with direct poisoning, and represented by the major part of the third peak on the sparrowhawk distribution diagrams. However, there is overlap between peaks 2 and 3, and the exact range of residues that each peak represents is subject to some interpretation. In the case of the peregrine data, two individuals had residues of 9.2 and 9.6 respectively, and these may well be cases of direct lethal toxicity. Assuming this to be the case, 25% of the sample represents cases of direct lethal poisoning, whilst a further 40% may have experienced sublethal effects. These proposals will be discussed further in relation to effects at the population level.

Like sparrowhawks, peregrines acquired substantial levels of pp' DDE in both tissues and eggs, and these were associated with marked eggshell thinning during the years following World War Two (Ratcliffe, 1980, 1993). With the later large-scale

Table 9.2. Distribution of peregrines between the residue peaks identified in the sparrowhawk distribution diagrams (*Figure 9.3*).

Concentration range (ppm)	Peak on sparrowhawk distribution diagram	No. individuals	%
0 – 1.9	First	7	35
2 – 9.9	Second	10	50
10 and above	Third	3	15

restrictions on DDT, however, eggshells have become thicker in more recent years (Ratcliffe, 1993).

Effects of OCs on populations

The following account will be restricted to effects on sparrowhawk and peregrine populations in Britain. It will be based, very largely, on the work of Newton & Haas (1984) and Newton (1986) for the sparrowhawk, and upon the work of Ratcliffe (1980, 1993) for the peregrine.

POPULATION CHANGES IN THE SPARROWHAWK

There were limited localised declines of sparrowhawks by around 1950 in some areas of south-east England where DDT was heavily used. However, no widespread decline was apparent until 1957–1960, despite the fact that DDE-related eggshell thinning occurred from 1947 onwards. There was a crash in sparrowhawk populations in agricultural areas of Britain in the late 1950s. This was related to the introduction of the cyclodiene insecticides and the lethal secondary poisoning they caused. As noted earlier, at least 50% of the birds found dead in the badly affected agricultural areas of eastern England contained residues of dieldrin plus heptachlor epoxide that were sufficiently high to have caused death. Thus, the eggshell thinning caused by DDE from 1947 did not bring any general decline of the population. The population crash of the late 1950s was related to the effects of cyclodienes (principally dieldrin). Whilst eggshell thinning caused by DDE was not in itself sufficient to cause decline, it may have accentuated the effect of the cyclodienes at the population level by reducing reproductive success. As mentioned earlier, sparrowhawk populations did not recover in eastern England until the geometric mean of the dieldrin concentration in livers fell below 1 ppm (Newton & Wyllie, 1992) (*Figures 9.1* and *9.4*). At this mean concentration, very few dead birds carried sufficiently high dieldrin levels to suggest direct lethal toxicity, and it seems very unlikely that such a small increase in adult mortality could explain the delay in recovery associated with 1 ppm or more of the insecticide. However, sublethal effects of dieldrin may have led to the death of 20% or more of the sample collected when the mean dieldrin concentration was close to 1 ppm (see also Walker & Newton, 1999). An effect of this magnitude might well be large enough to restrict or prevent the recovery of sparrowhawk populations. It is noteworthy that the recovery of the sparrowhawk in eastern England occurred despite the fact that DDE levels were still high enough to cause eggshell thinning, providing further evidence that this compound alone, at the residue levels occurring in Britain, was not responsible for population declines. In fact, DDE levels in sparrowhawks fell very little between 1963 and 1981 (Newton & Haas, 1984).

To test the possibility of sublethal effects of cyclodienes contributing to population declines, life history modelling was carried out using data for the sparrowhawk obtained in Britain (Sibly *et al.*, 2000). Scenarios were compared that attributed increase in instantaneous mortality rate to either lethal toxicity alone (9 ppm or more in the liver) or lethal toxicity plus sublethal effects (3 ppm or more). The latter scenario fitted the observed data better, and suggested that acute toxicity alone was insufficient to account for the population declines reported from field studies.

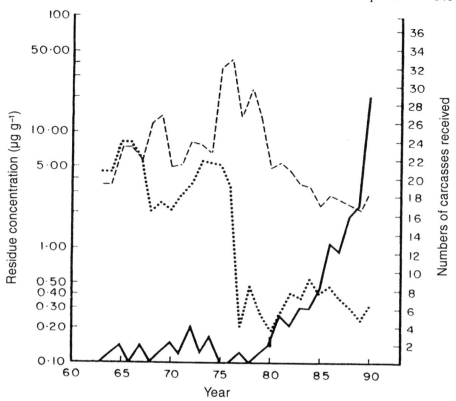

Figure 9.4. Numbers of sparrowhawk carcasses (solid line) received from the eastern area, 1963–1990, together with trends in the concentration of DDE (dashed line) and HEOD (dotted line) found in their livers. For both chemicals, lines show 5-year geometric means. On a regression analysis of individual $\log_{10}$ residue values (as the dependent variable) against year (as the independent variable), the downward trends in HEOD and DDE were statistically significant (HEOD, $b = -5.28$, $P <0.001$; DDE, $b = -2.33$, $P <0.02$). In contrast, there was no significant decline in the levels of PCBs (not shown, $b = +0.59$, $P >0.55$). From Newton & Wyllie (1992); reproduced from Walker *et al.* (2001) with permission.

The principal source of cyclodiene residues in sparrowhawks was evidently dressed seed (especially cereal seed). This conclusion is strengthened by the very sharp fall in liver dieldrin residues in eastern England following the ban on dieldrin seed dressings in 1975. DDE residues, on the other hand, do not originate from seed dressings. It appears that one of the main sources of them was from the soil. DDE is extremely persistent in soils (a half-life of 57 years has been estimated for one type of soil), and residues can be transferred to omnivorous birds, such as thrushes, by earthworms and other soil-dwelling invertebrates. Sparrowhawks prey on such omnivorous birds, and this route of transfer may explain why DDE levels in this predator have shown little tendency to fall following large restrictions in the use of DDT in Britain.

CHANGES IN PEREGRINE POPULATIONS

In a broad sense, changes in peregrine numbers have followed a similar trend to that just described for sparrowhawks (*Figure 9.5*). Although there was good evidence that

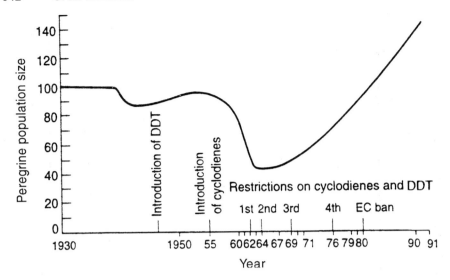

Figure 9.5. Peregrine population size in Britain (1930–1939 = 100) showing the 1961 population decline and subsequent recovery, together with an outline of pesticide usage from Ratcliffe (1993). Reproduced from Walker *et al.* (2001) with permission.

DDE-related eggshell thinning started around 1947, the crash of peregrine populations in agricultural areas did not occur until the late 1950s. This coincided with the widespread introduction of the cyclodiene insecticides, and was not found in remote areas, e.g. north-west Scotland inland sites, where there was little use of these chemicals. Recovery of peregrines followed the various bans and restrictions placed on cyclodienes, and the species has now returned to nearly all the coastal and inland breeding sites that were occupied around 1950. Indeed, the breeding population, now estimated at around 1,300 pairs over the country as a whole, exceeds estimates of numbers that were made before the crash.

As with the sparrowhawk, the population crashes of peregrines are attributed to the toxicity of the cyclodienes and not to the effects of DDT. The limited data on residues in peregrines suggest that sublethal effects of cyclodienes may have been important in addition to lethal ones (see earlier). DDE residues may have come from a number of sources, including racing pigeons dusted with DDT (Ratcliffe, 1993). This source may explain why DDE residues have declined more rapidly in the peregrine than in the sparrowhawk, following the various bans.

Conclusions

Large-scale mortality of grain-eating birds and their predators was reported from western Europe and North America following the introduction of the cyclodiene insecticides as seed dressings. It was soon apparent that birds were dying as a consequence of the acute toxicity of these compounds. A more difficult, and a more important question, however, was what effects the OCs were having on populations of birds. The studies just mentioned are amongst the very few which clearly established a link between the toxic effects of OCs and the decline of predatory birds.

A rapid decline of sparrowhawks as a consequence of cyclodiene toxicity also occurred in The Netherlands and Denmark between 1958 and 1960 (Newton, 1979). In North America, however, the decline of the peregrine and the bald eagle (*Haliaeetus leucocephalus*) occurred more slowly, and was associated with the effects of DDT rather than the effects of cyclodienes (Newton, 1979). Indeed, cyclodienes were used to a lesser extent in North America than in western European countries such as Great Britain, The Netherlands, and Denmark. Conversely, DDT was used on a larger scale in some parts of North America than in western Europe, and the decline of the peregrine there was related to eggshell thinning caused by DDE rather than by cyclodiene toxicity.

In Britain, the effects of cyclodienes were more evident upon bird-eating predators such as the sparrowhawk, the peregrine, and the merlin (*Falco columbarius*) (Newton *et al.*, 1978) than upon other predatory birds. The reasons for this deserve considera- tion. In the first place, the principal route of transfer of these compounds to predators is from dressed seed *via* grain-eating birds or small mammals. The detoxication of cyclodienes is dependent upon their conversion to water soluble and readily excretable products by the action of mono-oxygenases and associated conjugases (Walker, 1978, 1980). Mono-oxygenase activities tend to be higher in mammals than in birds of equivalent size (Walker, 1978; Ronis & Walker, 1989; Walker & Ronis, 1989), so birds may be more efficient 'vectors' of these compounds than mammals. In other words, small mammals may detoxify cyclodienes more rapidly than small birds, and therefore pass on smaller quantities of them to their prey. Another consideration is that the bird-eating sparrowhawk shows less hepatic microsomal mono-oxygenase activity towards OC substrates than does the less specialised kestrel, which commonly feeds on small rodents (Walker *et al.*, 1987). This agrees with a general tendency amongst birds, where specialised predators such as the sparrowhawk have substantially lower mono-oxygenase activities than do omnivores and herbivores (Walker, 1978; Ronis & Walker, 1989; Walker *et al.*, 2001), as might be anticipated because of their very limited requirement for the detoxication of naturally occurring xenobiotics in their food. Thus, bird-eating raptors may be particularly ineffective in the metabolic detoxication of cyclodienes. A further factor discussed earlier was the question of possible sublethal neurotoxic effects of cyclodienes. The marked sexual dimorphism of sparrowhawks and peregrines has been explained in terms of their dependence on sophisticated hunting skills to catch highly manoeuvrable prey (Newton, 1979). Predators feeding on less manoeuvrable prey (e.g. eagles, kites, and buzzards) have less dependence on such highly developed hunting skills, and show little sexual dimorphism. If cyclodienes produce sublethal neurotoxic effects that alter behaviour, it seems very probable that these will have a more serious effect on the feeding of bird-eaters than upon that of other types of predatory birds.

The OC insecticides have been studied in greater depth and detail than have other persistent lipophilic pollutants, especially in regard to their effects on populations of predatory birds. To what extent can they be regarded as useful models for other persistent lipophilic organic compounds? Polychlorinated biphenyls (PCBs), polychlorinated dibenzodioxins (PCDDs), organomercury and organotin compounds, and the 'superwarfarin' rodenticides are other important examples (Walker, 2001), and the list could be expanded considerably. In general terms, the study of the OCs has thrown a good deal of light on why such compounds have long biological half-lives and undergo biomagnification in food chains. The combination of high lipophilicity with metabolic recalcitrance usually leads to marked biological persistence (Walker et al., 2001). However, a particular feature of the OCs is their neurotoxicity, which raises questions about sublethal effects on behaviour (Walker, 2001). It is very difficult to study behavioural effects in the field, but this could be a very important issue when considering effects at the population level. Here, behavioural effects can be more important than direct toxicity. The evidence suggesting sublethal effects of cyclodienes on sparrowhawks, discussed here, raises wider issues about neurotoxic pollutants in general. All the major groups of insecticides, OCs, organophosphorus carbamate, and pyrethroid compounds are neurotoxic.

The detailed consideration of the distribution of residues in predatory birds has permitted the estimation of biological effects of OCs which can be incorporated into population models. The effects of DDE on eggshell thickness and fecundity in the peregrine (Peakall, 1993) and sparrowhawk (Newton, 1986) are cases in point. The distribution diagrams for cylodiene residues in sparrowhawks have allowed the estimation of lethal, and possible sublethal, effects (Sibly et al., 2000). These data have been incorporated into population models, to see how closely they can be related to changes of numbers in the field. In one sense, the OCs are very suitable for this kind of approach, because they are so persistent that their residues are relatively stable in carcasses and eggs. Biological effects can be estimated from residues. This is not possible with many other organic pollutants which are far less stable. With these, biomarkers of toxic action have a critical role to play in population modelling.

References

Eldefrawi, M. E. & Eldefrawi, A. T. (1990) Nervous-system-based insecticides. In: *Safer insecticides* (eds. E. Hodgson & R. J. Kuhr), pp. 155–208. Dekker, New York, USA.

Newton, I. (1979) *Population ecology of raptors*. Poyser, Berkhamsted, UK. 400 pp.

Newton, I. (1986) *The sparrowhawk*. Poyser, Calton, UK. 300 pp.

Newton, I. (1988) Determination of critical pollutant levels in wild populations, with examples from organochlorine insecticides in birds of prey. *Environmental Pollution* **55**, 29–40.

Newton, I. & Haas, M. B. (1984) The return of the sparrowhawk. *British Birds* **74**, 47–70.

Newton, I. & Wyllie, l. (1992) Recovery of a sparrowhawk population in relation to declining pesticide contamination. *Journal of Applied Ecology* **29**, 476–484.

Newton, I., Meek, E. & Little, B. (1978) Breeding ecology of the merlin in Northumberland. *British Birds* **71**, 376–398.

Peakall, D. B. (1993) DDE-induced eggshell thinning: an environmental detective story. *Environmental Reviews* **1**, 13–20.

Peakall, D. B., Lincer, J. L., Risebrough, R. W., Pritchard, J. B. & Kinter, W. B. (1973) DDE-induced shell thinning: structural and physiological effects in three species. *Comparative and General Pharmacology* **4**, 305–313.

Ratcliffe, D. A. (1967) Decrease in eggshell weight in certain birds of prey. *Nature, London* **215**, 208–210.

Ratcliffe, D. A. (1980) *The peregrine falcon (1st edition)*. Poyser, Calton. 410 pp.

Ratcliffe, D. A. (1993) *The peregrine falcon (2nd edition)*. Poyser, Calton. 460 pp.

Ronis, X. J. J. & Walker, C. H. (1989) The microsomal mono-oxygenases of birds. *Reviews in Biochemical Toxicology* **10**, 301–384.

Sibly, R. M., Newton, I. & Walker, C. H. (2000) Effects of dieldrin on population growth rates of sparrowhawks 1963–1986. *Journal of Applied Ecology* **37**, 540–546.

Walker, C. H. (1978) Species difference in microsomal mono-oxygenase activity and their relationship to biological half-lives. *Drug Metabolism Reviews* **7**, 295–323.

Walker, C. H. (1980) Species differences in some hepatic microsomal enzymes that metabolise xenobiotics. In: *Progress in drug metabolism, vol. 5* (eds. J. Bridges & L. Chasseaud), pp. 118–164. Wiley, Chichester, UK.

Walker, C. H. (2001) *Organic pollutants: an ecotoxicological perspective.* Taylor & Francis, London, UK. 282 pp.

Walker, C. H. & Newton, I. (1998) Effects of cyclodiene insecticides on the sparrowhawk (*Accipiter nisus*) in Britain – a reappraisal of the evidence. *Ecotoxicology* **7**, 185–189.

Walker, C. H. & Newton, I. (1999) Effects of cyclodiene insecticides on raptors in Britain – correction and updating of an earlier paper. *Ecotoxicology* **8**, 425–429.

Walker, C. H. & Ronis, X. J. J. (1989) Mono-oxygenases of birds, amphibians, and reptiles. *Xenobiotica* **19**, 1111–1121.

Walker, C. H., Newton, I., Hallam, S. & Ronis, X. J. J. (1987) Activities and toxicological significance of hepatic microsomal enzymes of the kestrel and sparrowhawk. *Comparative Biochemistry and Physiology* **86** (C), 379–382.

Walker, C. H., Hopkin, S. P., Sibly, R. M. & Peakall, D. B. (2001) *Principles of ecotoxicology (2nd edition).* Taylor & Francis, London, UK. 310 pp.

Wiemeyer, S. N. & Porter, R. D. (1970) DDE thins eggshells of captive American kestrels. *Nature, London* **227**, 737–738.

PART 3

Foraging Behaviour of Birds on Insects

10
Vision in Birds: The Avian Retina

JAMES K. BOWMAKER

Department of Visual Science, Institute of Ophthalmology, University College London, Bath Street, London, EC1V 9EL, UK

Introduction

A characteristic feature of almost all animals, excluding some subterranean, troglodytic and hypogean species, is prominent eyes. Vision is clearly a very important and ancient sense, allowing an animal to perceive an almost instantaneous and detailed picture of the surrounding environment. The visual systems of vertebrates have evolved and adapted to function in photic environments ranging from the broad spectrum of full sunlight to almost total darkness, including the restricted spectral ranges found in different coloured aquatic environments. Solar radiation is filtered by the earth's atmosphere so that, at sea level, about 80% of the energy is restricted to a narrow spectral band from about 300 nm in the ultraviolet to around 1,100 nm in the infra-red. Longer wavelengths are filtered out primarily by water vapour, whereas high energy, short wavelengths are absorbed by the ozone layer (for a full discussion, see Dartnall, 1975). This 800-nm daylight spectral range is theoretically available for vision, but in practice it is restricted even further at long wavelengths by the nature of the photoreceptor mechanism itself.

The key element in vision is the absorption of photons in the eye by photosensitive molecules, visual pigments. These are composed of a protein moiety, opsin, bound to a chromophore, the aldehyde of Vitamin A, retinal. The energy of an absorbed photon leads to a conformational change, a photo-isomerisation of the chromophore, from a twisted isomer, 11-*cis* retinal, to a straight isomer, all-*trans* retinal (for a review, see Knowles & Dartnall, 1977). However, the isomerisation requires an energy greater than that of light of wavelengths longer than about 800 nm, so that the effective daylight spectrum available for vision is limited to a range of approximately 500 nm, from 300 nm in the ultraviolet to about 800 nm in the far red.

It is only natural, as humans, to assume that what we perceive is the 'real' world, and that other animals see their surroundings either much as we do, or in some reduced or diminished way. However, this egocentric view could not be further from the truth. Our visual range is limited, extending only from about 400 nm in the violet to about 750 nm in the red. Wavelengths below 400 nm are absorbed by the ocular lens, which is pale yellow, whereas vision at long wavelengths is determined by the

Insect and Bird Interactions
© Intercept Ltd., PO Box 716, Andover, Hampshire, SP10 1YG, UK.

sensitivity of our most long-wave-sensitive visual pigment. Many vertebrates, including a variety of fish, frogs, lizards, and birds, have visual systems significantly different from that of humans, and are able to see in spectral regions outside our limited visual range (for a review, see Bowmaker, 1991a). The colourfulness of birds and the vivid and dramatic sexual displays of many species (peacocks and birds of paradise are just two obvious examples) strongly suggest that they must have excellent colour vision. But what is meant by 'excellent', and how does a bird's perception differ from ours?

The vertebrate eye

BASIC FEATURES

All vertebrate eyes are essentially similar, designed to focus an inverted image of the environment onto a photosensitive surface, the retina, at the back of the eye. The image is re-inverted by higher visual centres in the brain. The retina contains a complex array of photoreceptor cells differentiated into two classes; rods, that are specialised for vision in dim light and which operate at light levels less than that of bright moonlight (scotopic vision), and cones, that are specialised for vision in daylight (photopic vision). It is only the cone system that gives us the luxury of colour vision, whereas scotopic vision is a limited monochromatic, 'black and white' system. Behind the retina is a black pigment layer, the pigment epithelium, which is heavily involved in regulating the function of the retina, and in absorbing scattered and stray light within the eye.

Although rods and cones subserve different visual functions, they are built to a similar design. The cells consist of an inner segment that contains the basic cellular components (nucleus, concentrations of mitochondria, etc.), and a highly specialised photosensitive region, the outer segment, which contains a high concentration of visual pigment molecules built into the membranes of a stack of infoldings or discs (see Fein & Szuts, 1982). Since all the visual pigments contain the same chromophore, retinal, the spectral region in which the pigment is maximally sensitive must be determined by the nature of the opsin (Bowmaker & Hunt, 1999).

COLOUR VISION

The minimum requirement for colour vision, defined as an ability to distinguish wavelengths independently of brightness, is for two spectrally separated classes of cone. The two classes, each expressing a different visual pigment, are connected neurally by an opponent mechanism in which the quantum catch from one class of cone is compared with the quantum catch from the other. Theoretically, there is no upper limit to the number of spectral cone classes that may be involved in colour vision. However, because of the broad absorbance spectrum of visual pigments and the ever increasing complexity of the neural integration involved in increasing numbers of cone classes, there appears to be little advantage in increasing the number above four (Barlow, 1982; Bowmaker, 1983). Indeed, tetrachromacy, colour vision subserved by four spectrally distinct cone classes, probably arose very early in vertebrate evolution, and may have been present in ancestral fish about 450 million years ago (Goldsmith, 1991; Hisatomi *et al.*, 1994; Bowmaker, 1998).

In contrast, our human colour vision is only trichromatic, based on three classes of cone containing visual pigments with maximum sensitivities (λ_{max}) at long wavelengths (LWS) in the red at about 565 nm, at middle wavelengths (MWS) in the green close to 530 nm, and at short wavelengths (SWS) in the blue near 420 nm (Dartnall *et al.*, 1983; Nathans *et al.*, 1986; Schnapf *et al.*, 1987). These cones are not distributed evenly across the retina. The majority of the cones are concentrated in a small central region, the fovea, which is where our vision is centred to detect exquisite fine detail. In the fovea, more than 90% of the cones are LWS and MWS, whereas SWS cones are rare and relatively widely separated. The consequence of this is that our acuity at short wavelengths (in the blue) is poor, in comparison to our very high acuity at longer wavelengths (in the red/green).

The avian retina

What is the arrangement of cones in the avian retina? Diurnal birds probably have, at least at the retinal level, one of the most elaborate mechanisms for colour vision within the vertebrates. The retinas of these avian species contain a complex complement of photoreceptors, rods, double cones, and at least four classes of single cone. The cones are characterised by brightly coloured oil droplets (*Plate 1*), a feature restricted to birds and some reptiles (for a review, see Bowmaker, 1991b). The droplets are located in the distal ellipsoid region of the inner segment, and act as selective cut-off (long pass) filters interposed between the incident light and the visual pigment (Bowmaker & Knowles, 1977; Goldsmith *et al.*, 1984; Partridge, 1989). The four classes of single cone are spectrally distinct with maximum sensitivities extending from the near ultraviolet, around 360 nm, to the red, about 600 nm, and are thought to subserve tetrachromatic colour vision (Goldsmith, 1991; Maier & Bowmaker, 1993; Bowmaker *et al.*, 1997).

Although such general statements can be made concerning avian colour vision, in fact few species have been studied in any detail (for a recent review, see Hart, 2001). In the chicken (*Gallus gallus domesticus*), four cone visual pigments with wavelengths of maximum absorbance (λ_{max}) at about 418, 455, 505 and 569 nm have been identified by visual pigment extraction techniques (Fager & Fager, 1981; Yen & Fager, 1984; Okano *et al.*, 1989; Yoshizawa & Fukada, 1993) and electroretinography (Govardovskii & Zueva, 1977). Recently, the genes encoding these four cone visual pigments and the rod pigment have been isolated, sequenced, and expressed (Takao *et al.*, 1988; Kuwata *et al.*, 1990; Okano *et al.*, 1992; Wang *et al.*, 1992), confirming their presence in the retina and suggesting a close evolutionary relationship between the spectrally similar rod pigment and the green-sensitive cone pigment.

Detailed microspectrophotometric analysis of the photoreceptors of the chicken (Bowmaker & Knowles, 1977; Bowmaker *et al.*, 1997) and the closely related Japanese quail (*Coturnix coturnix japonica*) (Bowmaker *et al.*, 1993) and domestic turkey (*Meleagris gallopavo*) (Hart *et al.*, 1999), all members of the Galliformes, has established the location of the four cone pigments within distinct cone classes. The pattern identified in these species appears to be common amongst diurnal birds, and can be used as a model for other species (*Plate 1*, *Figures 10.1* and *10.3*). The red-

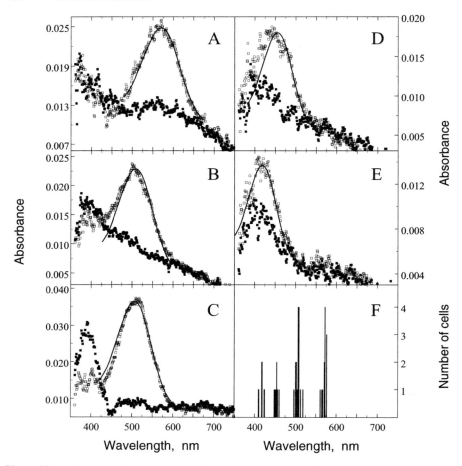

Figure 10.1. Mean absorbance spectra of visual pigments from the domestic chicken (*Gallus gallus domesticus*). Spectra were obtained by microspectrophotometry of individual isolated photoreceptors. Open symbols, before bleaching; filled symbols, after exposure to white light. (A) LWS (red) cones (both single and double cones), (B) MWS (green) single cones, (C) rods, (D) SWS (blue) single cones, (E) UVS (violet) single cones, (F) distribution histograms of the individual λ_{max} from all cones. The solid lines in (A) to (E) are visual pigment template curves with λ_{max} at 569, 507, 506, 453 and 418 nm respectively. Reproduced from Bowmaker *et al.* (1997) with permission.

sensitive pigment, with λ_{max} 569 nm (P569), dominates the retina, and is found in both members of the double cones and in a class of single cone containing a red (R-type) oil droplet. The filtering effect of the R-type droplet, with a cut-off at about 570 nm (*Figure 10.2*), narrows the spectral sensitivity of the single cone class by removing short wavelengths, and displaces the maximum sensitivity of the cell to longer wavelengths above 600 nm. In contrast, the oil droplet (P-type) of the principal member of the double cones has a cut-off at much shorter wavelengths, so that the spectral sensitivity of the double cones is broad, with a maximum close to 570 nm. The three remaining classes of single cone are identified as green-sensitive, with a P505 associated with a yellow (Y-type) droplet having a cut-off at about 510 nm, blue-sensitive with a P455 associated with a clear or colourless (C-type) droplet

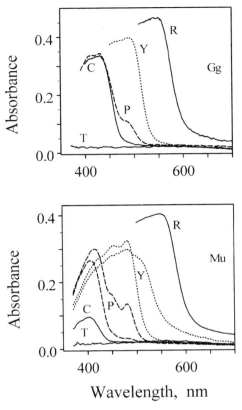

Figure 10.2. Mean absorbance spectra of cone oil droplets from the domestic chicken (Gg) and budgerigar (*Melopsittacus undulatus*, Mu). Letters indicate droplet type. **R**ed, **Y**ellow, **C**lear and **T**ransparent are all found in single cones, whereas **P**ale droplets are located in the Principal member of double cones. Variations in the absorbance of the Y- and P-type droplets occur in the budgerigar at different locations across the retina. The C-type droplets have low absorbance in species with UVS cones with λ_{max} around 365 nm, such as the budgerigar, but higher absorbance in species with UVS cones with λ_{max} above 400 nm, such as the chicken. Reproduced from Bowmaker *et al.* (1997) with permission.

cutting off at about 450 nm, and violet-sensitive with a P418 associated with a transparent (T-type) droplet that exhibits no significant absorbance throughout the spectrum (*Plate 1, Figures 10.1* and *10.3*).

Support for the role of the four classes of single cone in colour vision comes from the behaviourally determined increment spectral sensitivity function of the passeriform bird, the Pekin robin or red-billed Leiothrix (*Leiothrix lutea*) (Maier, 1992). *Leiothrix* shows high sensitivity in the near ultraviolet, with the sensitivity function exhibiting four marked peaks with maxima at about 370, 460, 520 and 620 nm (*Figure 10.4*). These sensitivity maxima closely match spectral sensitivity functions for the four classes of single cone derived from microspectrophotometric analysis of their visual pigments and oil droplets (Maier & Bowmaker, 1993). As was inferred from the behavioural data and confirmed by microspectrophotometry, the fourth class of single cone in *Leiothrix* contains an ultraviolet-sensitive visual pigment with λ_{max} at about 365 nm, and not at about 420 nm as in the galliform species.

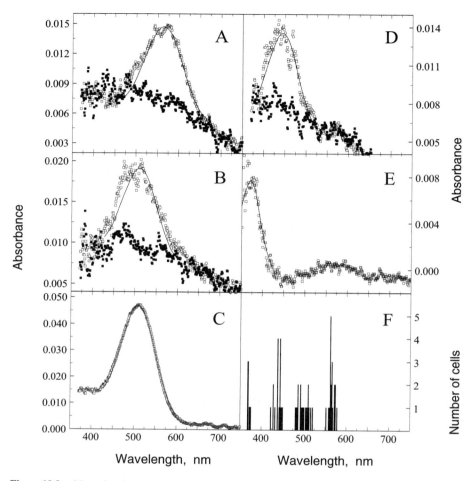

Figure 10.3. Mean absorbance spectra of visual pigments from the budgerigar (*Melopsittacus undulatus*). Open symbols, before bleaching; filled symbols, after exposure to white light. (A) LWS (red) cones (both single and double cones), (B) MWS (green) single cones, (C) rods, (D) SWS (blue) single cones, (E) UVS (ultraviolet) single cones, (F) distribution histograms of the individual λ_{max} from all cones. The solid lines in (A) to (E) are visual pigment template curves with λ_{max} at 564, 509, 509, 444 and 371 nm respectively. Reproduced from Bowmaker *et al.* (1997) with permission.

ULTRAVIOLET SENSITIVITY

Ultraviolet-sensitive visual pigments with λ_{max} close to 365 nm have been recorded by microspectrophotometry from a number of passeriforms, five species of finch (Bowmaker *et al.*, 1997; Hart *et al.*, 2000a), canary (*Serinus canaria*) (Das *et al.*, 1999), starling (*Sturnus vulgaris*) (Hart *et al.*, 1998), blackbird (*Turdus merula*), and blue tit (*Parus caeruleus*) (Hart *et al.*, 2000b), and the psittaciform, the budgerigar or shell parakeet (*Melopsittacus undulatus*) (*Figure 10.4*) (Bowmaker *et al.*, 1997; Wilkie *et al.*, 1998). In addition to species possessing a fourth cone visual pigment with λ_{max} at either 420 or 360 nm, other species such as the Humboldt penguin

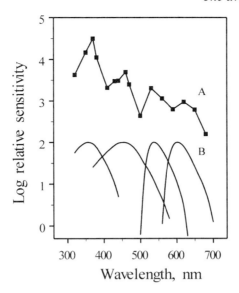

Figure 10.4. Comparison of the behaviourally determined increment threshold spectral sensitivity function of the Pekin robin (*Leiothrix lutea*) (A) and the derived effective spectral sensitivities of the four single cone classes (B). The functions have been arbitrarily displaced on the log. sensitivity axis and the curves in (B) have been normalized to log. 2. Visual pigment absorbance spectra were corrected for the filtering effects of the relevant oil droplets and for the absorbance of the ocular media (from Maier & Bowmaker, 1993). Note the correlation of the behavioural sensitivity peaks with the sensitivities of the single cones. Reproduced from Maier & Bowmaker (1993) with permission.

(*Spheniscus humboldti*) (Bowmaker & Martin, 1985), Manx shearwater (*Puffinus puffinus*) (Bowmaker *et al.*, 1997), and pigeon (*Columba livia*) (Bowmaker *et al.*, 1997; Yokoyama *et al.*, 1998; Wilkie *et al.*, 2000) have a violet/ultraviolet-sensitive pigment with λ_{max} close to 400 nm (*Table 10.1*).

Ultraviolet sensitivity is well established in avian species and has been most fully demonstrated in the pigeon, both behaviourally (Wright, 1972; Romeskie & Yager, 1976; Kreithen & Eisner, 1978; Emmerton & Delius, 1980; Emmerton, 1983; Emmerton & Remy, 1983; Remy & Emmerton, 1989) and electrophysiologically (Blough, 1957; Graf & Norren, 1974; Norren, 1975; Chen *et al.*, 1984; Wortel *et al.*, 1984; Chen & Goldsmith, 1986; Vos Hzn *et al.*, 1994). It has also been shown in three species of humming bird (Goldsmith, 1980; Goldsmith *et al.*, 1981), 13 species of passerines (Chen *et al.*, 1984; Chen & Goldsmith, 1986), two species of boobies (Pelecaniformes) (Reed, 1987), Pekin robin (Maier, 1992), kestrel (*Falco tinnunculus*) (Viitala *et al.*, 1995), zebra finch (*Taeniopygia guttata*) (Bennett *et al.*, 1996), bluethroat (*Luscinia svecica*) (Andersson & Amundsen, 1997), starling (Bennett *et al.*, 1997), and blue tit (Andersson *et al.*, 1998; Hunt *et al.*, 1998).

Although four spectrally distinct classes of single cone have been demonstrated in a number of birds with ultraviolet sensitivity subserved by a single class of cones with λ_{max} between about 360 nm and 420 nm, the possibility that the pigeon has five classes of single cone, with two classes maximally sensitive in the violet/ultraviolet range, cannot be ruled out. Microspectrophotometric analysis of pigeon photoreceptors

Table 10.1.   Summary of avian visual pigments, λ_{max} (nm) as determined from microspectrophotometry.

Species	uv	violet	blue	green	red	rod
Sphenisciformes:						
Humboldt penguin (*Spheniscus humboldti*)	–	403	450	–	543	504
Procellariformes:						
Manx shearwater (*Puffinus puffinus*)	–	402	452	P	P	505
Anseriformes:						
Mallard duck (*Anas platyrhynchos*)	–	420	452	502	570	505
Psittaciformes:						
budgerigar (*Melopsittacus undulates*)	371	–	444	508	564	509
Galliformes:						
domestic chicken (*Gallus gallus domesticus*)	–	418	455	507	569	506
Japanese quail (*Coturnix coturnix japonica*)	–	419	456	505	569	505
domestic turkey (*Meleagris gallopavo*)*	–	420	460	505	564	504
Columbiformes:						
pigeon (*Columba livia*)		409	453	507	568	506
	(366)	(410)				
Strigiformes:						
tawny owl (*Strix aluco*)	–	–	463	503	555	503
Passeriformes:						
Pekin robin (*Leiothrix lutea*)	355	–	453	501	567	500
zebra finch (*Taeniopygia guttata*)	ca 360–380	–	430	506	568	507
canary (*Serinus canaria*)	366	–	442	506	569	506
starling (*Sturnus vulgaris*)	362	–	449	504	563	503
Palaeognathous:**						
ostrich (*Struthio camelus*)		405	445	506	570	505
rhea (*Rhea americana*)		P	445	506	570	505

P, known to be present, but λ_{max} not determined; ( ), erg measurements from Vos Hzn *et al.* (1994); *, Hart *et al.* (1999); ** Wright & Bowmaker (2001); for all other references see Bowmaker *et al.* (1997).

(Bowmaker, 1977; Bowmaker *et al.*, 1997) has identified four cone pigments with λ_{max} at about 405, 460, 515 and 567 nm, and this is supported by recent studies of the molecular genetics of the cone opsins (Kawamura *et al.*, 1999; Wilkie *et al.*, 2000; Yokoyama *et al.*, 2000), but accumulating evidence from behaviourally determined wavelength discrimination functions (Wright, 1979; Emmerton & Delius, 1980), and from electrophysiological studies (Graf & Norren, 1974; Norren, 1975; Romeskie & Yager, 1976; Wortel *et al.*, 1984; Vos Hzn *et al.*, 1994) indicate that two spectrally distinct violet/ultraviolet cone mechanisms may be present with maxima at about 410 nm and 365 nm. If this is the case, then the pigeon may have the potential for pentachromatic colour vision, perhaps a recent evolutionary feature restricted to the Columbiformes.

FUNCTION OF OIL DROPLETS

Since the avian retina is characterised by brightly coloured oil droplets (*Plate 1*), the obvious question is what advantages, if any, do these droplets confer on birds (and turtles)? The effects of the oil droplets can be easily determined. The R-, Y- and C-types, acting as cut-off filters, will remove short wavelength light, displace the peak sensitivity of the cone to longer wavelengths than that of the visual pigment, and reduce the overall sensitivity of the cone (Bowmaker, 1991b; Kawamuro *et al.*, 1997). The reduction in sensitivity is probably of little consequence, since the cones will be functioning at high photopic light levels. In addition, the droplets act as individual lenses and will concentrate incident light into the cone outer segments (Baylor & Fettiplace, 1975). The displacement of the peak sensitivities to longer wavelengths must be of more significance. The LWS and MWS single cones will be maximally sensitive at about 620 and 530 nm respectively (*Figure 10.4*), and these may be ideal locations for discriminating objects (e.g. fruits, seeds, and insects) against a background of leaves and/or for discriminating between types of leaf. The spectral reflectance of leaves is, of course, dominated by chlorophyll, which has a spectral 'window' in its absorbance spectrum maximal at about 550 nm. The majority of variation in leaf colour is caused by additional pigments that are more red reflecting. The two peaks of the LWS and MWS cones of most birds lie either side of the 'window', with the LWS cone centred in the region of maximum leaf variation in the red (Bowmaker & Knowles, 1977). This situation is not unlike that in primates and humans, where it is believed that our red/green colour vision evolved recently (about 35 million years ago) specifically to enable ancestral primates to detect ripe (yellow/red) fruit and/or young, more palatable leaves against a highly variable green background (in terms of luminance and colour) of leaves (Mollon, 1989; Dominy & Lucas, 2001; Regan *et al.*, 2001).

A third effect of oil droplets is to narrow the spectral sensitivity functions of the cones, especially the LWS and MWS, and to remove short wavelength light (*Figure 10.4*). This will have the effect of increasing the rate of change of sensitivity with wavelength on the short-wave side of the sensitivity functions. Therefore, in an opponent neural process that compares the output of the LWS and MWS cones, in the spectral region where the two functions overlap, the sensitivity of one will be decreasing rapidly, whereas the other will be increasing rapidly. Theoretically, this should increase wavelength discrimination, but this has proved difficult to establish from behavioural experiments. Nevertheless, recent modelling of avian tetrachromacy with and without the filtering effects of oil droplets does suggest that the droplets should enhance wavelength discrimination (Vorobyev *et al.*, 1998). The removal of short-wave light and the focusing effect of the droplets may also enhance acuity by reducing chromatic aberration and light leakage into neighbouring cones.

CONE COMPLEMENT AND DISTRIBUTION

In diurnal birds, the complement of retinal photoreceptors is highly conserved across species. The retina is dominated by cones which make up about 80% of the total rod/cone population, and of these about 40–50% are double cones. The majority of single cones are LWS and MWS, which are present in about equal numbers (20–25% each), with the remaining 10% of cones comprising SWS and UVS cones (*Plate 1*) (Bowmaker, 1991b). The exact percentages of the different cone types may vary across the retina, often with dorso-ventral differentiation. These variations will affect the bird's perception of different regions of its visual field, and may perhaps be correlated with the different visual behaviours and visual ecology of the wide range of bird species (for a detailed discussion, see Hart, 2001).

Unlike the mammalian retina, in birds the cones are maintained at high levels right across the retina, though there is a central '*area centralis*' and/or a horizontal 'visual streak' around the eye's equator, where the density is significantly increased. Because of the lateral position of the eyes in birds, there may be two areas, one central and looking laterally, with the other, more posterior in the eye, used for forward binocular vision (Martin, 1985). In addition to the differences in sensitivity between cones and rods, cones function at a higher temporal resolution than rods, responding about three times faster than rods (Lamb & Pugh, 1990). This is important for an animal moving rapidly through its environment: a bird in flight needs to be aware of the scene moving past it (optical flow) and this can be achieved by having a high density of rapidly responding cones right across the retina to the periphery.

OWL, PENGUIN, AND OSTRICH

Although the arrangement of photoreceptors in the retina is highly conserved across avian species, there are significant differences in species that occupy photic environments other than 'normal' daylight. Nocturnal birds, notably owls, reverse the cone/rod ratio, with rods comprising about 80% of the photoreceptor complement, but nevertheless they still retain at least a residual trichromatic colour vision system (Bowmaker & Martin, 1978). However, the coloured oil droplets in the cones have reduced carotenoid concentrations (perhaps to increase overall sensitivity) and, for reasons that are not immediately apparent, the LWS cone pigment is displaced to shorter wavelengths from that in other birds, shifted from about 565–570 nm to 555 nm (*Table 10.1*). Low numbers of coloured oil droplets have also been reported in swallows and swifts, but in this case it is thought that this will increase relative sensitivity to short-wave light (blue), which should enhance the ability of the birds to detect flying insects as silhouettes against a relatively bright 'blue' sky (Lythgoe, 1979).

A further group of birds that have modified their retinal organisation are the penguins. They have the complication of requiring a visual system that will function both in air and under water (Martin & Young, 1984). This raises obvious problems of optical focusing, but also necessitates a colour vision system adapted for active hunting in a marine environment. Water acts as a monochromator and rapidly filters out long-wave light so that in clear oceanic water the daylight spectrum is dominated by blue/green light. Penguins appear to have adapted their colour vision to match the down-welling light by having their longest-wavelength-sensitive cones maximal at about 540–545 nm, and with an increased percentage of SWS and UVS cones (*Table 10.1*) (Bowmaker & Martin, 1985). This is somewhat similar to adaptations in oceanic fish, which have their cones maximally sensitive in the blue/green (for a review, see Bowmaker, 1995).

One group of birds that might be expected to have a visual system somewhat different from the 'normal' avian pattern are the paleognathous birds. This group of large, flightless birds includes the ostriches (Struthioidae) from Africa, the emus (Dromicienae) from Australia, and the rheas (Rheidae) from South America. They are probably monophyletic, originating from a common ancestor about 50 million years ago, and most likely represent an early offshoot of the avian stem (Härlid *et al.*, 1998). Their visual world is distinctly different from that of a small passeriform but, somewhat surprisingly, the organisation of their retinal photoreceptors follows the typical avian pattern: they are, in effect, 'big chickens' (Wright & Bowmaker, 2001). The retina is cone dominated, with double cones and four classes of spectrally distinct single cones, though the UVS cone has a peak sensitivity around 405 nm, rather than 420 nm as in the chicken (*Table 10.1*). It is noteworthy that it is the larger, ground-feeding birds that generally lack true ultraviolet sensitivity.

ULTRAVIOLET VISION: A SPECIAL SENSE?

The perception of ultraviolet light by birds was probably first appreciated about twenty years ago when it was shown that humming birds could detect 'ultraviolet' nectar guides in flowers (e.g. Goldsmith, 1980), but, perhaps rather late in the day, behaviourists have only recently realised that birds perceive the world in a different manner from humans. Not only are avian species able to see considerably further into the near ultraviolet than humans, they will also divide the spectrum into different 'colours' (Wright, 1979). The consequences of this are many. First, species that do not appear sexually dimorphic to humans may be distinctly dimorphic to themselves, either because of feather colouring, reflectance and/or patterning in the ultraviolet, or because of 'colour name' differences in the human visible spectrum. This has been clearly demonstrated recently in blue tits, where differences in ultraviolet coloration distinguish the sexes (Andersson *et al.*, 1998; Hunt *et al.*, 1998).

Sexual dimorphism, visible to human observers, may also be enhanced for birds by differences in ultraviolet plumage reflectance (Burkhardt, 1989, 1996; Andersson, 1996; Bennett *et al.*, 1996; Andersson & Amundsen, 1997; Bennett *et al.*, 1997). Further, predator and/or prey mimicry may also be perceptually significantly different between human vision and avian vision (Church *et al.*, 1998). Ultraviolet sensitivity and tetrachromatic colour vision will also cause more general perceptual differences in the visual environment that relate to aspects such as the localisation of prey (e.g. scent markings, Viitala *et al.*, 1995), and colour preferences (Derim-Oglu & Maximov, 1994; Hunt *et al.*, 1997).

The visual environment offers a wealth of information in the ultraviolet, and many animals, both invertebrate and vertebrate, take full advantage of this. Indeed, ultraviolet sensitivity is probably common amongst not only birds, but many diurnal vertebrates outside of the mammals, and is retained even in some nocturnal rodents (Jacobs *et al.*, 1991; Jacobs, 1993). Because it is so common, rather than asking what are the functions of ultraviolet perception, perhaps a more interesting point for consideration is why many mammals, including primates and humans, have lost the ability to detect these short wavelengths.

References

Andersson, S. (1996) Bright ultraviolet coloration in the Asian whistling thrushes (*Myiophonus* spp.). *Proceedings of the Royal Society of London, B* **263**, 843–848.
Andersson, S. & Amundsen, T. (1997) Ultraviolet colour vision and ornamentation in bluethroats. *Proceedings of the Royal Society of London, B* **264**, 1587–1591.
Andersson, S., Örnborg, J. & Andersson, M. (1998) Ultraviolet sexual dimorphism and assortative mating in blue tits. *Proceedings of the Royal Society of London, B* **265**, 445–450.
Barlow, H. B. (1982) What causes trichromacy? A theoretical analysis using comb-filtered spectra. *Vision Research* **22**, 635–643.
Baylor, D. A. & Fettiplace, R. (1975) Light path and photon capture in turtle photoreceptors. *Journal of Physiology* **248**, 433–464.

Bennett, A. T. D., Cuthill, I. C., Partridge, J. C. & Maier, E. J. (1996) Ultraviolet vision and mate choice in zebra finches. *Nature, London* **380**, 433–435.

Bennett, A. T. D., Cuthill, I. C., Partridge, J. C. & Lunau, K. (1997) Ultraviolet plumage colors predict mate preferences in starlings. *Proceedings of the National Academy of Sciences* **94**, 8618–8621.

Blough, D. S. (1957) Spectral sensitivity in the pigeon. *Journal of the Optical Society of America* **47**, 827–833.

Bowmaker, J. K. (1977) The visual pigments, oil droplets and spectral sensitivity of the pigeon (*Columba livia*). *Vision Research* **17**, 1129–1138.

Bowmaker, J. K. (1983) Trichromatic colour vision: why only three receptor channels? *Trends in Neuroscience* **6**, 41–43.

Bowmaker, J. K. (1991a) Evolution of visual pigments and photoreceptors. In: *Vision and visual dysfunction, vol. 2, evolution of the eye and visual system* (eds. R. L. Gregory & J. R. Cronly-Dillon), pp. 63–81. Macmillan, London, UK.

Bowmaker, J. K. (1991b) Photoreceptors, photopigments and oil droplets. In: *Vision and visual dysfunction, vol. 6, the perception of colour* (ed. P. Gouras), pp. 108–127. Macmillan, London, UK.

Bowmaker, J. K. (1995) The visual pigments of fish. *Progress in Retinal and Eye Research* **15**, 1–31.

Bowmaker, J. K. (1998) Evolution of colour vision in vertebrates. *Eye* **12**, 541–547.

Bowmaker, J. K. & Hunt, D. M. (1999) Molecular biology of photoreceptor spectral sensitivity. In: *Adaptive mechanisms in the ecology of vision* (eds. S. N. Archer, M. B. A. Djamgoz, E. R. Loew, J. C. Partridge & S. Valerga), pp. 439–462. Kluwer, London, UK.

Bowmaker, J. K. & Knowles, A. (1977) The visual pigments and oil droplets of the chicken *Gallus gallus*. *Vision Research* **17**, 755–764.

Bowmaker, J. K. & Martin, G. R. (1978) Visual pigments and colour vision in a nocturnal bird *Strix aluco* (tawny owl). *Vision Research* **18**, 1125–1130.

Bowmaker, J. K. & Martin, G. R. (1985) Visual pigments and oil droplets in the penguin *Spheniscus humboldti*. *Journal of Comparative Physiology, A* **156**, 71–77.

Bowmaker, J. K., Kovach, J. K., Whitmore, A. V. & Loew, E. R. (1993) Visual pigments and oil droplets in genetically manipulated and carotenoid deprived quail: a microspectrophotometric study. *Vision Research* **33**, 571–578.

Bowmaker, J. K., Heath, L. A., Wilkie, S. E. & Hunt, D. M. (1997) Visual pigments and oil droplets from six classes of photoreceptor in the retinas of birds. *Vision Research* **37**, 2183–2194.

Burkhardt, D. (1996) Ultraviolet perception by bird eyes and some implications. *Naturwissenschaften* **83**, 492–497.

Burkhardt, D. A. (1989) UV vision: a bird's eye view of feathers. *Journal of Comparative Physiology, A* **164**, 787–796.

Chen, D. M. & Goldsmith, T. H. (1986) Four spectral classes of cones in the retinas of birds. *Journal of Comparative Physiology, A* **159**, 473–479.

Chen, D. M., Collins, J. S. & Goldsmith, T. H. (1984) The ultraviolet receptor of bird retinas. *Science, New York* **225**, 337–340.

Church, S. C., Bennett, A. T. D., Cuthill, I. C. & Partridge, J. C. (1998) Ultraviolet cues affect the foraging behaviour of blue tits. *Proceedings of the Royal Society of London, B* **265**, 1509–1514.

Dartnall, H. J. A. (1975) Assessing the fitness of visual pigments for their photic environments. In: *Vision in fishes* (ed. M. Ali), pp. 543–563. Plenum Press, New York, USA.

Dartnall, H. J. A., Bowmaker, J. K. & Mollon, J. D. (1983). Human visual pigments: microspectrophotometric results from the eyes of seven persons. *Proceedings of the Royal Society of London, B* **220**, 115–130.

Das, D., Wilkie, S. E., Hunt, D. M. & Bowmaker, J. K. (1999) Visual pigments and oil droplets in the retina of a passerine bird, the canary *Serinus canaria*: microspectrophotometry and opsin sequences. *Vision Research* **39**, 2801–2815.

Derim-Oglu, E. N. & Maximov, V. V. (1994) Small passerines can discriminate ultraviolet surface colors. *Vision Research* **34**, 1535–1539.

Dominy, N. J. & Lucas, P. W. (2001) Ecological importance of trichromatic vision to primates. *Nature, London* **410**, 363–366.

Emmerton, J. (1983) Pattern discrimination in the near ultraviolet by pigeons. *Perception and Psychophysics* **34**, 555–559.

Emmerton, J. & Delius, J. D. (1980) Wavelength discrimination in the visible and ultraviolet spectrum by pigeons. *Journal of Comparative Physiology, A* **141**, 47–52.

Emmerton, J. & Remy, M. (1983) The pigeon's sensitivity to ultraviolet and 'visible' light. *Experientia* **39**, 1161–1163.

Fager, L. Y. & Fager, R. S. (1981) Chicken blue and chicken violet, short wavelength sensitive visual pigments. *Vision Research* **21**, 581–586.

Fein, A. & Szuts, E. Z. (1982) *Photoreceptors: their role in vision.* Cambridge University Press, Cambridge, UK. 212 pp.

Goldsmith, T. H. (1980) Hummingbirds see near ultraviolet light. *Science, New York* **207**, 786–788.

Goldsmith, T. H. (1991) The evolution of visual pigments and colour vision. In: *Vision and visual dysfunction, vol. 6, the perception of colour* (ed. P. Gouras), pp. 62–89. Macmillan, London, UK.

Goldsmith, T. H., Collins, J. S. & Perlman, D. L. (1981) A wavelength discrimination function for the hummingbird *Archilochus alexandri. Journal of Comparative Physiology, A* **143**, 103–110.

Goldsmith, T. H., Collins, J. S. & Licht, S. (1984) The cone oil droplets of avian retinas. *Vision Research* **24**, 1661–1671.

Govardovskii, V. I. & Zueva, L. V. (1977) Visual pigments of chicken and pigeon. *Vision Research* **17**, 537–543.

Graf, V. & Norren, D. V. (1974) A blue sensitive mechanism in the pigeon retina: λ_{max} 400 nm. *Vision Research* **14**, 1203–1209.

Härlid, A., Janke, A. & Arnason, U. (1998) The complete mitochondrial genome of *Rhea americana* and early avian divergences. *Journal of Molecular Evolution* **46**, 669–679.

Hart, N. S. (2001) The visual ecology of avian photoreceptors. *Progress in Retinal and Eye Research* **20** (5), 675–703.

Hart, N. S., Partridge, J. C. & Cuthill, I. C. (1998) Visual pigments, oil droplets and cone photoreceptor distribution in the European starling (*Sturnus vulgaris*). *Journal of Experimental Biology* **201**, 1433–1446.

Hart, N. S., Partridge, J. C. & Cuthill, I. C. (1999) Visual pigments, cone oil droplets, ocular media and predicted spectral sensitivity in the domestic turkey (*Meleagris gallopavo*). *Vision Research* **39**, 3321–3328.

Hart, N. S., Partridge, J. C., Bennett, A. T. D. & Cuthill, I. C. (2000a) Visual pigments, cone oil droplets and ocular media in four species of estrildid finch. *Journal of Comparative Physiology, A* **186**, 681–694.

Hart, N. S., Partridge, J. C., Cuthill, I. C. & Bennett, T. A. D. (2000b) Visual pigments, oil droplets, ocular media and cone photoreceptor distribution in two species of passerine birds: the blue tit (*Parus caeruleus* L.) and the blackbird (*Turdus merula* L.). *Journal of Comparative Physiology, A* **186**, 375–387.

Hisatomi, O., Kayada, S., Aoki, Y., Iwasa, T. & Tokunaga, F. (1994) Phylogenetic relationships among vertebrate visual pigments. *Vision Research* **34**, 3097–3102.

Hunt, S., Cuthill, I. C., Swaddle, J. P. & Bennett, A. T. D. (1997) Ultraviolet vision and band-colour preferences in female zebra finches *Taeniopygia guttata. Animal Behaviour* **54**, 1383–1392.

Hunt, S., Bennett, A. T. D., Cuthill, I. C. & Griffiths, R. (1998) Blue tits are ultraviolet tits. *Proceedings of the Royal Society of London, B* **265**, 451–455.

Jacobs, G. H. (1993) The distribution and nature of colour vision among the mammals. *Biological Reviews* **68**, 413–471.

Jacobs, G. H., Neitz, J. & Deegan, J. F. (1991) Retinal receptors in rodents maximally sensitive to ultraviolet light. *Nature, London* **353**, 655–656.

Kawamura, S., Blow, N. S. & Yokoyama, S. (1999) Genetic analyses of visual pigments of the pigeon (*Columba livia*). *Genetics* **153**, 1839–1850.

Kawamuro, K., Irie, T. & Nakamura, T. (1997) Filtering effect of cone oil droplets detected in the P-III response spectra of Japanese quail. *Vision Research* **37**, 2829–2834.

Knowles, A. & Dartnall, H. J. A. (1977) The photobiology of vision. In: *The eye, vol. 2B* (ed. H. Davson), pp. 1–689. Academic Press, New York, USA.

Kreithen, M. L. & Eisner, T. (1978) Ultraviolet light detection by the homing pigeon. *Nature, London* **272**, 347–348.

Kuwata, O., Imamoto, Y., Okano, T., Kokame, K., Kojima, D., Matsumoto, H., Morodome, A., Fukada, Y., Shichida, Y., Yasuda, K., Shimura, Y. & Yoshizawa, T. (1990) The primary structure of iodopsin, a chicken red-sensitive cone pigment. *FEBS Letters* **272**, 128–132.

Lamb, T. D. & Pugh, E. N. (1990) Physiology of transduction and adaptation in rod and cone photoreceptors. *Seminars in the Neurosciences* **2**, 3–13.

Lythgoe, J. N. (1979) *The ecology of vision.* Oxford University Press, Oxford, UK. 169 pp.

Maier, E. J. (1992) Spectral sensitivities including the ultraviolet of the passeriform bird *Leiothrix lutea. Journal of Comparative Physiology, A* **170**, 709–714.

Maier, E. J. & Bowmaker, J. K. (1993) Colour vision in a passeriform bird *Leiothrix lutea*: correlation of visual pigment absorbance and oil droplet transmission with spectral sensitivity. *Journal of Comparative Physiology, A* **172**, 295–301.

Martin, G. R. (1985) Eye. In: *Form and function in birds, vol. 3* (eds. A. S. King & J. McLelland), pp. 311–373. Academic Press, London, UK.

Martin, G. R. & Young, S. R. (1984) The eye of the Humboldt penguin *Spheniscus humboldti*: visual fields and schematic optics. *Proceedings of the Royal Society of London, B* **223**, 197–222.

Mollon, J. D. (1989) "Tho' she kneel'd in that place where they grew…". *Journal of Experimental Biology* **146**, 21–38.

Nathans, J., Thomas, D. & Hogness, D. S. (1986) Molecular genetics of human color vision: the genes encoding blue, green, and red pigments. *Science, New York* **232**, 193–202.

Norren, D. V. (1975) Two short wavelength sensitive cone systems in pigeon, chicken and daw. *Vision Research* **15**, 1164–1166.

Okano, T., Fukada, Y., Artamonov, I. D. & Yoshizawa, T. (1989) Purification of cone visual pigments from chicken retina. *Biochemistry* **28**, 8848–8856.

Okano, T., Kojima, D., Fukada, Y., Shichida, Y. & Yoshizawa, T. (1992) Primary structures of chicken cone visual pigments: vertebrate rhodopsins have evolved out of cone visual pigments. *Proceedings of the National Academy of Sciences* **89**, 5932–5936.

Partridge, J. C. (1989) The visual ecology of avian cone oil droplets. *Journal of Comparative Physiology, A* **165**, 415–426.

Reed, J. R. (1987) Scotopic and photopic spectral sensitivities of boobies. *Ethology* **76**, 33–55.

Regan, B. C., Julliot, C., Simmen, B., Vienot, F., Charles-Dominique, P. & Mollon, J. D. (2001) Fruits, foliage and the evolution of primate colour vision. *Philosophical Transactions of the Royal Society, B* **356**, 229–283.

Remy, M. & Emmerton, J. (1989) Behavioral spectral sensitivities of different retinal areas in pigeons. *Behavioral Neuroscience* **103**, 170–177.

Romeskie, M. & Yager, D. (1976) Psychophysical studies of pigeon color vision – I. Photopic spectral sensitivity. *Vision Research* **16**, 501–505.

Schnapf, J. L., Kraft, T. W. & Baylor, D. A. (1987) Spectral sensitivity of human cone photoreceptors. *Nature, London* **325**, 439–441.

Takao, M., Yasui, A. & Tokunaga, F. (1988) Isolation and sequence determination of the chicken rhodopsin gene. *Vision Research* **28**, 471–480.

Viitala, J., Korpimäki, E., Palokangas, P. & Koivula, M. (1995) Attraction of kestrels to vole scent marks visible in ultraviolet light. *Nature, London* **373**, 425–427.

Vorobyev, M., Osorio, D., Bennett, A. T. D., Marshall, N. J. & Cuthill, I. C. (1998) Tetrachromacy, oil droplets and bird plumage colours. *Journal of Comparative Physiology, A* **183**, 621–633.

Vos Hzn, J. J., Coemans, M. A. J. M. & Nuboer, J. F. W. (1994) The photopic sensitivity of the yellow field of the pigeon's retina to ultraviolet light. *Vision Research* **34**, 1419–1425.

Wang, S.-Z., Adler, R. & Nathans, J. (1992) A visual pigment from chicken that resembles rhodopsin: amino acid sequence, gene structure, and functional expression. *Biochemistry* **31**, 3309–3315.

Wilkie, S. E., Vissers, P. M. A. M., Das, D., DeGrip, W. J., Bowmaker, J. K. & Hunt, D. M. (1998) The molecular basis for UV vision in birds: spectral characteristics, cDNA sequence and retinal localization of the UV-sensitive visual pigment of the budgerigar (*Melopsittacus undulatus*). *Biochemical Journal* **330**, 541–547.

Wilkie, S. E., Robinson, P. R., Cronin, T. W., Poopalasundaram, S., Bowmaker, J. K. & Hunt, D. M. (2000) Spectral tuning of avian violet- and ultraviolet-sensitive visual pigments. *Biochemistry* **39**, 7895–7901.

Wortel, J. F., Wubbels, R. J. & Nuboer, J. F. W. (1984) Photopic spectral sensitivities of the red and the yellow field of the pigeon retina. *Vision Research* **24**, 1107–1113.

Wright, A. A. (1972) The influence of ultraviolet radiation on the pigeon's color discrimination. *Journal of the Experimental Analysis of Behavior* **17**, 325–327.

Wright, A. A. (1979) Color-vision psychophysics: a comparison of pigeon and human. In: *Neural mechanisms of behavior in the pigeon* (eds. A. M. Granda & J. H. Maxwell), pp. 89–127. Plenum, New York, USA.

Wright, M. W. & Bowmaker, J. K. (2001) Retinal photoreceptors of paleognathous birds: the ostrich (*Strutho camelus*) and rhea (*Rhea americana*). *Vision Research* **41**, 1–12.

Yen, L. & Fager, R. S. (1984) Chromatographic resolution of the rod pigment from the four cone pigments of the chicken retina. *Vision Research* **24**, 1555–1562.

Yokoyama, S., Radlwimmer, F. B. & Kawamura, S. (1998) Regeneration of ultraviolet pigments of vertebrates. *FEBS Letters* **423**, 155–158.

Yokoyama, S., Blow, N. S. & Radlwimmer, F. B. (2000) Molecular evolution of colour vision of zebra finch. *Gene* **259**, 17–24.

Yoshizawa, T. & Fukada, Y. (1993) Preparation and characterization of chicken rod and cone pigments. In: *Methods in neurosciences, vol.15, photoreceptor cells* (ed. P. A. Hargrave), pp. 161–179. Academic Press, San Diego, USA.

11
Avian Ultraviolet Vision and its Implications for Insect Defensive Coloration

STUART C. CHURCH, INNES C. CUTHILL, ANDREW T. D. BENNETT AND JULIAN C. PARTRIDGE

School of Biological Sciences, University of Bristol, Woodland Road, Bristol, BS8 1UG, UK

Introduction

The protective colour patterns of animals, such as those involved in camouflage, warning coloration, and mimicry, are often viewed as some of the most convincing examples of natural selection in the wild, and have become classic examples in evolutionary biology and ecology (Poulton, 1890; Cott, 1940; Wickler, 1968; Kettlewell, 1973). For example, Cott (1940, p. xii) opined that "…it is no exaggeration to say that the modification of outward appearance by visual characters, directed towards a seeing public, and serving to either facilitate recognition or to frustrate it, has been one of the main results attained in the evolution of the higher animals; and such characters comprise some of the most outstanding examples of adaptation in the whole field of biology." Similarly, Fisher (1958) described mimicry as "… the greatest Neo-Darwinian application of Natural Selection."

The essence of understanding protective coloration in animals lies in our ability to quantify and classify colour patterns accurately and relate these classifications to their selective consequences. As humans, we consider ourselves to be primarily visual animals, and attach huge importance to colour perception *via* our sense of aesthetics and its implications for our emotional state (Gregory, 1997). Perhaps more than any other sense, we feel that we have an intuitive understanding and appreciation of colour patterns in the natural world. From a human perceptual viewpoint, it is easy to accept, on purely intuitive grounds, that, for example, a palatable 'green' insect resting on a 'green' leaf, which looks camouflaged to us, is also camouflaged to its visually-searching natural predators, and will therefore suffer less predation than, say, a palatable 'red' insect on the same leaf. Indeed, more than a hundred years of research on animal coloration, including research on protective coloration, has been based on the (usually implicit) assumption that other animals perceive and classify colour patterns in the same way as humans (Bennett *et al.*, 1994). However, it is

Insect and Bird Interactions
© Intercept Ltd., PO Box 716, Andover, Hampshire, SP10 1YG, UK.

becoming increasingly clear that this anthropocentric view is very limited. The main point to consider is that 'colour' is not an inherent property of an object, but rather a property of the neural machinery of the animal perceiving the light reflected from the object (Endler, 1978, 1990; Lythgoe, 1979; Chittka, 1992; Thompson *et al.*, 1992; Bennett *et al.*, 1994). An object reflecting light at particular wavelengths may result in different sensations of colour to different animal species, or even different individuals within a species (human colour blindness being an obvious example). Since protective colour patterns evolve in response to selection by visually-guided predators, it is predicted that colour patterns will evolve which are tuned to the visual systems of those predators. Since humans have not been the major selective agent in the evolution of virtually all examples of protective coloration, it is important to consider the visual systems of natural predators.

From the perspective of insect protective coloration, avian predators are likely to be a major selective force (Dempster, 1984), although other vertebrate predators and insect predators or parasitoids may also impose selection. Birds have long been the primary experimental animal used in field and laboratory tests of theories involving crypsis, mimicry, and aposematism (e.g. Pietrewicz & Kamil, 1977; Roper & Redston, 1987; Roper & Cook, 1989; Dittrich *et al.*, 1993; Alatalo & Mappes, 1996; Rowe & Guilford, 1996; Mappes & Alatalo, 1997a,b; Lindström *et al.*, 1999). During the past two decades, though, it has become apparent that the majority of avian predators possess a visual system that is very different from that of humans (Burkhardt, 1989; Bennett & Cuthill, 1994; Bennett *et al.*, 1994; Cuthill *et al.*, 2000a). Birds have probably the most sophisticated visual system of any vertebrate (Goldsmith, 1990) and, importantly, see a range of hues to which humans are blind. This fact has been ignored in virtually all studies of protective coloration among insects, yet has clear implications for our understanding of these phenomena (Bennett *et al.*, 1994).

Avian and human colour vision

In all animals, differential stimulation of interacting photoreceptor types forms the basis of colour vision, with at least two spectrally distinct photoreceptor types required for wavelength (hue) discrimination. Normal human colour vision (see also Bowmaker, 2003) is based on three photoreceptor types; a long-wave-sensitive ('red') cone, a medium-wave-sensitive ('green') cone, and a short-wave-sensitive ('blue') cone. These cones have peak sensitivities (λ_{max}) at wavelengths of approximately 560, 530 and 420 nm respectively (Dartnall *et al.*, 1983), and human spectral sensitivity ranges from approximately 400 to 700 nm. Humans are said to be 'trichromatic' in that all three cones are neurally integrated, and the myriad hues which we perceive are derived from differential stimulation of these three cone types.

In contrast to humans, behavioural and physiological studies of more than 35 species suggest that diurnal birds usually possess four spectrally distinct single cone types (reviewed by Cuthill *et al.*, 2000a; Hart, 2001). Microspectrophotometric analyses of the avian retina (e.g. Bowmaker *et al.*, 1997; Hart *et al.*, 1998; Hart, 2001) have revealed the presence of long-wave-sensitive (LWS; λ_{max} 543–571 nm), medium-wave-sensitive (MWS; λ_{max} 497–509 nm), short-wave-sensitive (SWS; λ_{max} 430–463 nm), and either violet-sensitive (VS: λ_{max} 402–426 nm) or ultraviolet-sensitive (UVS; λ_{max} 355–376 nm) cone types (*Figure 11.1*). Birds, therefore, are likely to possess

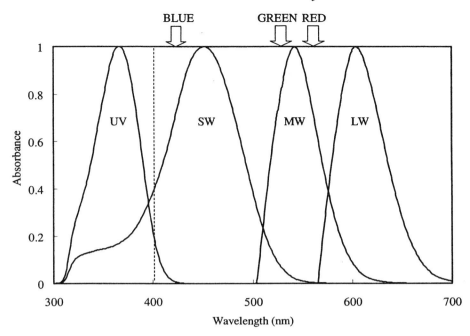

Figure 11.1.   Spectral sensitivities of single cones of the European starling (*Sturnus vulgaris*) based on data derived from microspectrophotometry of the cone photopigments and oil droplets (Hart *et al.*, 1998). Mean values for the wavelengths of maximum absorbance (λ_{max}) for the different visual pigment types are: UVS = 362 nm, SWS = 449 nm, MWS = 504 nm, LWS = 563 nm. For comparison, human λ_{max} values (VS = 419 nm, MWS = 531 nm, LWS = 558 nm; Dartnall *et al.*, 1983) are indicated by the positions of the arrows.

four-dimensional ('tetrachromatic') colour vision (Thompson *et al.*, 1992). It is important to note that simply possessing four cone types does not guarantee that colour vision is tetrachromatic, since not all cone types may contribute to colour vision (Neumeyer, 1992). However, all four cones have been shown to be involved in opponent channels (i.e. neural comparison) in birds (Osorio *et al.*, 1999a,b). Further-more, goldfish, which also possess four classes of visual pigments (Yokoyama *et al.*, 1998; Bowmaker & Hunt, 1999), have been shown to be tetrachromatic via colour-mixing experiments (Neumeyer, 1992). The VS/UVS cones of birds, coupled with a UV-transmitting lens and ocular media (e.g. Hart *et al.*, 1998) allows diurnal birds to detect wavelengths in the UV-A region (320–400 nm). Diurnal birds are therefore able to detect wavelengths in a range of the spectrum to which humans are blind. This is a fundamental difference in the manner in which birds and humans perceive the world.

Although the focus of this chapter is how differences in UV sensitivity affect our understanding of insect protective colour patterns, there are other crucial differences between avian and human vision which may also cause birds to perceive colours differently from humans. First, each type of avian cone cell is associated with a characteristic oil droplet (not found in humans) through which light is filtered prior to reaching the photopigment-containing cone outer segments (Partridge, 1989; Bowmaker, 1991; Hart, 2001). These oil droplets act as short wavelength cut-off filters, and narrow the range of sensitivity of each cone type. The suggested advantages of narrowed spectral sensitivity and reduced overlap in waveband is increased colour saturation and enhanced discrimination of certain classes of spectra (Govardovskii, 1983; Goldsmith, 1990; Vorobyev *et al.*, 1998; Dyer, 2001), as well as improved colour constancy (the phenomenon whereby objects are perceived as the same colour despite changes in the illuminant; Osorio *et al.*, 1997; Vorobyev *et al.*, 1998). Second, an important consequence of visual systems with more than two dimensions is the existence of 'non-spectral' colours. These are colours that are not part of the rainbow and cannot be simulated by any monochromatic light source. They arise from simultaneous stimulation of two or more photoreceptors that are sensitive to *non-adjacent* wavebands. In humans, stimulation of the red and blue cones results in the perception of purple (the only non-spectral colour we can perceive). An avian tetrachromat, with four single cone types, might have five non-spectral colours. These would correspond to the three possible two-way combinations (UVS/VS+MWS, UVS/VS+LWS, SWS+LWS) and two possible three-way combinations (UVS/VS+SWS+LWS, UVS/VS+MWS+LWS) of receptors sensitive to non-adjacent wavebands of light (see Burkhardt, 1989; Goldsmith, 1990, 1994; Thompson *et al.*, 1992). If birds are truly tetrachromatic, then it is the non-spectral colours that are one of the main reasons it is particularly hard to get a subjective impression of a bird's view of objects' hues. We can, however, describe both the reflectance spectra and their likely processing by the bird's visual system in ways that can be analysed quantitatively (Endler, 1990; Endler & Théry, 1996; Vorobyev *et al.*, 1998; Cuthill *et al.*, 1999).

Ultraviolet wavelengths: foraging and function

In birds, it has been hypothesised that UV wavelengths may be important in orientation, regulation of circadian rhythms, mate choice (reviewed by Bennett & Cuthill, 1994; Cuthill *et al.*, 2000a) and, most importantly from the perspective of protective coloration, foraging. Unfortunately, our understanding of the role that UV plays in the foraging behaviour of birds is still in its infancy. As an integral part of the visual system of birds, it is probable that UV wavelengths provide information which birds are able to use in the detection and classification of prey items. This seems likely, given that many of the invertebrates, fruits, seeds, and flowers on which birds feed reflect in the UV (Silberglied, 1979; Burkhardt, 1982; Willson & Whelan, 1989; see next section). Indeed, experimental evidence has slowly accumulated that birds are able to utilise UV wavelengths when foraging.

Some of the earliest demonstrations of UV vision in birds were in hummingbirds (Huth & Burkhardt, 1972; Goldsmith, 1980). This was perhaps a natural choice, given the importance of UV vision for bee pollinators (Chittka & Menzel, 1992; Chittka *et al.*, 1994; Giurfa *et al.*, 1995; Kevan *et al.*, 1996; Lunau *et al.*, 1996; Vorobyev *et al.*, 1997). Surprisingly, though, the role of UV cues in natural foraging by hummingbirds has not been investigated. This may be because red coloration is commonly found in bird-pollinated flowers, and it has been postulated (Raven, 1972) that red represents a 'private channel' for flowers to signal to birds but not bees (see also Porch, 1931, cited in Lunau & Maier, 1995).

In an early psychophysical experiment, Emmerton & Remy (1983) measured the performance of pigeons in a seed detection task. Trained pigeons (*Columba livia*) were given access to 20 grains of corn on a diffusing plate for 30 seconds. The plate was illuminated from below with a monochromatic light source, which was varied from 320 to 640 nm. On the basis of the number of seeds consumed in 30 seconds, Emmerton & Remy (1983) demonstrated that the pigeons foraged most effectively at longer wavelengths (peak sensitivity at approximately 580 nm), yet were still able to perform the task in the UV. The distribution of cone cells is uneven in the pigeon retina, and the retinal image of seeds in this foraging task would have fallen on the pigeon's dorso-temporal 'red' field (as opposed to its ventral 'yellow' field), which is likely to have relatively low UV sensitivity. The red field is so called because of the high concentration of cones with red oil droplets in this area; these are the LWS and double cones (Bowmaker, 1991). The area of the retina which is looking down and forward is therefore apparently more suited to longer wavelength-based discriminations. Emmerton & Remy (1983) therefore suggested that UV information might be of little value in grain detection tasks. However, it does not necessarily follow that these conclusions hold for all birds, since other species do not have the same spatial distribution of photoreceptors as pigeons. For example, in contrast to the pigeon, the starling (*Sturnus vulgaris*) has more UV cones in its dorsal retina (Hart *et al.*, 1998), which would make it particularly suited to using UV cues in foraging tasks. Furthermore, the pigeon has a violet-sensitive, rather than UV-sensitive cone, with peak sensitivity at around 410 nm (Bowmaker *et al.*, 1997).

More recently, Viitala *et al.* (1995) found that kestrels (*Falco tinnunculus*) used UV cues to locate areas containing vole trails. Using reflectance spectrophotometry, they showed that the scent marks of male voles contrasted with their background in the UV, but less so in the human-visible spectrum. In field manipulations, Viitala *et al.* (1995) demonstrated that kestrels hunted in areas containing artificial scent marks rather than in areas containing water-treated trails, or no trails at all. Furthermore, in laboratory preference experiments, kestrels spent more time looking at, and more time in the vicinity of, arenas which had added vole scent marks, but only when under UV illumination (*Figure 11.2*). Overall, Viitala *et al.*'s (1995) experiments suggest

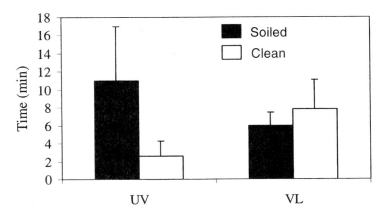

Figure 11.2.   The effect of UV wavelengths on kestrel (*Falco tinnunculus*) foraging behaviour, measured as the mean time (minutes, mean + s.d.) spent by kestrels above arenas differing in illumination and presence or absence of vole scent marks. The four treatments were: dry vole trails in ultraviolet (UV) light, clean arena in ultraviolet light, dry vole trails in visible (VL) light, and clean arena in visible light. Redrawn from Viitala *et al.* (1995) with permission.

that vole trails, and therefore areas of high vole abundance, are more conspicuous to kestrels when UV cues are present. The same appears to be true for another diurnally hunting raptor, the rough-legged buzzard (*Buteo lagopus*; Koivula & Viitala, 1999). In contrast, however, similar laboratory experiments on adults and juveniles of another major predator of voles, Tengmalm's owl (*Aegolius funereus*), revealed no preference for arenas containing vole scent marks under UV illumination (Koivula *et al.*, 1997). This is not altogether surprising, since *A. funereus* is a nocturnal predator, and owls may lack UV cones (Bowmaker & Martin, 1978).

Birds foraging for fruits may also use UV wavelengths. Many fruits, particularly those with waxy layers or 'blooms', reflect UV wavelengths to some degree (Burkhardt, 1982; Cuthill *et al.*, 2000a). Siitari *et al.* (1999) carried out experiments on redwings (*Turdus iliacus*) foraging for bilberries (*Vaccinium myrtillus*). Under full-spectrum illumination, adult redwings showed a significant preference for bilberries in which the waxy, UV-reflecting layer was intact over berries with the wax rubbed off. This preference disappeared when UV cues were experimentally removed. Since juvenile redwings exhibited no preference under UV+ or UV− conditions, Siitari *et al.* (1999) concluded that the preference for UV-reflecting berries was probably learned.

The effect of UV wavelengths on avian food preferences is not necessarily simple, though. Maddocks *et al.* (2001) found that removing UV had no effect on the preferences of zebra finches (*Taeniopygia guttata*) foraging for red and white millet seeds which were presented in equal numbers. Instead, removal of long wavelengths (approx. 600–700 nm) was found to have the greatest effect on the finches' seed selection. In contrast, Church *et al.* (2001), using a similar experimental paradigm, showed that UV wavelengths had a significant effect on the direction of *frequency-dependent* seed preferences (i.e. preferences depended on the relative availabilities of the two seed types; Allen, 1988) for the same seed types, yet found no effect of long wavelengths. Thus, UV may only be important in affecting avian prey preferences under certain environmental conditions.

Insects are important prey items for birds, and many insects, particularly adult Lepidoptera, are known to reflect UV to some extent (Silberglied, 1979). However, whether birds actually utilise UV reflectance patterns in prey detection or recognition is unknown. In the first experimental study to investigate the importance of UV to birds hunting for cryptic insect prey, Church *et al.* (1998b) looked at the behaviour of blue tits (*Parus caeruleus*) searching for cabbage moth (*Mamestra brassicae*) or winter moth (*Operophtera brumata*) caterpillars in a laboratory arena. The latency to find the first prey item was increased in trials where UV wavelengths were removed from the illuminating light. This effect was most pronounced when the UV contrast between prey and background was greatest, namely *M. brassicae* against a cabbage leaf (*Figure 11.3*). However, the reduction in performance was temporary, presumably due to the birds learning to attend to other salient visual cues (Church *et al.*, 1998b). It is unwise to interpret this experiment as conclusively demonstrating that blue tits locate prey using 'UV cues'. The detrimental effect of removing UV could be *via* alteration of the hue of both prey and background, and it is the combined colour change which might render the task more difficult. However, subsequent experiments using a similar protocol with blue tits foraging for cryptic winter moth caterpillars and sunflower seeds (S. C. Church, I. C. Cuthill, A. T. D. Bennett & J. C. Partridge, *unpublished*) failed to reveal any significant effect of UV cues on latency to find prey.

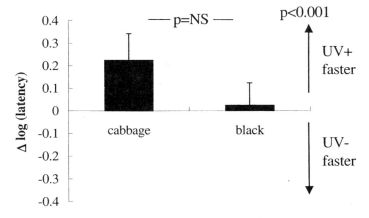

Figure 11.3.   Difference in latencies, Δ log (latency), to find first prey item under UV+ (human-visible plus UV) and UV- (human-visible only) illumination by blue tits searching for cabbage moth (*Mamestra brassicae*) caterpillars on cabbage and black paint backgrounds. Positive values indicate that the blue tits find prey faster when ultraviolet cues are present. (Data derived from Church *et al.*, 1998b). Reproduced from Cuthill *et al.* (2000a) with permission.

These experiments differed from the previous study in that birds were pre-exposed to the lighting conditions for ten minutes prior to testing. Thus, both sets of experiments suggest that UV cues offer little benefit in detecting cryptic prey if the bird can visually adapt to the light environment in which it is foraging. Since these experiments were carried out on cryptic prey, they are likely to represent conservative tests of the effects of UV on foraging because most natural visual backgrounds (e.g. leaves, bark, soil) reflect relatively little UV. As it is unlikely that any natural prey will differ from its visual background *only* in the UV, there are always likely to be additional cues which birds can learn to attend to (e.g. human-visible wavelengths, shape, size, visual texture). Ultimately, experiments using precisely constructed artificial prey are required to examine the importance of UV from a behavioural viewpoint, and to determine the capabilities of avian visual systems.

Insect UV reflectance patterns

UV reflectance patterns appear to be common among insects (Silberglied, 1979), with the most numerous examples of UV reflectance being found among adult Lepidoptera (e.g. Mazokhin-Porshnyakov, 1957; Silberglied, 1979; Eguchi & Meyer-Rochow, 1983; Meyer-Rochow & Jarvilheto, 1997). Although this chapter focuses on the role of the avian visual system in the evolution of insect protective colour patterns, it is worth stressing that many insect colour patterns represent trade-offs between defensive, thermoregulatory, and sexual functions (Scoble, 1992). Certainly, UV patterns seem to be used widely by Lepidoptera in mate assessment (e.g. Silberglied & Taylor, 1973; Silberglied, 1979, 1984; Meyer-Rochow, 1991; Brunton & Majerus, 1995), and there may even be greater sexual dimorphism in the UV than in the human-visible spectrum (Silberglied, 1984). It has also been suggested that the white, UV-reflective elytra of some Tenebrionidae (Coleoptera) play an important role in thermoregulation (Pope & Hinton, 1977). Thus, a complete understanding of a given insect colour pattern can only be achieved by considering all three factors (i.e. defensive, thermoregulatory, and sexual functions) in both the UV and human-visible spectrum.

Insect coloration is most commonly achieved by means of structural and/or pigment colours (Hinton, 1973; Silberglied, 1979). Insect structural colours are produced either by means of diffusion (scattering), diffraction, or interference (Scoble, 1992; Vulinec, 1997), with the type of structural colour produced depending upon the microstructure of the cuticle. Although simple scattering by the wing scales can produce UV reflectance (Siblerglied, 1979), some of the most impressive colour patterns result from interference patterns. These interference colours generally result in iridescent patterns in which the observed colour depends on the angle of the viewer to the insect, and can produce strong UV reflectance (e.g. in the Pieridae; Ghiradella *et al.*, 1972). Interference colours can be combined with pigment colours in Lepidoptera, allowing any human-visible colour potentially to be combined with UV (Silberglied, 1979).

Among the insects, UV reflectance appears to be particularly common in adult Lepidoptera. As Silberglied (1979) states: "The wings of butterflies bear a wide variety of UV reflection patterns, including the most intense such reflection found in nature." Several studies have explored the UV reflectance over a variety of Lepidopteran families. In a pioneering study, Crane (1954) used a series of narrow band-pass filters to photograph 41 species of Trinidadian butterflies at human-visible and UV wavelengths. She found that UV reflectance was low in the majority of species, with no 'unexpected' areas discovered which might be used for intra-specific signalling. Ultimately, Crane concluded that UV was unimportant, and that "ultraviolet sensitivity is a mere by-product of the physiological processes of the insect eye." However, during the following thirty years, a large body of evidence of high UV reflectance of butterflies accumulated (Mazokhin-Porshnyakov, 1957; Silberglied & Taylor, 1973; Roland, 1978; Silberglied, 1979, 1984; Eguchi & Meyer-Rochow, 1983; Meyer-Rochow & Eguchi, 1983; Brunton & Majerus, 1995). Eguchi & Meyer-Rochow (1983) have provided the largest survey of UV patterns in Lepidoptera. They looked at 43 species of Lepidoptera from 10 different families using UV photography, and ranked species according to whether UV reflectance was 'strongly present', 'present', or 'not present'. Their results indicated that UV reflectance was widespread, with all human colour classes being associated with some degree of UV reflectance. Interestingly, while UV cues were, predictably, associated with human 'blue' and 'blue-green', they were also strongly associated with 'red-orange'. These counter-intuitive associations may result from the combination of structural and pigment colours, as suggested by Silberglied (1979). In butterfly colour patches which were white, various shades of brown, or black, UV reflectance declined with increasing darkness, presumably due to increasing levels of melanin, which strongly absorbs UV (Majerus, 1998). Indeed, blackflies (*Simulium vittatum*) raised in a UV-rich environment tend to be more melanic (Zettler *et al.*, 1998), presumably so that UV wavelengths can be absorbed before they can cause genetic damage.

Although research has focused primarily on the Lepidoptera, strong UV reflectance has also been found in other insect orders. In a survey of UV reflectance among Coleoptera, Pope & Hinton (1977) demonstrated, using UV photography, strong UV patterns among the families Carabidae, Scarabaeidae, Buprestidae, Elateridae,

Tenebrionidae, Cerambycidae, Anthribidae, and Curculionidae. High UV reflectance has also been found in Diptera (Hinton, 1973; Deonier, 1974; Steinly *et al.*, 1978) and Odonata (Robey, 1975; Silberglied, 1979). However, among insect orders other than the Lepidoptera, the prevalence and functional significance of UV reflectance is generally poorly understood. We are therefore blind to much of the sensory information which may be available to avian predators.

Consequences of avian UV vision for insect protective coloration

SIGNAL DESIGN AND DEFENSIVE COLORATION

Avian UV vision has implications for our understanding of insect protective coloration. In general, "a colour pattern must resemble a random sample of the background in order to be cryptic, and must deviate from the background in one or more ways in order to be conspicuous" (Endler, 1978). In terms of signal design, cryptic prey will be selected to minimise signal–to–noise ratios in the visual systems of potential predators, while conspicuous (e.g. aposematic) prey ought to be selected to maximise signal–to–noise ratios in their natural habitat (Vane-Wright & Boppré, 1993). It is therefore important to assess the spectral properties of both the prey item and its visual background if one is to comprehend the design and protective value of a colour pattern (Endler, 1978). All other things being equal, a highly UV-reflective insect may look very conspicuous to an avian predator on a UV-absorbing background, but quite cryptic on a UV-reflecting background. The degree of UV reflectance from the insect alone provides little information regarding any possible defensive function of its coloration. It is not sufficient to disregard the UV when considering protective coloration simply because the species in question has low UV reflectance, irrespective of its background. While it is true that many natural backgrounds of insects (e.g. soil, bark, leaves) do not usually reflect strongly in the UV (Kevan *et al.*, 1996; Church *et al.*, 1998a), other natural backgrounds, such as the sky (Silberglied, 1979), clouds (Mazokhin-Porshniakov, 1957), and some fruits (Burkhardt, 1982; Willson & Whelan, 1989), flowers (Silberglied, 1979; Chittka *et al.*, 1994; Dyer, 1996), and sands (Pope & Hinton, 1977) have a strong UV component.

UV AND CRYPSIS

In order to be cryptic to birds and not just to the human eye, insects ought to match their visual backgrounds throughout the avian-visible spectrum. In cases where reflectance spectra of cryptic caterpillars have been measured over this range, this is often, but not always, the case. Majerus *et al.* (2000) demonstrated that the degree of colour to background matching of the melanic and non-melanic forms of the peppered moth (*Biston betularia*) depends, to a large extent, on the amount of UV reflected by different types of lichens on tree bark. The black scales of *B. betularia* (i.e. melanic form) are highly UV-absorbing, while white scales (non-melanic) are UV-reflecting. Majerus *et al.* (2000) showed that the reflectance spectra of non-melanic *B. betularia* were more cryptic than the melanic form on a background of crustose lichens, which sometimes reflected UV, than on foliose lichens, which reflected little UV. It may therefore be the case that the preferred resting sites of non-

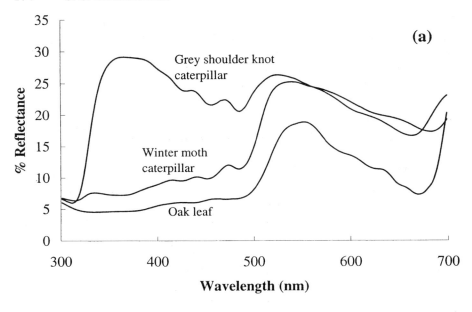

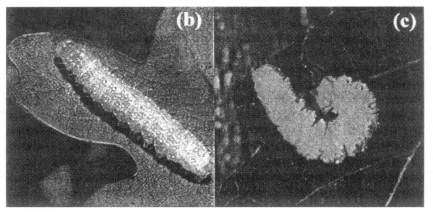

Figure 11.4.   (a) Reflectance spectra in the wavelength range 300–700 nm of the caterpillars of the grey shoulder knot (*Lithophane ornitopus*) and the winter moth (*Operophtera brumata*) and the underside of a common oak (*Quercus robur*). While the reflectance characteristics of the winter moth, like most 'green' caterpillars of oak, are very similar to those of its oak leaf background, the grey shoulder knot contrasts strongly, particularly in the UV (data derived from Church *et al.*, 1998a), as illustrated by the images of the grey shoulder knot (b) under UV illumination (300–400 nm), and (c) in the human-visible spectrum (400–700 nm). Adapted from Cuthill *et al.* (2000a) with permission.

melanic *B. betularia* are on the undersides of branches, where crustose lichens are predominantly found. However, while this is certainly an intriguing new twist on a classic evolutionary study, one must be careful when ascribing function to a colour pattern in the absence of definite knowledge of the background on which the animal rests.

Church *et al.* (1998a) demonstrated that the spectral reflectance of a number of Lepidopteran larvae from oak trees tends to match closely those of either leaves or twigs over the whole range of wavelengths to which birds are sensitive. However, in one species, the grey shoulder knot (*Lithophane ornitopus*), which appears moderately cryptic to the human eye, maximum reflectance is in the UV (*Figure 11.4*). It is feasible that the grey shoulder knot is aposematic rather than cryptic, and is advertising distastefulness to birds via a UV communication channel (Church *et al.*, 1998a). However, the crucial experimental tests which could confirm this hypothesis have yet to be carried out. It is also conceivable that the caterpillar hides in places where there is little UV illumination, so matching in this waveband is not important (and the UV reflectance is an incidental by-product of structural features).

In a study of the UV reflectance of beetles, Pope & Hinton (1977) showed that some desert Tenebrionidae (e.g. *Onymacris* spp.) have strong UV reflectance. They suggested that this might render the beetles cryptic to UV-detecting predators on UV-reflecting sand. At the time of publication, Pope & Hinton (1977) thought this only applied to invertebrate predators, since the ability of birds to detect UV was not fully appreciated. They also suggested that the colours of these beetles could be aposematic, since they also possess sharply contrasting black and white areas. This illustrates the difficulty of interpreting the function of colour patterns, even when UV reflectance characteristics are known. Ultimately, behavioural experiments are required to ascertain whether a particular prey species is cryptic or conspicuous, and what the selective consequences of such coloration are likely to be.

While UV cues should not be considered 'special' compared to other wavelengths to which birds are sensitive, it is possible that UV cues will be useful to birds foraging for cryptic prey in some ambient light environments. Endler (1993) used field spectrophotometry to demonstrate how the spectral composition of ambient light in forests depends largely on forest geometry (for example, the size of gaps in the canopy through which the sky is visible), time of day, and weather. In particular, Endler's category of 'woodland shade' (shaded areas illuminated primarily by light reflected from leaves and the sky) and light at dawn/dusk are particularly rich in short wavelengths. As such, a significant proportion of the visual information available to birds for detecting cryptic prey under these conditions is likely to be in the UV. Furthermore, since light availability may limit the ability of birds to forage (Kacelnik & Krebs, 1982), possession of a UV receptor may allow birds to extend the time period over which they are able to forage at dawn and dusk.

In general, prey species that are cryptic to the human eye are also likely to be cryptic to birds. Vorobyev *et al.* (1998) used a modelling approach to demonstrate that perception of colour patterns by humans and birds (bird plumage spectra in their case) is positively correlated. It therefore seems reasonable to assume that cryptic species will, in many cases, also be perceived similarly. However, Vorobyev *et al.* (1998) warn that the greatest discrepancies will arise when spectra vary in the UV, or when they are very complex (e.g. multi-peaked, as in the case of non-spectral colours). Thus, although human and avian perception may sometimes be closely correlated, it is still unwise to ignore the UV.

UV, APOSEMATISM, AND MIMICRY

In recent years, there has been a surge of interest in the design features of aposematic and mimetic colour patterns (e.g. Guilford, 1986; Guilford & Dawkins, 1991; Schuler

& Roper, 1992; Marples & Roper, 1996; Rowe & Guilford, 1996; Roper & Marples, 1997a,b), yet these patterns have only ever been assessed from a human perspective (but see Cherry & Bennett, 2001, for a notable exception). In general, it seems likely that aposematic patterns that are conspicuous and highly contrasting to us (e.g. red/ yellow and black) will be at least as conspicuous to birds, as the avian eye is well designed for long wavelength discriminations (Emmerton & Delius, 1980; Vorobyev & Osorio, 1998; Vorobyev *et al.*, 1998; Osorio *et al.*, 1999a). However, the possibility that UV wavelengths might be used to signal unpalatability to avian predators (Church *et al.*, 1998a) has only just begun to be explored. Lyytinen *et al.* (2001) found that great tits (*Parus major*) were poor at learning to use UV presence/absence as a discriminative stimulus for unpalatable, artificial prey. However, the colours they used (pale green, with or without UV reflection) were quite unsaturated, unlike normal aposematic colours. Some aposematic butterflies that are seen as 'red' by humans often have a strong UV component (Eguchi & Meyer-Rochow, 1983; Meyer-Rochow & Eguchi, 1983). Although UV was once thought to represent a 'private channel' for intra-specific communication among insects (e.g. Silberglied, 1979, 1984; Meyer-Rochow & Eguchi, 1983) under the assumption that birds were UV blind, avian predators are likely to see 'UV-red' aposematic coloration as a non-spectral colour, to which humans have no perceptual equivalent. This 'UV-red' combination could be an effective signal to avian predators since it is essentially the inverse of 'green', and any 'UV-red' insect would therefore produce a high degree of chromatic contrast while resting on leaves. In terms of signal design, such a high degree of contrast could aid the detectability and memorability of the warning signal (Gittleman & Harvey, 1980; Guilford & Dawkins, 1991). Thus, as in the case of crypsis, consideration of UV wavelengths has the potential to change our understanding of the ways in which protective colour patterns function.

Despite the potential for non-spectral aposematic coloration, some aposematic and mimetic species have 'red' or 'red/orange' colour pattern elements which reflect very little UV. For example, the unpalatable African Danaid butterfly (*Danaus chrysippus*) is polymorphic, with the 'red' dorsal coloration in each morph reflecting virtually no UV (S. C. Church, D. A. S. Smith, I. C. Cuthill & A. T. D. Bennett, *unpublished*). Furthermore, for each colour morph of *D. chrysippus*, there are several Batesian and Mullerian mimics (Smith *et al.*, 1993; Owen *et al.*, 1994). For these mimics to be effective, their reflectance properties ought to match those of *D. chrysippus* in the UV as well as in human-visible wavelengths. Simple visual comparison of the reflectance spectra suggests, reassuringly, that the colours of the model and its mimics would, at the very least, be perceived as reasonably similar by avian predators (*Figure 11.5*). Similarly, in an early study, Lutz (1933) used UV photography to demonstrate that mimetic resemblances could extend into the UV. Although he was unaware of the ability of birds to detect UV wavelengths, Lutz prophetically stated that "Apparently there is...more to be known about the visual powers of vertebrates..." While these cases suggest there may be similarities in the UV, human-perceived mimicry does not inevitably extend into the UV. Other studies based on UV photography suggest that the colour match in the UV in Lepidopteran (Remington, 1973) and Coleopteran (Hinton, 1973) mimicry systems may be quite variable. For example, the day-flying Scarabaeid beetles *Glycyphana horsfieldi*, *Glycosia tricolor*, and *Glycyphana binotata* are Mullerian mimics which look very similar to the human observer, being

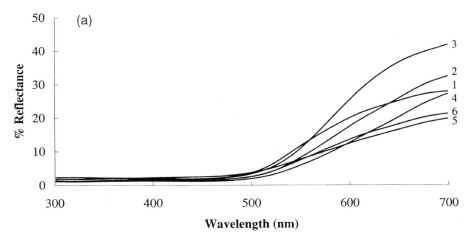

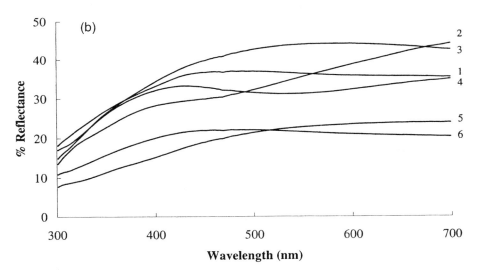

Figure 11.5. Reflectance spectra of (a) the main 'red' coloration, and (b) the 'white' spots from the dorsal surface of the forewings of the unpalatable butterfly *Danaus chrysippus* f. *aegyptius* (1) and the corresponding mimic morphs of *Hypolimnas misippus* (2), *Mimacraea marshalli* (3), *Pseudacraea poggei* (4), *Acraea encedon* (5) and *Acraea encedana* (6). Colour mimicry appears to extend into the ultraviolet; in all species, the 'white' spots reflect strongly in the near-UV (300–400 nm) while the 'red' areas have no UV reflectance.

predominantly black with yellow spots on the elytra (Hinton, 1973). However, the yellow spots on *G. horsfieldi* strongly reflect UV, while the other two species absorb UV. There are also more subtle markings on the side of the abdomen that can be distinguished in both the UV and human-visible region of the spectrum. If birds are the main predators of the beetles, it is possible that these visual cues could help them to discriminate between the model and mimics. In general, though, we possess very little information regarding the extent of model-mimic similarity in the UV. As a result, we cannot be certain that current colour patterns that have been classified as mimetic are, in fact, mimetic from an avian perspective.

The differences between the perceptual abilities of humans and birds also have consequences for our understanding of some theories associated with mimetic coloration. For example, much recent attention has focused on explaining why mimicry is sometimes imperfect (e.g. Dittrich *et al.*, 1993; Howse & Allen, 1994; Lindström *et al.*, 1997). However, all discussion of 'imperfect mimicry' to date has considered only the mismatch between model and mimic colour patterns as perceived by humans, and does not consider similarity as perceived by birds (or any other predator, for that matter). Depending on the exact nature of the visual pattern being viewed, it is conceivable that an avian predator might perceive the similarity of a given model to a given mimic as more or less perfect than would a human. Classification of mimetic coloration in species subject to high avian predation can only be achieved successfully if colour is interpreted from the perspective of a bird (Cuthill & Bennett, 1993), either *via* models of avian perception (e.g. Vorobyev & Osorio, 1998) or *via* precise behavioural experiments. When studying mimicry systems, it is also important to appreciate the exact nature of the visual task faced by potential predators. For example, Azmeh *et al.* (1998), when attempting to quantify the similarity of hover fly mimics to wasp models, justify overlooking UV reflectance by stating "UV photography of the relevant hover fly species shows that there are no extra UV components of the colour patterns". While this may well be true, this approach is of limited use unless the wasp models have similarly low UV reflectance. The only way to determine this is actually to measure the UV reflectance of both model and mimic. In the absence of UV reflectance data, there is no simple way around this problem.

Conclusions

If the function and design of protective colour patterns are to be fully comprehended, it is necessary to be able to assess the conspicuousness (in the case of crypsis and warning coloration) and similarity (in the case of mimicry) of such patterns in an objective, reliable, and ecologically meaningful manner. Here, we have focused on how considering UV wavelengths, which have been overlooked consistently in previous studies of protective coloration, can enhance our understanding of such processes. Our emphasis on UV wavelengths does not imply in any way that UV wavelengths are somehow 'special' compared to other wavelengths; it is merely the fact that these wavelengths have been overlooked that has prompted the need to redress the balance. However, UV wavelengths do present special problems for researchers studying animal colour patterns. In particular, the majority of video and still cameras, as well as image analysis software, is specifically tailored to represent the world through human rather than avian eyes. Although such human-based approaches can be misleading (D'Eath, 1998; Fleishman *et al.*, 1998; Cuthill *et al.*, 2000b; Fleishman & Endler, 2000; Oliveira *et al.*, 2000), they have nevertheless been applied freely in studies of animal coloration (e.g. Dittrich *et al.*, 1993; Azmeh *et al.*, 1998; Villafuerte & Negro, 1998). In general, attempts to quantify colour patterns should, at the very least, encompass the range of wavelengths to which a predator is sensitive. In the case of birds, this can be achieved by reflectance spectrophotometry (e.g. Endler, 1990; Cuthill *et al.*, 1999), or video/still imaging using a combination of human visible and UV-sensitive cameras (Eisner *et al.*, 1969). Both approaches unfortunately require a reasonable amount of background knowledge and specialist equipment, although relatively cheap and field-portable equipment is becoming increasingly available.

We have attempted to highlight the paucity of knowledge concerning the distribution of UV reflectance among the natural prey of avian predators. This lack of knowledge also extends to other visual tasks (e.g. mate choice) and other taxa (e.g. insects, fish, amphibians), and represents something of a barrier to our understanding of protective colour patterns, as well as of coloration in a much broader context. There is clearly a need for researchers to 'prospect' for UV information contained in insect colour patterns and their natural backgrounds. Until we know this basic information for enough species, we will not be able to begin to appreciate the role which UV wavelengths could play in avian foraging behaviour. However, simply collecting UV reflectance data is only the beginning of the story. Although considerable progress in understanding the function of defensive coloration can be made by considering reflectance spectra alone, a full understanding of the visual signals perceived by animals also depends on (a) the visual system of the signal receiver, and (b) the spectral properties of the environment in which the object is viewed (Lythgoe, 1979; Endler, 1990, 1993; Endler & Théry, 1996). These issues are not restricted solely to the consideration of UV wavelengths, but encompass all wavelengths to which signal receivers are sensitive. Although our understanding of avian vision at the retinal level has increased enormously during the past decade (e.g. Bowmaker *et al.*, 1997; Hart *et al.*, 1998, 1999, 2000; Bowmaker & Hunt, 1999; Hart, 2001), we still lack important information regarding the functioning of the visual pathways in the avian brain, as well as basic psychophysical data to determine what the avian colour vision system is actually capable of. We also lack understanding about the spectral composition of the light environments in which organisms are viewed in nature. This is an important component of signal design, since organisms may choose to display in specific light environments in order to enhance signal quality (e.g. Endler & Théry, 1996, Fleishman *et al.*, 1997). However, detailed spectral irradiance data for terrestrial systems only exists for tropical forests (Endler, 1993), and even then do not fully cover UV wavelengths.

Concluding remarks

In summary, if we are to comprehend the true nature of protective coloration, or even animal colour signals in general, we can benefit enormously by considering the nature of the visual task from the predator's (or receiver's) perspective, and in an appropriate ecological context. In the case of insect protective coloration that has evolved in response to selection by avian predators, this means at least considering UV wavelengths and, ideally, other differences between avian and human vision, such as differences in peak sensitivity (λ_{max}) of visual pigments. Relying solely upon our own limited visual system can help us to find answers, but only with limited success.

References

Alatalo, R. V. & Mappes, J. (1996) Tracking the evolution of warning signals. *Nature, London* **382**, 708–710.

Allen, J. A. (1988) Frequency-dependent selection by predators. *Philosophical Transactions of the Royal Scoiety of London,* B 319, 485–503.

Azmeh, S., Owen, J., Sorensen, K., Grewcock, D. & Gilbert, F. (1998) Mimicry profiles are affected by human-induced habitat changes. *Proceedings of the Royal Society of London,* B 265, 2285–2290.

Bennett, A. T. D. & Cuthill, I. C. (1994) Ultraviolet vision in birds: what is its function? *Vision Research* 34, 1471–1478.

Bennett, A. T. D., Cuthill, I. C. & Norris, K. J. (1994) Sexual selection and the mismeasure of color. *American Naturalist* 144, 848–860.

Bowmaker, J. K. (1991) Visual pigments, oil droplets and photoreceptors. In: *Vision and visual dysfunction, vol. 6* (ed. P. Gouras), pp. 108–127. CRC Press, Boston, USA.

Bowmaker, J. K. (2003) Vision in birds: the avian retina *(this volume).*

Bowmaker, J. K. & Hunt, D. M. (1999) Molecular biology of photoreceptor spectral sensitivity. In: *Adaptive mechanisms in the ecology of vision* (eds. S. N. Archer, M. B. A. Djamgoz, E. R. Loew, J. C. Partridge & S. Valerga), pp. 439–464. Chapman & Hall, London, UK.

Bowmaker, J. K. & Martin, G. R. (1978) Visual pigments and colour vision in a nocturnal bird *Strix aluco* (tawny owl). *Vision Research* 18, 1125–1130.

Bowmaker, J. K., Heath, L. A., Wilkie, S. E. & Hunt, D. M. (1997) Visual pigments and oil droplets from six classes of photoreceptor in the retinas of birds. *Vision Research* 37, 2183–2194.

Brunton, C. F. A. & Majerus, M. E. N. (1995) Ultraviolet colours in butterflies – intraspecific or interspecific communication? *Proceedings of the Royal Society of London,* B 260, 199–204.

Burkhardt, D. (1982) Birds, berries and UV. *Naturwissenschaften* 69, 153–157.

Burkhardt, D. (1989) UV vision: a bird's eye view of feathers. *Journal of Comparative Physiology, A* 164, 787–796.

Cherry, M. I. & Bennett, A. T. D. (2001) Egg colour matching in an African cuckoo, as revealed by ultraviolet-visible reflectance spectrophotometry. *Proceedings of the Royal Society of London,* B 268, 565–571.

Chittka, L. (1992) The color hexagon: a chromaticity diagram based on photoreceptor excitations as a generalized representation of color opponency. *Journal of Comparative Physiology, A* 170, 533–543.

Chittka, L. & Menzel, R. (1992) The evolutionary adaptation of flower colours and the insect pollinators' colour vision. *Journal of Comparative Physiology, A* 171, 171–181.

Chittka, L., Shmida, A., Troje, N. & Menzel, R. (1994) Ultraviolet as a component of flower reflections, and the colour perception of Hymenoptera. *Vision Research* 34, 1489–1508.

Church, S. C., Bennett, A. T. D., Cuthill, I. C., Hunt, S., Hart, N. S. & Partridge, J. C. (1998a) Does lepidopteran larval crypsis extend into the ultraviolet? *Naturwissenschaften* 85, 189–192.

Church, S. C., Bennett, A. T. D., Cuthill, I. C. & Partridge, J. C. (1998b) Ultraviolet cues affect the foraging behaviour of blue tits. *Proceedings of the Royal Society of London,* B 265, 1509–1514.

Church, S. C., Merrison, A. S. L. & Chamberlain, T. M. M. (2001) Avian ultraviolet vision and frequency-dependent seed preferences. *Journal of Experimental Biology* 204, 2491–2498.

Cott, H. B. (1940) *Adaptive coloration in animals.* Methuen, London, UK. 508 pp.

Crane, J. (1954) Spectral reflectance characteristics of butterflies (Lepidoptera) from Trinidad, B.W.I. *Zoologica, New York* 39, 85–115.

Cuthill, I. C. & Bennett, A. T. D. (1993) Mimicry and the eye of the beholder. *Proceedings of the Royal Society of London,* B 253, 203–204.

Cuthill, I. C., Bennett, A. T. D., Partridge, J. C. & Maier, E. H. (1999) Plumage reflectance and the objective assessment of avian sexual dichromatism. *American Naturalist* 153, 183–200.

Cuthill, I. C., Partridge, J. C., Bennett, A. T. D., Church, S. C., Hart, N. S. & Hunt, S. (2000a) Ultraviolet vision in birds. *Advances in the Study of Behavior* 29, 159–214.

Cuthill, I. C., Partridge, J. C., Bennett, A. T. D., Hart, N. S., Hunt, S. & Church, S. C. (2000b) Avian colour vision and avian video playback experiments. *Acta Ethologica* 3, 29–37.

Dartnall, H. J. A., Bowmaker, J. K. & Mollon, J. D. (1983) Microspectrophotometry of human photoreceptors. In: *Colour vision: physiology & psychophysics* (eds. J. D. Mollon & L. T. Sharpe), pp. 69–80. Academic Press, London, UK.

D'Eath, R. B. (1998) Can video images imitate real stimuli in animal behaviour experiments? *Biological Reviews of the Cambridge Philosophical Society* **73**, 267–292.

Dempster, J. P. (1984) The natural enemies of butterflies. In: *The biology of butterflies* (eds. R. I. Vane-Wright, & P. R. Ackery), pp. 97–104. Academic Press, London, UK.

Deonier, D. L. (1974) Ultraviolet-reflective surfaces on *Ochthera mantis mantis* (DeGeer) (Diptera: Ephydridae). Preliminary Report. *Entomologists News* **85**, 193–201.

Dittrich, W., Gilbert, F., Green, P., McGregor, P. & Grewcock, D. (1993) Imperfect mimicry: a pigeon's perspective. *Proceedings of the Royal Society of London, B* **251**, 195–200.

Dyer, A. G. (1996) Reflection of near-ultraviolet radiation from flowers of Australian native plants. *Australian Journal of Botany* **44**, 473–488.

Dyer, A. G. (2001) Ocular filtering of ultraviolet radiation and the spectral spacing of photoreceptors benefit von Kries colour constancy. *Journal of Experimental Biology* **204**, 2391–2399.

Eguchi, E. & Meyer-Rochow, V. B. (1983) Ultraviolet photography of forty-three species of Lepidoptera representing ten families. *Annotationes Zoologicae Japonenses* **56**, 10–18.

Eisner, T., Silberglied, R. E., Aneshansley, D., Carrell, J. E. & Howland, H. C. (1969) Ultraviolet video viewing: the television camera as an insect eye. *Science, New York* **166**, 1172–1174.

Emmerton, J. & Delius, J. D. (1980) Wavelength discrimination in the 'visible' and UV spectrum by pigeons. *Journal of Comparative Physiology, A* **141**, 47–52.

Emmerton, J. & Remy, M. (1983) The pigeon's sensitivity to ultraviolet and 'visible' light. *Experientia* **39**, 1161–1163.

Endler, J. A. (1978) A predator's view of animal colour patterns. *Evoutionary Biology* **11**, 319–364.

Endler, J. A. (1990) On the measurement and classification of colour in studies of animal colour patterns. *Biological Journal of the Linnean Society* **41**, 315–352.

Endler, J. A. (1993) The color of light in forests and its implications. *Ecological Monographs* **63**, 1–27.

Endler, J. A. & Théry, M. (1996). Interacting effects of lek placement, display behavior, ambient light, and color patterns in three neotropical forest-dwelling birds. *American Naturalist* **148**, 421–452.

Fisher, R. A. (1958). *The genetical theory of natural selection (2nd edition)*. Dover, London, UK. 291 pp.

Fleishman, L. J. & Endler, J. A. (2000) Some comments on visual perception and the use of video playback in animal behavior studies. *Acta Ethologica* **3**, 15–27.

Fleishman, L. J., Bowman, M., Saunders, D., Miller, W. E., Rury, M. J. & Loew, E. R. (1997) The visual ecology of Puerto Rican anoline lizards: habitat light and spectral sensitivity. *Journal of Comparative Physiology, A* **181**, 446–460.

Fleishman, L. J., McClintock, W. J., D'Eath, R. B., Brainards, D. H. & Endler, J. A. (1998) Colour perception and the use of video playback experiments in animal behaviour. *Animal Behaviour* **56**, 1035–1040.

Ghiradella, H., Aneshansley, D., Eisner, T., Silberglied, R. E. & Hinton, H. E. (1972) Ultraviolet reflection of a male butterfly: interference colour caused by thin layer elaboration of wing scales. *Science, New York* **178**, 1214–1217.

Gittleman, J. L. & Harvey, P. H. (1980) Why are distasteful prey not cryptic? *Nature, London* **286**, 149–150.

Giurfa, M., Nunez, J., Chittka, L. & Menzel, R. (1995) Color preferences of flower-naïve honeybees. *Journal of Comparative Physiology, A* **177**, 247–259.

Goldsmith, T. H. (1980) Hummingbirds see near ultraviolet light. *Science, New York* **207**, 786–788.

Goldsmith, T. H. (1990) Optimization, constraint, and history in the evolution of eyes. *Quarterly Review of Biology* **65**, 281–322.

Goldsmith, T. H. (1994) Ultraviolet receptors and color vision: evolutionary implications and a dissonance of paradigms. *Vision Research* **34**, 1479–1487.

Govardovskii, V. I. (1983) On the role of oil drops in colour vision. *Vision Research* **23**, 1739–1740.

Gregory, R. L. (1997) *Eye and brain.* Oxford University Press, Oxford, UK. 256 pp.

Guilford, T. (1986) How do warning colours work? Conspicuousness may reduce recognition errors in experienced predators. *Animal Behaviour* **34**, 286–288.

Guilford, T. & Dawkins, M. S. (1991). Receiver psychology and the evolution of animal signals. *Animal Behaviour* **42**, 1–14.

Hart, N. S. (2001) The visual ecology of avian photoreceptors. *Progress in Retinal and Eye Research* **20**, 675–703.

Hart, N. S., Partridge, J. C. & Cuthill, I. C. (1998). Visual pigments, oil droplets and cone photoreceptor distribution in the European starling (*Sturnus vulgaris*). *Journal of Experimental Biology* **201**, 1433–1446.

Hart, N. S., Partridge, J. C. & Cuthill, I. C. (1999) Visual pigments, cone oil droplets, ocular media and predicted spectral sensitivity in the domestic turkey (*Meleagris gallopavo*). *Vision Research* **39**, 3321–3328.

Hart, N. S., Partridge, J. C., Cuthill, I. C. & Bennett, A. T . D. (2000) Visual pigments, oil droplets, ocular media and cone photoreceptor distribution in two species of passerine: the blue tit (*Parus caeruleus* L.) and the blackbird (*Turdus merula* L.). *Journal of Comparative Physiology, A* **186**, 375–387.

Hinton, H. E. (1973) Natural deception. In: *Illusion in nature and art* (eds. R. L. Gregory & E. H. Gombrich), pp. 161–191. Duckworth, London, UK.

Howse, P. E. & Allen, J. A. (1994) Satyric mimicry – the evolution of apparent imperfection. *Proceedings of the Royal Society of London, B* **257**, 111–114.

Huth, H. H. & Burkhardt, D. (1972) Der spektrale Sehbereich eines Violetta Kolibris. *Naturwissenschaften* **59**, 650.

Kacelnik, A. & Krebs, J. R. (1982) The dawn chorus in the great tit (*Parus major*): proximate and ultimate causes. *Behaviour* **83**, 287–309.

Kettlewell, H. B. D. (1973) *The evolution of melanism.* Oxford University Press, Oxford, UK. 423 pp.

Kevan, P., Giurfa, M. & Chittka, L. (1996) Why are there so many and so few white flowers? *TIPS* **1**, 280–284.

Koivula, M. & Viitala, J. (1999) Rough-legged buzzards use vole scent marks to assess hunting areas. *Journal of Avian Biology* **30**, 329–332.

Koivula, M., Korpimaki, E. & Viitala, J. (1997) Do Tengmalm's owls see vole scent marks visible in ultraviolet light? *Animal Behaviour* **54**, 873–877.

Lindström, L., Alatalo, R. V. & Mappes, J. (1997) Imperfect Batesian mimicry – the effects of the frequency and the distastefulness of the model. *Proceedings of the Royal Society of London, B* **264**, 149–153.

Lindström, L., Alatalo, R. V., Mappes, J., Riipi, M. & Vertainen, L. (1999) Can aposematic signals evolve by gradual change? *Nature, London* **397**, 249–251.

Lunau, K. & Maier, E. J. (1995) Innate color preferences of flower visitors. *Journal of Comparative Physiology, A* **177**, 1–19.

Lunau, K., Wacht, S. & Chittka, L. (1996) Colour choices of naïve bumble bees and their implications for colour perception. *Journal of Comparative Physiology, A* **178**, 477–489.

Lutz, F. E. (1933) 'Invisible' colors of flowers and butterflies. *Proceedings of the American Academy of Arts and Science* **74**, 281–285.

Lythgoe, J. N. (1979) *The ecology of vision.* Oxford University Press, Oxford, UK. 244 pp.

Lyytinen, A., Alatalo, R. V., Lindström, L. & Mappes, J. (2001) Can ultraviolet cues function as aposematic signals? *Behavioral Ecology* **12**, 65–70.

Maddocks, S. A., Church, S. C. & Cuthill, I. C. (2001) The effects of the light environment on prey choice by zebra finches. *Journal of Experimental Biology* **204**, 2509–2515.

Majerus, M. (1998) *Melanism: evolution in action.* Oxford University Press, Oxford, UK. 352 pp.

Majerus, M. E. N., Brunton, C. F. A. & Stalker, J. (2000) A bird's eye view of the peppered moth. *Journal of Evolutionary Biology* **13**, 155–159.

Mappes, J. & Alatalo, R. V. (1997a) Batesian mimicry and signal accuracy. *Evolution* **51**, 2050–2053.

Mappes, J. & Alatalo, R. V. (1997b) Effects of novelty and gregariousness in survival of aposematic prey. *Behavioral Ecology* **8**, 174–177.

Marples, N. M. & Roper, T. J. (1996) Effects of novel colour and smell on the response of naïve chicks towards food and water. *Animal Behaviour* **51**, 1417–1424.

Mazokhin-Porshnyakov, G. A. (1957) Reflecting properties of butterfly wings and the role of ultra-violet rays in the vision of insects. *Biophysics* **2**, 285–296.

Meyer-Rochow, V. B. (1991) Differences in ultraviolet wing patterns in the New Zealand lycaenid butterflies *Lycaena salustius*, *L. rauparaha*, and *L. feredayi* as a likely isolating mechanism. *Journal of the Royal Society of New Zealand* **21**, 169–177.

Meyer-Rochow, V. B. & Eguchi, E. (1983) "Flugelfarben, wie sie Falter sehen": a study of UV and other colour patterns in Lepidoptera. *Annotationes Zoologicae Japonenses* **56**, 85–99.

Meyer-Rochow, V. B. & Jarvilheto, M. (1997) Ultraviolet colours in *Pieris napi* from northern and southern Finland: Arctic females are the brightest! *Naturwissenschaften* **84**, 165–168.

Neumeyer, C. (1992) Tetrachromatic color vision in goldfish: evidence from color mixture experiments. *Journal of Comparative Physiology, A* **171**, 639–649.

Oliveira, R. F., Rosenthal, G. G., Schlupp, I., McGregor, P. K., Cuthill, I. C., Endler, J. A., Fleishman, L. J., Zeil, J., Barata, E., Burford, F., Gonçalves, D., Haley, M., Jakobsson, J., Jennions, M. D., Körner, K. E., Lindström, L., Peake, T., Pilastro, A., Pope, D. S., Roberts, A., Rowe, C., Smith, J. & Waas, J. R. (2000) Considerations on the use of video playbacks as visual stimuli: the Lisbon workshop consensus. *Acta Ethologica* **3**, 61–65.

Osorio, D., Marshall, N. J. & Cronin, T. W. (1997) Stomatopod photoreceptor tuning as an adaptation to colour constancy underwater. *Vision Research* **37**, 3299–3309.

Osorio, D., Jones, C. D. & Vorobyev, M. (1999a) Accurate memory for colour but not pattern contrast in chicks. *Current Biology* **9**, 199–202.

Osorio, D., Vorobyev, M. & Jones, C. D. (1999b) Colour vision of domestic chicks. *Journal of Experimental Biology* **202**, 2951–2959.

Owen, D. F., Smith, D. A. S., Gordon, I. J. & Owiny, A. M. (1994) Polymorphic mimicry in a group of African butterflies: a re-assessment of the relationship between *Danaus chrysippus*, *Acraea encedon* and *Acraea encedana* (Lepidoptera: Nymphalidae). *Journal of Zoology* **232**, 93–108.

Partridge, J. C. (1989) The visual ecology of avian cone oil droplets. *Journal of Comparative Physiology, A* **165**, 415–426.

Pietrewicz, A. T. & Kamil, A. C. (1977) Visual detection of cryptic prey by blue jays (*Cyanocitta cristata*). *Science, New York* **195**, 580–582.

Pope, R. D. & Hinton, H. E. (1977) A preliminary survey of ultraviolet reflectance in beetles. *Biological Journal of the Linnean Society* **9**, 331–348.

Poulton, E. B. (1890). *The colours of animals, their meaning and use, especially considered in the case of insects.* Kegan Paul, Trench & Trubner, London, UK. 360 pp.

Raven, P. H. (1972) Why are bird-visited flowers predominately red? *Evolution* **76**, 674.

Remington, C. L. (1973) Ultraviolet reflectance in mimicry and sexual signals in Lepidopetra. *Journal of the New York Entomological Society* **81**, 124–125.

Robey, C. W. (1975) Observations on the breeding behaviour of *Pachydiplax longipennis* (Odonata: Libellulidae). *Psyche* **82**, 89–96.

Roland, J. (1978) Variation in spectral reflectance of alpine and arctic *Colias* (Lepidoptera: Pieridae). *Canadian Journal of Zoology* **56**, 1447–1453.

Roper, T. J. & Cook, S. E. (1989) Responses of chicks to brightly coloured insect prey. *Behaviour* **110**, 276–293.

Roper, T. J. & Marples, N. M. (1997a) Colour preferences of domestic chicks in relation to food and water presentation. *Applied Animal Behaviour Science* **54**, 207–213.

Roper, T. J. & Marples, N. M. (1997b) Odour and colour as cues for taste-avoidance learning in domestic chicks. *Animal Behaviour* **53**, 1241–1250.

Roper, T. J. & Redston, S. (1987) Conspicuousness of distasteful prey affects the strength and durability of one-trial avoidance learning. *Animal Behaviour* **35**, 739–747.

Rowe, C. & Guilford, T. (1996) Hidden colour aversions in domestic chicks triggered by pyrazine odours of insect warning displays. *Nature, London* **383**, 520–522.

Schuler, W. & Roper, T. J. (1992) Responses to warning coloration in avian predators. *Advances in the Study of Behaviour* **21**, 111–144.

Scoble, M. J. (1992) *The lepidoptera: form, function and diversity.* Oxford University Press, Oxford, UK. 404 pp.

Siitari, H., Honkavaara, J. & Viitala, J. (1999) Ultraviolet reflection of berries attracts foraging birds. A laboratory study with redwings (*Turdus iliacus*) and bilberries (*Vaccinium myrtillus*). *Proceedings of the Royal Society of London, B* **266**, 2125–2129.

Silberglied, R. E. (1979) Communication in the ultraviolet. *Annual Review of Ecology and Systematics* **10**, 373–398.

Silberglied, R. E. (1984) Visual communication and sexual selection among butterflies. In: *The biology of butterflies* (eds. R. I. Vane-Wright & P. E. Ackery), pp. 207–223. Academic Press, London, UK.

Silberglied, R. E. & Taylor, O. R. (1973) Ultraviolet differences between the sulphur butterflies, *Colias eurytheme* and *C. philodice*, and a possible isolating mechanism. *Nature, London* **241**, 406–408.

Smith, D. A. S., Owen, D. F., Gordon, I. J. & Owiny, A. M. (1993) Polymorphism and evolution in the butterfly *Danaus chrysippus* (L.) (Lepidoptera: Danainae). *Heredity* **71**, 242–251.

Steinly, B. A., Deonier, D. L. & Regensburg, J. T. (1978) Scanning electron microscopy of ultraviolet-reflecting pruinosity in species of *Ochthera* (Diptera: Ephydridae). *Entomologist's News* **89**, 117–124.

Thompson, E., Palacios, A. & Varela, F. J. (1992) Ways of coloring: comparative color vision as a case study for cognitive science. *Behavior and Brain Science* **15**, 1–74.

Vane-Wright, R. I. & Boppré, M. (1993) Visual and chemical signalling in butterflies – functional and phylogenetic perspectives. *Philosophical Transactions of the Royal Society of London, B* **340**, 197–205.

Viitala, J., Korpimaki, E., Palokangas, P. & Koivula, M. (1995) Attraction of kestrels to vole scent marks visible in ultraviolet light. *Nature, London* **373**, 425–427.

Villafuerte, R. & Negro, J. J. (1998) Digitial imaging for colour measurement in ecological research. *Ecology Letters* **1**, 151–154.

Vorobyev, M. & Osorio, D. (1998) Receptor noise as a determinant of colour thresholds. *Proceedings of the Royal Society of London, B* **265**, 351–358.

Vorobyev, M., Gumbert, A., Kunze, J., Giurfa, M. & Menzel, R. (1997) Flowers through insect eyes. *Israel Journal of Plant Science* **45**, 93–101.

Vorobyev, M., Osorio, D., Bennett, A. T. D., Marshall, N. J. & Cuthill, I. C. (1998) Tetrachromacy, oil droplets and bird plumage colours. *Journal of Comparative Physiology, A* **183**, 621–633.

Vulinec, K. (1997) Iridescent dung beetles: a different angle. *Florida Entomologist* **80**, 132–141.

Wickler, W. (1968) *Mimicry.* Weidenfeld & Nicholson, London, UK. 253 pp.

Willson, M. F. & Whelan, C. J. (1989) Ultraviolet reflectance of fruits of vertebrate-dispersed plants. *Oikos* **55**, 341–348.

Yokoyama, S., Radlwimmer, F. B. & Kawamura, S. (1998) Regeneration of ultraviolet pigments of vertebrates. *FEBS Letters* **423**, 155–158.

Zettler, J. A., Adler, P. H. & McCreadie, J. W. (1998) Factors influencing larval color in the *Simulium vittatum* complex (Diptera: Simuliidae). *Invertebrate Biology* **117**, 245–252.

12

Warning Colours and Warning Smells: How Birds Learn to Avoid Aposematic Insects

NICOLA M. MARPLES[1] AND TIM J. ROPER[2]

[1]*Department of Zoology, Trinity College, Dublin D2, Ireland and* [2]*School of Biological Sciences, Sussex University, Brighton, East Sussex, BN1 9QG, UK*

Introduction

The term 'warning coloration' (or 'aposematism') refers both to a fact of nature and to a hypothesis:

The *fact of nature* is that prey that are chemically defended are usually brightly coloured; for example, the cinnabar moth caterpillar (*Tyria jacobaeae*) has black and orange stripes and it is also toxic. This association between bright coloration and chemical defence is best known in insects, but not confined to them. It is also found in other invertebrates such as the sea-slug, and in some vertebrates such as the South American poison-dart frog (*Dendrobates pumilio*).

The *hypothesis* is that conspicuous coloration sends a signal to the predator that the prey has some kind of defence mechanism and had better be avoided. So, we are dealing with a prey–predator communication system from which both partners benefit: the prey because it does not get eaten, and the predator because it does not get poisoned.

However, this fails to explain why chemically defended prey need to be conspicuously coloured. There is no inevitable association between chemical defence and conspicuous coloration since some chemically defended prey, such as the buff-tip moth (*Phalera bucephala*), are cryptic. Presumably, then, conspicuous coloration must confer some special advantage from the prey's point of view. What could that advantage be?

The most generally accepted answer to this question is that conspicuous coloration somehow facilitates avoidance learning by the predator. Chemical defence is protective; if a predator samples a chemically defended prey it gets sick, and therefore it learns to avoid the prey in question. It has been suggested that this learning process is perhaps more rapid if the prey is conspicuously coloured than if it is cryptic. In the language of learning theory, the hypothesis is that conspicuous colours constitute more salient stimuli for learning.

Insect and Bird Interactions
© Intercept Ltd., PO Box 716, Andover, Hampshire, SP10 1YG, UK.

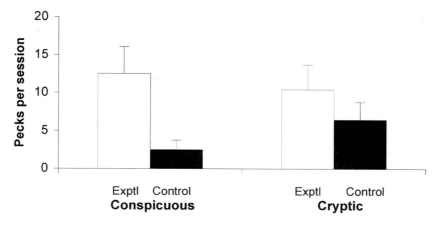

Figure 12.1.   Mean (+/–s.d.) of the number of pecks at pins made by chicks in each group in a 4-minute test trial. In the 'conspicuous' groups, the pins were red and the background white, while in the 'cryptic' groups the pins were white on the same background. In the experimental groups (exptl), the chicks had been trained with distasteful pins, while in the 'control' groups, training pins were neutral tasting. After Roper & Redston (1987).

This idea, that warning coloration facilitates avoidance learning, was first suggested just over a century ago in a letter from Wallace to Darwin (Wallace, 1867), and it is so generally accepted that it has actually come to be regarded not just as a property of warning coloration, but as a definition of warning coloration (Wickler, 1968; Guilford, 1990). It is strange, however, that until a few years ago, no attempt had been made to test this idea. When Wickler wrote his book on mimicry (Wickler, 1968), there was no evidence whatsoever that warning coloration acted in the manner that he suggests.

Warning colours as significant stimuli for learning

So, the question arises – do conspicuous colours facilitate avoidance learning, as claimed by Wickler, Guilford, and others?

Redston and Roper tested this idea using domestic chicks (*Gallus gallus domesticus*) in a single trial, passive-avoidance, bead-pecking task (Roper & Redston, 1987). The design of the experiment was as follows: chicks were put two at a time into an arena containing four coloured beads attached to the floor. The beads were either red or white and the background was either red or white, so the beads either matched the background (i.e. were cryptic) or contrasted with the background (i.e. were conspicuous). For half the chicks (the experimental birds), the beads had been painted with dimethyl anthranilate, which chicks find extremely distasteful. They were allowed to peck once at these nasty-tasting beads and were then removed from the arena. For the other birds (the controls), the beads tasted of nothing. Again, the chicks were allowed to peck once and were then removed.

The next day, the chicks were put back into the arena for a four-minute test session with beads the same colour as those they had experienced in the training trial. For this test session, none of the beads was distasteful. The authors simply measured the number of pecks that the chicks made to the beads. The prediction was that if conspicuous stimuli support more effective avoidance learning, chicks should have learnt better about the contrasting beads than about matching beads. So, the chicks given contrasting beads should peck less, in the test trial, than the chicks given matching beads. The results (*Figure 12.1*) show a greater avoidance learning by the experimental birds when the beads were conspicuous than when they were cryptic. Therefore, conspicuousness, or at least contrast between prey and background, does facilitate avoidance learning.

Providing conspicuousness is not the only thing that warning colours do, because there is also evidence that at least some warning colours and patterns induce avoidance in naïve birds – that is, birds that have never before been exposed to the prey in question. For example, there is good evidence that domestic chicks and various other species of birds tend to avoid black and yellow striped, or red insects, even if they have been hand-reared and have never had previous experience of warningly coloured insect prey (Schuler & Hesse, 1985; Roper & Cook, 1989).

So, Wickler and Guilford were right, but only partly right. Warning coloration does facilitate avoidance learning, but that is not the only thing it does; it also triggers innate avoidance in naïve birds.

Aposematic odours

Many insects that are warningly coloured also produce a distinctive odour, usually when they are attacked. A good example is the 7-spot ladybird which, when disturbed, exudes a yellowish fluid from the joints of its legs. The fluid is toxic, so it is part of the insect's primary defence mechanism: but it also has a strong smell, rather like new-mown hay, but more acrid and persistent. Rothschild and her colleagues have shown that the odour comes from two members of a family of compounds called the 2-methoxy-3alkylpyrazines, and that these same compounds are produced by many other aposematic insects (Rothschild & Moore, 1987; Moore *et al.*, 1990).

Given their association with warning coloration, it is tempting to suggest that these odours play a role in the signalling system between prey and predator. There are two ways in which they might do this:

(1) The simplest possibility is that the odours are warning signals in their own right: that is, they act in the same way as warning colours, but *via* a different modality. As we have seen, warning colours have two basic effects: they elicit neophobia in naïve birds, and they act as a cue for learned avoidance in experienced birds. So, if aposematic odours act as warning stimuli, we predict that they should have these two properties.

(2) Rothschild, however, has suggested something slightly different, which is that these odours act as 'alerting stimuli' (Rothschild & Moore, 1987). The idea of an alerting stimulus is that it is a stimulus that does not contain detailed information itself, but that it attracts the receiver's attention for detailed information that follows. A lecturer noisily clearing the throat at the start of a lecture is producing an alerting stimulus. Rothschild's hypothesis, then, is that these odours, that are associated with aposematism, are not warning signals *per se*; rather, they alert the predator's attention to the real warning signal, which is the colour pattern. This hypothesis predicts that aposematic odours should enhance the response of the predator towards warning coloration.

The experiments below were aimed at investigating some of these odours, with these two hypotheses in mind. The experiments were done with domestic chicks, using either food or water as the ingestive stimulus. In the case of food, we put the food on the floor of the test cage and put a ring of olfactant, in solution, around it. In the case of water, we added the olfactant to a ring of filter paper around the rim of the jar.

EXPERIMENT 1. NEOPHOBIC RESPONSES TO ODOURS

Our first experiment looked at the unlearned responses of domestic chicks towards a variety of odours (Marples & Roper, 1996). We gave naïve chicks access to food or water that was either accompanied, or not accompanied, by an odour. Also, because aposematic insects are often both colourful and smelly, we added the odour cue to food or water that was either familiar or novel in colour. We were therefore able to look at the interaction between novel odour and novel colour. As an index of neophobia, we measured the latency with which the chicks ate the food or drank the water.

We did this experiment with five different odours. Two of them were pyrazines that are commonly produced by warningly coloured insects. We also chose bitter almond, because it is associated with toxicity in plants, and as a control we used two odours that are not naturally associated with toxicity, namely vanilla and thiazole.

The results were very clear (*Figure 12.2*). When pyrazine odour was added to familiar-looking food or water, it elicited no neophobia. Chicks ate or drank with short latencies, regardless of whether or not pyrazine was present. As expected, novel colour produced an increase in latency to eat or drink, but presence of an odour produced an even larger increase. Presence of odour interacted synergistically with a novel colour cue, so as to produce a heightened neophobic response.

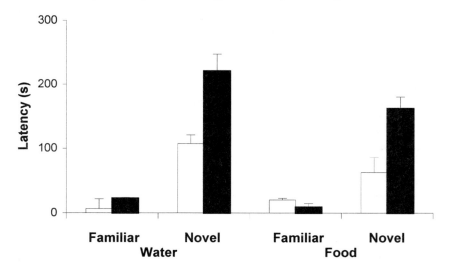

Figure 12.2.   Mean latencies (+/–s.e.) for chicks to ingest water or food that were either familiar or novel in appearance, in the presence (dark bars) or absence (pale bars) of pyrazine odour. After Marples & Roper (1996).

This looks very much like what one would expect of an alerting stimulus. If the odour were a warning signal, we would expect it to elicit neophobia in its own right, but it does not do so. Rather, it enhances the neophobic response to novel colour. The 'Rothschild hypothesis' seems to be rather strikingly confirmed.

Similar effects occurred with both the pyrazines and with almond. With vanilla, on the other hand, the effect was weaker, and with thiazole it was not statistically significant (Marples & Roper, 1996). So, the effect is not elicited by just any smell: it is only elicited, or is at least elicited more strongly, by smells that are naturally associated with toxicity. Since the chicks were naïve, this seems to suggest that there is innate recognition of certain classes of warning odour. We do have evidence from other experiments that chicks can detect the odours of vanilla (Roper & Marples, 1997) and thiazole (T. J. Roper & N. M. Marples, *unpublished*), so the lack of response to these odours was not because the birds could not smell them.

EXPERIMENT 2. COLOUR AND ODOUR AS COMPOUND AVOIDANCE CUES

Some odours have one of the properties expected of an alerting stimulus: they enhance the neophobic response to a novel colour. What about their ability to enhance avoidance learning? If a predator is presented with a prey object that is conspicuously coloured and smells nasty, does the nasty smell enhance the predator's ability to learn to avoid the colour?

To test this, we trained chicks to avoid bitter-tasting fluid that was novel in both colour and odour (note that this experiment and subsequent ones were done with fluid, not food, because it is easier to control). We then tested chicks in four groups:

*   Group 1 with fluid that was the same colour *and* odour as the training stimulus
*   Group 2 with the same colour but a different odour
*   Group 3 with the same odour but a different colour
*   Group 4 with fluid which was different in both colour and odour than the training stimulus (Roper & Marples, 1997).

If chicks had learned to associate the colour with the bitter taste, they should avoid 1 and 2; if they had learned to associate the smell with the bitter taste, they should avoid 1 and 3; if they had learned to associate both colour and smell, they should avoid 1, 2 and 3. We carried out the experiment with both almond and vanilla as the odours.

With almond, the results were clear (*Figure 12.3a*). As expected, the chicks avoided the stimulus that was the same colour and odour as the training stimulus (SCO), and they readily drank the fluid that was a different colour and odour (DCO). They also avoided the stimulus that was the same odour but a different colour (SO), and they readily drank the stimulus that was the same colour but a different odour (SC). So, when an odour cue and a colour cue were presented simultaneously, the odour cue took precedence: the chicks learned little or nothing about the colour cue.

This is exactly the opposite of what we expect if the odour were acting as an alerting stimulus. Far from improving the predator's ability to learn about the colour cue, the odour completely blocks learning of the colour cue. In this case, then, it is not acting as an alerting stimulus, it is acting as a powerful warning stimulus in its own right.

With vanilla, the picture was again completely different (*Figure 12.3b*). The birds readily drank all the fluids: they showed no evidence of having learned anything at all.

(a)

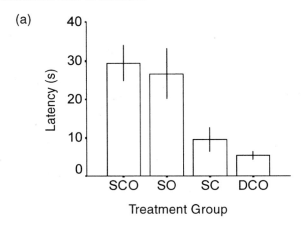

(b)

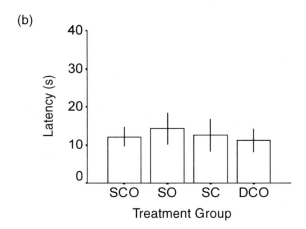

Figure 12.3.   Mean (+/–s.e.) latency to drink palatable fluid, in a test trial, in chicks who had been trained to avoid coloured unpalatable fluid accompanied by a novel odour. The odour in question was (a) almond, or (b) vanilla. The test fluid was either: SCO, the same odour and colour; SO, a different colour from that used in training but the same odour; SC, a different odour from that used in training but the same colour; or DCO, different in both colour and odour from that used in training .

This was a strange result because, if the birds were simply ignoring the vanilla odour, they should have learned about the colour cue. We therefore postulate that the birds had, in fact, learned to avoid the vanilla odour, but had forgotten about it by the time they were tested 24 hours later.

EXPERIMENT 3. MEMORY FOR ALMOND AND VANILLA

We tested this idea that memory for almond odour was more persistent than memory for vanilla (Experiment 4 in Roper & Marples, 1997). We gave chicks a series of trials in which they encountered either a bitter-tasting fluid that smelled of almond or vanilla, or an innocuous-tasting fluid that was odourless. They experienced five trials with each fluid, in random order. Twenty-four hours after the last training trial, they were given a memory test to see whether they would avoid the odorous fluid.

(a)

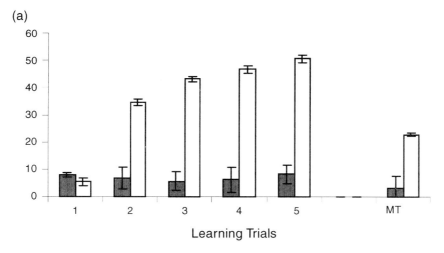

(b)

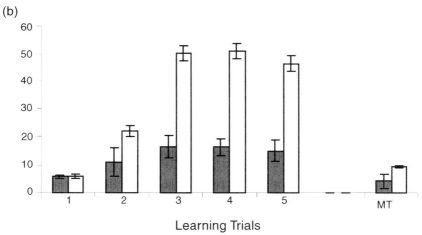

Figure 12.4. Mean (+/– s.e.) latencies to drink fluid that was either palatable (solid grey) or unpalatable (pale bars) during five discrimination learning trials. The unpalatable fluid was accompanied by odour of either (a) almond, or (b) vanilla; the palatable fluid was odourless. MT, mean latency to drink in a memory test, when chicks were presented with palatable fluid which was either accompanied by the odour used during training (pale bars) or odourless (solid grey). After Roper & Marples (1997).

The results showed that, with either almond or vanilla, the chicks learned to avoid the odorous fluids (*Figure 12.4*). So, they were learning to associate the odour with the bitter taste. When they were tested 24 hours later, they still showed significant avoidance of almond odour, but they did not significantly avoid vanilla. Therefore, memory for almond was indeed more persistent than memory for vanilla.

Conclusions

These experiments show three things:

• First, they show that odours can function for purposes of anti-predator protection:

they can elicit neophobia when paired with a novel visual cue, and they can constitute cues for avoidance learning. Whether they do so primarily by acting as alerting stimuli, or by acting as warning signals in their own right, is unclear. At the moment, they seem to have some of the properties of both types of signal.

• Secondly, the results show that not all odours are equally effective in this respect: odours that are associated with toxicity in the real world seem to be more effective than odours that are benign. This suggests that birds may have innate recognition of the kinds of odours that are likely to constitute warning stimuli.

• Thirdly, the fact that odour cues overshadow colour cues suggests the possibility of olfactory mimicry that is independent of visual mimicry. In other words, two species may be olfactory mimics if they smell the same, even though they may look different. Given the high degree of convergence that there is in the chemicals used as warning stimuli (Rothschild & Moore, 1987), olfactory mimicry may be extremely widespread.

References

Guilford, T. C. (1990) The evolution of aposematism. In: *Insect defenses: adaptive mechanisms and strategies of prey and predators* (eds. D. L. Evans & J. O. Schmidt), pp. 23–61. State University of New York Press, New York, USA.

Marples, N. M. & Roper, T. J. (1996) Effects of novel colour and smell on the response of naïve chicks towards food and water. *Animal Behaviour* 51, 1417–1424.

Moore, B. P., Brown, W. V. & Rothschild, M. (1990) Methylalkylpyrazines in aposematic insects, their host plants and mimics. *Chemoecology* 1, 43–51.

Roper, T. J. & Cook, S. E. (1989). Responses of chicks to aposematic prey: effects of prey colour and early experience. *Behaviour* 100, 276–293.

Roper, T. J. & Marples, N. M. (1997) Odour and colour as cues for taste-avoidance learning in domestic chicks. *Animal Behaviour* 53, 1241–1250.

Rothschild, M. & Moore, B. P. (1987) Pyrazines as alerting signals in toxic plants and insects. In: *Insects–plants* (eds. V. Labeyrie, G. Fabres & D. Lachaise), pp. 97–101. Junk, Dordrecht, The Netherlands.

Roper, T. J. & Redston, S. (1987) Conspicuousness of distasteful prey affects the strength and durability of one-trial avoidance learning. *Animal Behaviour* 35, 739–747.

Schuler, W. & Hesse, E. (1985) On the function of warning coloration: a black and yellow pattern inhibits prey-attack by naïve domestic chicks. *Behavioral Ecology and Sociobiology* 16, 249–255.

Wallace, A. R. (1867) [Untitled]. *Proceedings of the Entomological Socety of London* 5, 80–81.

Wickler, W. (1968) *Mimicry in plants and animals*. Weidenfeld & Nicholson, London, UK. 225 pp.

13
The Chemical Defences of an Aposematic Complex: A Case Study of Two-Spot and Seven-Spot Ladybirds

NICOLA M. MARPLES

Department of Zoology, Trinity College, Dublin D2, Ireland

Introduction

Whenever we come across classically aposematic animals, we home in on the colour pattern as the most striking feature. Indeed, the alternative term for aposematism, 'warning coloration', all but states that colour is the only relevant cue. This hides the underlying truth that insect defences are far more complex and far more interesting than simple flags of toxicity. In fact, colour is only one of a whole battery of defences which come into play as a bird approaches an insect. Some of these defences are detectable at some distance from the insect, such as colour, odour, size, and shape; some are behavioural responses, such as freezing, reflex bleeding, or dropping off the leaf; some are only discovered by the predator in retrospect, such as palatability, and toxicity. Some are even features of the predator in relation to that particular prey type, such as its perception of the insect as novel, or its memory of previous encounters. All these, and many more, act to alter the likelihood of the insect being attacked, and so all should be considered together if we are trying to look at insect survival and bird predation. That is a daunting prospect. Research has tended to focus on the simplest questions so far, and relatively few of the theoretical predictions of how aposematism works have been tested empirically. In this chapter, I will use an example of an aposematic complex, that of the 2-spot *(Adalia bipunctata)* and 7-spot *(Coccinella septempunctata)* ladybirds, to present some of these tests. I will show that, although some results support the theoretical work, some contradict the accepted views of how aposematism works.

The rival hypotheses

Brakefield (1985) hypothesised that the 2-spot ladybird might be a Batesian mimic of the 7-spot ladybird; that is, 7-spots would be chemically protected and 2-spots gain protection by resemblance to the 7-spot, and not by their own toxicity. Both these species synthesise different alkaloids: the 7-spot contains coccinelline, and the 2-spot

Insect and Bird Interactions
© Intercept Ltd., PO Box 716, Andover, Hampshire, SP10 1YG, UK.

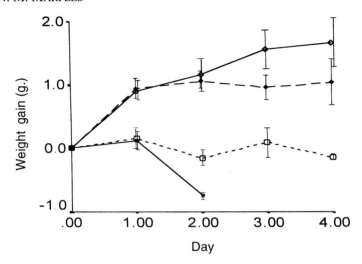

Figure 13.1.   The gain in weight of nestlings on diets supplemented by mealworm (—◇—); 2-spot ladybird (- - - ◆- - -); high level of 7-spot ladybird (— ■—); or low level of 7-spot ladybird (- - -□- - -).

adaline (Pasteels *et al.*, 1973). However, alkaloids are a huge family of chemicals, which range in toxicity from strychnine to caffeine; thus, the presence of alkaloids does not ensure toxicity.

Both ladybird species are quoted in the literature as being inedible to birds, and are very often assumed to be Mullerian mimics, although little direct proof of this view has been available. Fraser & Rothschild (1960) demonstrated that ground-up 7-spot material was fatal if injected into guinea pigs. Kristin (1988), however, surveyed the food brought to nestlings of insectivorous birds and found that, of twelve species bringing in ladybirds, nine species ate 7-spots and only three ate 2-spots. de Jong *et al.* (1991) and Holloway *et al.* (1991) showed that the amount of alkaloid present in the 2-spot is about twice as much as in the 7-spot, supporting the view that maybe the 2-spot is better protected than the 7-spot.

Toxicity assay

I tested the toxicity of both these ladybird species directly (Marples *et al.*, 1989) using nestling blue tits (*Parus caeruleus*) as my predator, as I had evidence from camera nest boxes that blue tits brought ladybirds to their young. Five-day-old nestlings were taken into the laboratory and fed diets incorporating small amounts of 2-spot or 7-spot ladybird, or a control diet incorporating mealworm. Two levels of 7-spot treatment were tested. The nestlings' weights were monitored at each hourly feed (*Figure 13.1*), and they were kept on the diet for four days, unless they became ill. After these four days, they were given a ladybird-free diet for a day, then returned to their home nest boxes at 10 days of age. All returned nestlings fledged successfully.

The 2-spot, high-level 7-spot, and control food pellets contained respectively one 2-spot ladybird, or $\frac{1}{3}$ of a 7-spot or $\frac{1}{3}$ of a mealworm (approximately equivalent weights) added to an insectivore diet. One of these pellets was provided per hour, along with pure nestling diet pellets *ad libitum*, and water. *Figure 13.1* shows that there was no difference between the weight gains of nestlings on control or 2-spot treatments, but the high level 7-spot treatment (lowest line on the graph) quickly became ill and had to be taken off this diet before the end of the third day. In the low-level 7-spot treatment, half the amount of ladybird was added to the food pellets, and even here the birds showed a large and significant reduction in growth rate compared to control or to 2-spot diet. Tests on purified coccinelline showed that this was the active agent in the 7-spot defence (Marples, 1993a).

So, we have evidence that the 7-spot is strongly chemically protected, but despite their high alkaloid content, the 2-spots are not toxic. This clearly supports the view that, if mimicry is involved in the defence at all, it must be Batesian in nature. However, we are left with two questions:

(1) Why does the 2-spot ladybird synthesise alkaloid?
(2) Does Batesian mimicry of the 7-spot protect the 2-spot from bird predation?

Why does the 2-spot ladybird synthesise alkaloid?

TASTE AS A DEFENCE

The alkaloid in 2-spots may not be toxic, but it carries a bitter taste, and it is possible that this alone would prevent bird predation. Experiments done with Japanese quail (*Coturnix coturnix japonica*) (Marples, 1990) tend to refute this. Adult quail were offered 2-spot or 7-spot ladybirds or a palatable 'control' beetle (*Alphitobius diaperinus*), and their latencies to attack the prey were recorded over 9 trials (*Figure 13.2*). They took longer to accept the 2-spot than they did the palatable control beetles, but they eventually did so. By the end of the experiment, they were eating 2-spots as fast as the control beetles. Only in the birds offered the toxic 7-spot ladybird was learned avoidance evident. Distastefulness is clearly not a sufficient deterrent to maintain the defence of 2-spots, although it may raise it initially.

This experiment demonstrates that distastefulness should not be confused with toxicity. With extensive experience of nasty tastes, birds may learn to accept them, as long as there is no toxic effect. Non-toxic chemicals in aposematic insects should therefore be regarded as taste mimicry or, at most, as a short-term defence against naïve predators, not an effective long term defence.

Defence against other predator groups

Although birds are assumed to be major predators of small insects, arthropod predators are also important, and parasitoids may be much more so. Larval ladybirds may be attacked and killed by ants, and even adult ladybirds are attacked. In an experiment to test the effectiveness of ladybird alkaloids against ants (Marples, 1993b), the ants clearly avoided food pellets containing 2-spot or 7-spot ladybird if offered alternative pellets containing mealworm. However, they avoided the 7-spot more strongly than the 2-spot when given only the two ladybird species as food. Even over a 12-day study, the ants did not stop eating the 2-spot ladybird altogether, even though there was food provided *ad libitum* outside the trial period. So, the 2-spot chemical defence is partially effective against ants, but not dramatically so.

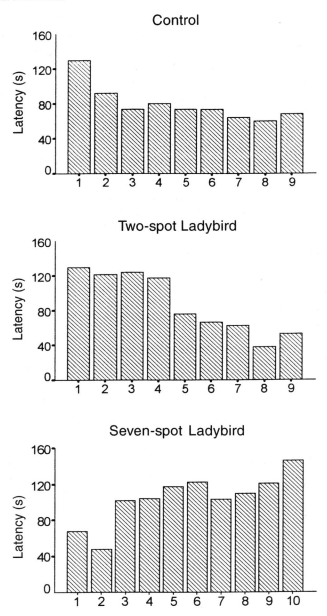

Figure 13.2.   Latencies in successive trials for quail to attack two species of ladybird or a palatable beetle *Alphitobius diaperinus* (control treatment).

Another important selection pressure on ladybird survival comes from parasitoids (Majerus & Kearns, 1989). *Perilitus* wasps parasitise both 7-spots and 2-spots (Cartwright *et al.*, 1982), but while the 7-spot ends up being killed by the parasitoid larva, the 2-spot appears to encapsulate the egg, and suffers no obvious harm. Although the 2-spot is smaller than the 7-spot, this probably is not the reason for their

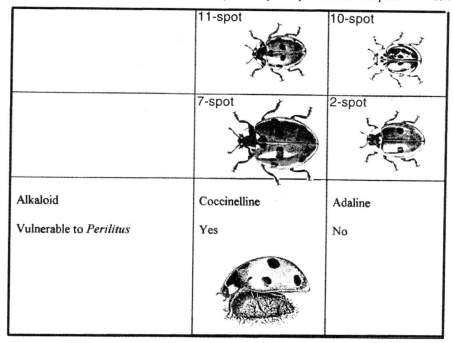

Figure 13.3.   Four species of ladybirds (4× life size), their alkaloid content, and vulnerability to the parasitoid *Perilitus.*

different survival when parasitised, since the 11-spot is much the same size as the 2-spot and is parasitised by *Perilitus* very successfully (Cartwright *et al.*, 1982) (*Figure 13.3*). The 10-spot, which is the same small size as the 2-spot and the 11-spot, is not vulnerable to this parasitoid. There is the possibility that the alkaloid content of these ladybirds is relevant to these differences in susceptibility since the 11- and 7-spots both contain coccinelline (Pasteels *et al.*, 1973) and are vulnerable, while the 2- and 10-spots both contain adaline and are not. There is no strong evidence that adaline protects the 2-spot from this parasitoid, but the circumstantial evidence looks suggestive, so it would be well worth studying.

Does Batesian mimicry of the 7-spot protect the 2-spot from bird predation?

THE SIZE PROBLEM

When considering the proposed Batesian defence of 2-spot ladybirds, the size difference between the 2-spot and the 7-spot poses a bit of a problem. The 2-spot is about a third smaller, and the size ranges do not overlap. With such a simple cue available to the predators, why do not the birds learn to eat the mimic and avoid the model? Are they unable to use size to select small objects as food?

To investigate this, I offered wild birds balls of pastry of three sizes, the middle size being 3 mm in diameter and the others being 1.5 times bigger and smaller. This size difference is approximately that of the 2- and 7-spot ladybirds (Marples, 1993b).

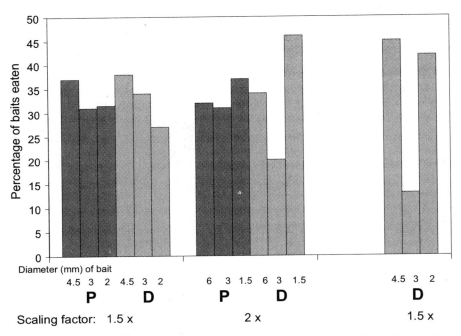

Figure 13.4.   Percentage of 3 different sizes of baits eaten by foraging birds when all are: P, palatable, or D, the middle-sized baits are unpalatable for two size differences, 1.5 times and 2 times the size of the middle-sized bait. The baits were presented in sets of trials ordered from left to right on the graph (see text).

After a day, about half the baits had been taken, and the rest were collected and counted. For the first four days, the baits were all kept palatable, then for the next four days, the middle-sized baits were made unpalatable by the addition of quinine hydrochloride to the pastry mixture. The birds preferred the larger baits even when all baits were palatable (*Figure 13.4*, first set of columns), but the discrimination needed for selection of 2-spots by size would require selection of the smaller of two, so it is only when both big and small baits were preferred over middle-sized that we would feel confident that the birds could discriminate mimic from model. This was not the case when the size difference was a factor of 1.5, but when the size difference was increased to a factor of 2 (*Figure 13.4*, second set of columns), the birds could do the discrimination, selecting against the middle-sized baits when they were unpalatable. In the final stage of the experiment (*Figure 13.4*, third set of columns), the bait sizes were again a factor of 1.5 in size difference and, having learned on the larger size difference, the birds could now do the discrimination

So, birds could use size differences to discriminate model and mimic, but they do not necessarily always do this until they are prompted to do so on an easier task. The size difference between the 2-spot and the 7-spot seems to be close to the limit of effective mimicry, with experienced birds able to distinguish between them using size.

MULTIFACETED DEFENCE

So far, we have been considering one aspect of the ladybird defence at a time but, as

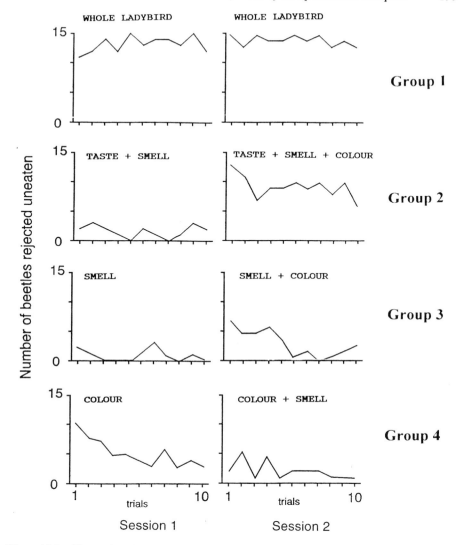

Figure 13.5.   The number of beetles rejected by adult quail when presented with beetles protected by one or more aspects of the 7-spot ladybird defence system.

stated above, there are a large number of elements involved. It is likely that these interact in their effects on a predator. As several of them are shared by the 2-spot and 7-spot, looking at these interactions may aid our understanding of the proposed mimicry system acting between these species.

An experiment was designed to look at the interaction between four of the aspects of the ladybird defence: colour, taste, smell, and toxicity (Marples *et al.*, 1994). Sixty adult quail experienced with (and averse to) 7-spot ladybirds were offered either whole 7-spot (Group 1), a palatable beetle dipped in reflex blood and so carrying only the taste and odour of a 7-spot (Group 2), a palatable beetle in the presence of the odour of 7-spot (Group 3), or a palatable beetle with washed 7-spot elytra stuck over its own elytra, presenting the colour pattern alone (Group 4). The survival of the prey was recorded over 10 trials (*Figure 13.5*, session 1). These initial treatments showed that, while whole ladybird was avoided throughout, none of the cues afforded much protection when presented alone. Only colour caused even initial avoidance, and this soon waned.

The birds were then offered combined cues (*Figure 13.5*, session 2), with groups 2 and 3 gaining the colour pattern cue, and group 4 gaining the odour. The results show that only when the colour was novel was any aversion seen, but even when odour was present as well, avoidance did not last. When all three non-toxic cues were present (Group 2), avoidance was initially as high as that caused by whole ladybird (Group 1), but soon fell as the birds began to accept the beetles in later trials. This strengthens the view that toxicity is necessary for a full and sustained defence. This experiment also suggests that the interaction of taste, smell, and colour are more effective than any of them alone, so the multifaceted approach to protection appears to defend the ladybird better than if it just had one aspect to its defence. Experiments in which chicks were offered combinations of novel colour and odour confirm this synergistic effect (Marples & Roper, 1996).

THE IMPORTANCE OF NOVELTY

Both the above experiments show that novelty is an important aspect of a defence. The degree to which this is true may have been underestimated because most work has been carried out in the laboratory, where birds are more prone to try new foods than are wild birds. In recent work on wild birds, the extremely conservative feeding habits of some individuals became apparent (Marples *et al.*, 1998). Colour-ringed blackbirds (*Turdus merula*) and robins (*Erithacus rubecula*) were trained to eat green pastry baits, then were offered a choice between these and novel red baits. They showed astonishing variation in their willingness to accept the novel baits (*Table 13.1*). Some accepted the new colour without hesitation on the first trial, while others waited up to 125 trials (four months). This avoidance of the novel colour was not restricted to red, which might be argued to be a warning colour to which they had an innate aversion, or to which the birds might have become averse due to experience of aposematic insects in the area. Yellow and green were also tested as novel colours, and they were similarly avoided by some and accepted by others. Green was significantly more avoided than either of the classic 'warning colours'.

Such hesitancy on the part of certain birds is hard to explain. Birds in a pair often differed widely in their responses. One male watched his mate eating the red baits and taking them back to the young for many weeks, but refused to eat them himself. It was only when he fell ill with an infected injury that he accepted the red baits. All birds picked up and presumably tasted the baits very early in this period, so they knew the baits were not unpalatable, but they persisted in their refusal of the novel colour, selecting only the familiar one. It does not seem likely that the birds were extremely afraid of the baits; it seems better described as a refusal to incorporate anything new into the diet, however palatable they knew it to be.

This 'dietary conservatism' is also present in quail, and a selective breeding experiment showed that the trait is heritable and almost certainly a single gene (Marples & Brakefield, 1995). If such dietary conservatism is as common in all bird

Table 13.1. Time taken by two species of wild bird (blackbird (*Turdus merula*) and robin (*Erithacus rubecula*)) to contact (pick up or peck) or recruit into the diet (eat three times) unfamiliar colours of palatable baits.

Novel colour	Bird species	Individual bird	Trials to Contact	Trials to Recruit
Red	Blackbird	BR	87	125
		WDBL	2	49
		WNR	2	114
		MNL	1	3
		WDRL	12	>15
		RL	2	10
		NGL	1	3
		M6	3	>10
		PI	11	14
		FYEW	1	6
		FHUT	1	3
		BC	6	24
		YRR	4	>6
	Robin	DWYL	2	32
		GWDL	>68	>68
		EER	10	18
Yellow	Blackbird	BR	4	16
		WDBL	2	4
		WNR	1	3
		MNL	>50	>50
		WDRL	1	19
		YGL	4	6
		ELB	1	3
		FYEW	1	3
		FHUT	1	14
		BC	3	9
		SN	6	8
	Robin	DWYL	13	26
		GWDL	51	>59
		OMR	3	20
		PPR	11	>41
		EWDL	7	41
		RHUT	3	>13
Green	Blackbird	MM	>42	>42
		MOL	>23	>23
		MF	25	58
	Robin	RER	5	>6

populations as it was in my populations of blackbirds and robins, then we need to reassess our view of aposematic animals and predation on them by birds.

For at least part of the year, food is common enough that birds can turn down possible food items in favour of known safe ones. In situations where food is abundant, it may be that odd-looking food is not under much threat of predation. At other times of the year, conditions are so tough that perhaps birds dare not risk sampling novel, and potentially dangerous, food items. Being made sick when you are near starvation may be too disastrous. At such times, any novel appearance, taste, or smell may be sufficient to prevent attack. Until we know what proportion of the time the predators spend in one of these two states, we cannot guess how strong a selection pressure bird predators actually exert on palatable but conspicuous prey. It may be that the answer to our questions about the 2-spot ladybird, and other insect species like it, is that they get away with their lack of toxicity and lack of similarity to their supposed models by only being abundant at times when most birds are unwilling to attack new prey. Many insect species, including the 2-spot, become scarce or hidden during winter, when the birds may be more hungry, and early spring, when the birds are feeding young.

Although there are cases of Batesian mimicry which are not open to dispute, there is the danger of assuming mimicry is happening in cases where other mechanisms, such as dietary conservatism, may be more important. Direct tests of wild predators' responses to aposematic defences are badly needed.

Acknowledgements

I would like to thank all those who helped with the practical work mentioned in this paper, particularly Paul Brakefield, Willem van Veelen, Tim Roper, David Harper, and Rob Thomas. I would also like to thank those who provided raw materials or data for these experiments, including Mike Majerus, Richard Cowie, and Shelly Hinsley. I am grateful for the support provided by NERC, Royal Society, and the AFRC.

In preparation of the text, I am grateful for useful comments from Mike Speed.

References

Brakefield, P. M. (1985) Polymorphic Mullerian mimicry and interactions with thermal melanism in ladybirds and a soldier beetle: a hypothesis. *Biological Journal of the Linnean Society* **26**, 243–267.

Cartwright, B., Eikenbary, R. D. & Augalet, G. W. (1982) Parasitism by *Parilitus coccinellae* (Hymenoptera: Braconidae) of indigenous coccinellid hosts and the introduced *Coccinella septempunctata* (Coleoptera: Coccinellidae) with notes on winter mortality. *Entomophaga* **27**, 237–244.

de Jong, P. W., Holloway, G. J., Brakefield, P. M. & de Vos, H. (1991) Chemical defence in ladybird beetles (Coccinellidae). II Amount of reflex fluid, the alkaloid adaline and individual variation in defence in the 2-spot ladybird (*Adalia bipunctata*). *Chemoecology* **2**, 15–19.

Fraser, J. F. D. & Rothschild, M. (1960) Defence mechanism in warningly-coloured moths and other insects. *Proceedings of the 11th International Congress of Entomology, Vienna, 1960* **3**, 249–256.

Holloway, G. J., de Jong, P. W., Brakefield, P. M. & de Vos, H. (1991) Chemical defence in ladybird beetles (Coccinellidae). I Distribution of coccinelline and individual variation in defence in 7-spot ladybirds (*Coccinella septempunctata*). *Chemoecology* **2**, 7–14.

Kristin, A. (1988) Coccinellidae and Syrphidae in the food of some birds. In: *Ecology and the effectiveness of aphidophaga* (eds. E. Niemczyk & A. F. G. Dixon), pp. 321–324. Academic Publishers, The Hague.

Majerus, M. & Kearns, P. (1989) *Ladybirds*. Richmond, Slough, UK. 103 pp.

Marples, N. M. (1990) *The influences of predation on ladybird colour patterns*. PhD Thesis, University of Wales, Cardiff, UK.

Marples N. M. (1993a) Toxicity assays of ladybirds using natural predators. *Chemoecology* **4**, 33–38.

Marples, N. M. (1993b) Is the alkaloid in two-spot ladybirds (*Adalia bipunctata*) a defence against ant predation? *Chemoecology* **4**, 29–32.

Marples, N. M. & Brakefield, P. M. (1995) Genetic variation for the rate of recruitment of novel insect prey into the diet of a bird. *Biological Journal of the Linnean Society* **55**, 17–27.

Marples, N. M. & Roper, T. J. (1996) Effects of novel colour and smell on the response of naïve chicks towards food and water. *Animal Behaviour* **51**, 1417–1424.

Marples, N. M., Brakefield, P. M. & Cowie, R. J. (1989) Differences between the 7-spot and 2-spot ladybird beetles (*Coccinellidae*) in their toxic effects on a bird predator. *Ecological Entomology* **14**, 79–84.

Marples, N. M., van Veelen, W. & Brakefield, P. M. (1994) The relative importance of colour, taste and smell in the protection of an aposematic insect *Coccinella septempunctata*. *Animal Behaviour* **48**, 967–974.

Marples, N. M., Roper, T. J. & Harper, D. C. G. (1998) Responses of wild birds to novel prey: evidence of dietary conservatism. *Oikos* **83**, 161–165.

Pasteels, J. M., Deroe, C., Tursch, B., Braekman, J. C., Daloze, D. & Hootele, C. (1973) Distribution et activites des alcaloides defensifs des coccinellidae. *Journal of Insect Physiology* **19**, 1771–1784.

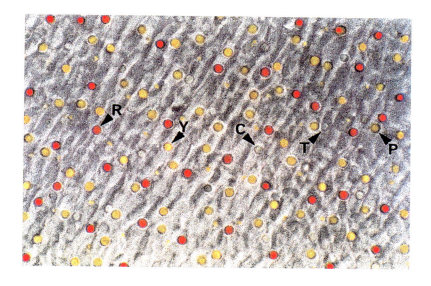

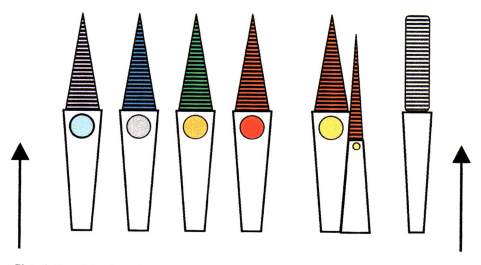

Plate 1. Top: light photomicrograph of an area of fresh retina from the ostrich (*Struthio camelus*) illustrating the five different types of oil droplet, **R**ed, **Y**ellow, **C**lear, **T**ransparent and **P**ale found in the inner segments of cones. This arrangement is typical of most diurnal birds. Reproduced from Wright & Bowmaker (2001) with permission.

Bottom: Schematic diagram illustrating the combinations of visual pigments and oil droplets in the five classes of cones and rods in a diurnal avian retina. Double cones contain the LWS pigment (red) in both members of the outer segments in combination with a P-type droplet in the Principal member and, depending on species, a small **A**ccessory droplet in the smaller Accessory member. The four classes of single cone consist of an LWS pigment (red) plus an R-type droplet, an MWS pigment (green) with a Y-type droplet, an SWS pigment (blue) with a C-type droplet and a UVS pigment (violet) with a T-type droplet. The colours of the droplets in the diagram have been matched to those in the photomicrograph above. The numbers are the percentages of each cone class in a typical passerine retina. Rods (extreme right) do not contain an oil droplet. The direction of the incident light is indicated by the large arrows and is filtered by the oil droplets before being absorbed by the visual pigments.

Plate 2. Some aposematic species of Lepidoptera and Coleoptera, which excrete the odour of 2-methoxy-3-isobutyl pyrazine. Top row, left to right: Atala hairstreak (*Eumaeus atala*), *Pseudolycus naemopterus*, southern festoon (*Zerynthia polyxena*); centre: *Papilio rumansovia* with either side, top to bottom: seven-spot ladybird (*Coccinella septempunctata*), white ermine (*Spilosoma lubricipeda*), five-spot burnet (*Zygaena lonicerae*); bottom row, left to right; garden tiger (*Arctia caja*), cinnabar (*Tyria jacobaeae*), cream-spot tiger (*Arctia villica*).

Plate 3. Great tit nest with 6-day-old chicks fitted with neck collars.

Plate 4. Moths collected from the gut of great tit chicks using neck collars. The parents prepare these moths before feeding the chicks by removing the wings, legs, or even the whole thorax and head (Barba *et al.*, 1996). The complete moths are about 23 mm long.

Plate 5. Retrieving a spider from the gut of a 6-day-old great tit chick fitted with a neck collar.

14

The Effect of the Odour of Pyrazine and Colours on Recall of Past Events and Learning in Domestic Chicks (*Gallus gallus domesticus*)

ANAT BARNEA[1], GADI GVARYAHU[2] AND MIRIAM ROTHSCHILD[3]

[1]*Department of Natural and Life Sciences, The Open University of Israel, Ramat-Aviv, Tel Aviv 61392, Israel,* [2]*Department of Animal Sciences, Hebrew University of Jerusalem, Rehovot 76100, Israel and* [3]*Ashton Wold, Peterborough, PE8 5LZ, UK*

Introduction

Until quite recently, the role played by odours as alerting signals, especially in the relationship between insectivorous bird predators and their prey, has been somewhat neglected. Attention is usually centred on warning coloration (Cott, 1940; Schuler & Roper, 1992) and behaviour, rather than smell. Bitter or sweet tastes have also, rather dubiously, been classified as alerting signals, since they can deter or encourage the final ingestion of captured insect prey or tempting fruit or seeds.

The aromatic pyrazines are exceptionally widely distributed in nature (Woolfson & Rothschild, 1990), and are one of the most successful alerting signals. They cut across the man-made classifications, utilised by both plants and vertebrate and invertebrate animals, and range from continent to continent and ocean to ocean. *Plate 2* illustrates a range of aposematic Lepidoptera and Coleoptera known to excrete the odour of pyrazine. They are not in themselves harmful or beneficial, and their message is essentially 'come' or 'go'.

The pyrazine nucleus comprises a six-membered aromatic ring containing two para-oriented tertiary nitrogen atoms. Over 100 pyrazine compounds have been identified, some of which are scentless, while the methoxyalkylpyrazines produce one of the most powerful and persistent odours known to man (Moore & Brown, 1981; Moore *et al.*, 1990). The nucleus is a stable one, which is readily derived biosynthetically from amino acids and sugar degradation products. Pyrazines are consequently widespread, not only in natural bouquets and stimulating attractive flavours, but in the decomposition or pyrolysis products of biological tissues, and by fungal decay. As odoriferous signals, they possess four especially important characteristics: (1) a low olfactory threshold (Moore *et al.*, 1990); (2) an arresting impact;

Insect and Bird Interactions
© Intercept Ltd., PO Box 716, Andover, Hampshire, SP10 1YG, UK.

(3) an evocative quality difficult to define, akin to the sensation of *déja vu*, which is known to be associated in man with the limbic system (Penfield & Jasper, 1954); and (4) perception at a distance, not only on contact (Guilford *et al.*, 1987).

The frequent conjugation of the odour of pyrazine with the alerting colour red was first noted in ladybird beetles (Coccinellidae) and tiger moths (Arctiidae) (Rothschild, 1961). Both characteristics are likely to be selected, since they are advantageous to both potential predator and potential insect prey, which may only become available to the former at relatively long intervals. The bird thus avoids unnecessary expenditure of energy culminating in an unpleasant experience, and the toxic insect avoids an attack which is likely to inflict injury, even if eventually futile.

Both signals occur 'right across the board' in different Orders and even Phyla; and their ability to either attract or deter is well illustrated. For example, the colour red lures birds to edible berries and fruit or nectar-laden flowers, but warns them against attacking toxic Coleoptera or Lepidoptera. The crimson wattles of turkeys can, as circumstances dictate, demonstrate either sexual attraction or masculine aggression, while baboons may charm their mates or warn rivals by displaying their bright red, swollen posteriors. We ourselves use red as a 'come' or 'go' signal, on the one hand by the coloration of children's delicious candies and, on the other, by the dye of dangerous sleeping pills. It is worth noting that the red and green cone system of primates occupies about 80 per cent of their total colour system (Rothschild & Moore, 1987).

We have assumed that the reaction of birds to so-called 'warning' colours presented in water are learned, and are not innate. The few experiments which indicate inherited aversion to certain colours in naïve chicks have been carried out with solid food and prey, not water (Schuler & Roper, 1992), and it is well known that birds may react differently to food or drink (Roper & Marples, 1997a).

In our experiments, we have tried to examine the possibility that pyrazine not only functions as a powerful alerting signal for birds, but can stimulate the vitally important recall of past events, as it is known to do for mammals (Kaye *et al.*, 1989). We are also interested in the mutual relationship between pyrazine odour and the alerting or warning colour red. Although a simple association is clearly advantageous as a means of protection, in some circumstances the relationship is synergistic and more complex, and the interactions of different colours and odours remain, at the present time, somewhat mysterious.

In a previous publication, one of us (Rothschild *et al.*, 1984) suggested that the odour of pyrazine might "assist or hasten the process of learning", and we have also attempted to test this possibility in naïve chicks.

Methods

Male domestic chicks (*Gallus gallus domesticus*), marked individually, were housed in white wooden cages (120 × 120 cm), heated, and illuminated for 12 hours light/12 hours dark. Chicks were fed *ad libitum*, and provided with plain tap water in four plastic tubes (14 cm length, 3.5 cm diameter), each attached to one of the cage walls.

Training and testing of the chicks was carried out in an experimental cage, made of a white painted, wooden waiting chamber (32 × 22 cm) connected to a testing chamber (85 × 62 cm) by a guillotine door. Both chambers opened from above. In the testing chamber, two drinking tubes were attached to each side of the wall opposite the guillotine door.

Water tubes were painted red, yellow, green, or indigo with paints manufactured by the Tambor paint factory (Israel), matched respectively to No. 45A, No. 9B, No. 133A and No. 116A on the RHS Colour Chart (Royal Horticultural Society, 1965).

Experiments involving bitter taste and pyrazine odour were carried out in a different room, to avoid contamination of the room which was used for the colour-only experiments. For bitter taste, we used quinine hydrochloride (Sigma, 0.03 M aqueous solution). The tested odour was 2-methoxy-3-isobutyl pyrazine (Pyrazine Specialities Inc., Georgia, USA), diluted by dissolving 10 µl in 100 ml distilled water. This concentration of the pyrazine solution is based on values of human olfactory thresholds (Woolfson & Rothschild, 1990). A rectangular filter paper (15 × 50 mm) was placed on each side of the water tubes. Five drops (250 µl) of odour solution or water were added to the filter papers of the test or control tubes. Odour solutions and water evaporated, and after a certain time (equivalent to about eight sessions within an experiment), filter papers were almost dry. Therefore, to ensure constant concentrations throughout the experiment, water and odour solutions were renewed after every eighth session within an experiment, by adding five drops to the filter papers.

Experimental design

Table 14.1 summarises four sets of experiments. Each set included two experiments testing effects of either colours, or colours combined with pryazine odour, on recall and learning. Each experiment consisted of five or six replicates with 10–13 chicks in each. Only chicks which completed the training process (see below for details) were included in the analysis. Ninety-one per cent of the total number of chicks which were used successfully completed the learning process and, accordingly, the total number of chicks in each experiment ranged between 40 and 61. Overall, 393 chicks were tested and the results analysed.

Our previous experience with chicks has shown that a total of 30 chicks is a manageable number to handle at one time, as it allows completion of all learning sessions of DAYS 5 and 6 (see below) within a reasonable time. If experiments involve more chicks, the learning process lasts too long, and chicks do not co-operate as they do at the beginning of the day. Each replicate consisted of about 10 chicks, and therefore three replicates were tested each time. The three replicates were randomly chosen from all eight experiments presented in *Table 14.1*, so that sets of experiments were broken down and, as a result, different replicates of the same experiment were tested at different times. This design obtained internal validity (Keppel, 1982, p. 339)

Table 14.1. Experiments designed to investigate the effect of pyrazine odour and colours (red, yellow, green, and indigo) on recall and learning (see text).

Experimental set	Colours with/without pyrazine odour (no. replicates; total no. chicks tested)	
1	red (5;47)	red + pyrazine (6;54)
2	yellow (6;44)	yellow + pyrazine (5;43)
3	green (5;50)	green + pyrazine (5;49)
4	indigo (5;45)	indigo + pyrazine (6;61)

and allowed inter-experiment comparisons, as it took into consideration the arbitrariness of the assays, which might be due to the use of different hatchlings, different times of the year, and other unknown factors between the runs.

The general frame of the experiments

DAYS 1–3   Day-old chicks were kept in housing cages where food and water were available *ad libitum*. Water was provided in five water tubes, each of a different colour (red, yellow, green, indigo, and white), with a daily random change in their location.

DAY 4   After overnight water deprivation, chicks were individually trained to drink from tubes in the experimental cage. Each chick was given five training sessions, in random order. During each session, the chick stayed in the waiting chamber for 15 seconds, and was then moved through the guillotine door into the testing chamber, which had two water tubes placed on the far wall, coloured alike, and filled with tap water (referred to as 'neutral colour'). The chick was allowed 30 seconds to drink from either tube before being returned to the housing cage.

DAYS 5 and 6   After overnight water deprivation, the chicks were taught the association between:

(1) a specific colour and taste of water,
(2) a colour, pyrazine odour, and taste of water.
One of the water tubes was permanently white, containing plain tap water (referred to as the 'white tube'). The other tube contained bitter water, coloured in one of the tested colours, either with or without pyrazine odour (referred to as the 'test tube'). On DAY 5, each chick had five learning sessions similar to the ones on DAY 4, but now it was given the choice of drinking either from the white or test tube. The location of the tubes was changed randomly after every session. From the end of DAY 5, chicks were given water *ad libitum* in neutral coloured tubes in the housing cage until the end of the experiment on DAY 21.

On DAY 6, after overnight water deprivation, each chick was given as many sessions as needed to reach the 'learning criterion', defined as drinking from the white tube (the 'correct' one) on three consecutive sessions. Chicks which only drank from the white tube in all sessions were removed from the experiment.

DAYS 7–10   Chicks were given water *ad libitum* in the housing cage.

DAY 11   After overnight water deprivation, chicks were first tested for recall. They were randomly chosen and individually introduced to the experimental cage, where they were exposed to the same situation as the one presented to them on DAYS 5 and 6, except that now both tubes (white and test) were filled with plain tap water. This testing regime resembled the natural situation, in which a predator re-encounters a specific combination of visual and odoriferous signals already known to be disagreeable. Each chick had only one session (described above) and had to choose between the two tubes. Chicks that drank from the white tube (the 'correct' one, originally containing plain tap water) were considered to recall the association with colour and bitter water correctly, and were put back into the housing cage for future recalls.

Chicks which drank from the test tube (the 'wrong' one, which originally contained bitter water) were removed from the experiment. Length of recall was therefore defined as the number of days from the learning day during which the chick remembered to avoid the tested tube.

DAYS 13, 14, 15, 18 and 21   The same recall procedure was carried out as on DAY 11. The location of tubes in the experimental cage was randomly changed between recalls. The experiment ended on DAY 21.

Data analysis

To test the effect of different colours and pyrazine odour, we used two-way Analysis of Variance. Survival analysis (SAS, 1990) was used to test the effect of different colours and pyrazine odour on recall. We used the LIFETEST procedure, which tests for equality of survival curves across strata (Cox & Oakes, 1984), for survival analysis.

Results

EFFECTS OF COLOURS (WITHOUT PYRAZINE ODOUR) ON RECALL

Figure 14.1 presents recall curves when bitter water was associated with coloured tubes (red, yellow, green, and indigo). Statistical analysis showed that three colours (red, yellow, and green), out of the four tested, did not differ significantly in their effect on recall. Indigo was the only colour which differed significantly, by improving recall (comparison of indigo v. red ($\chi^2 = 4.6093$, df = 1, $P = 0.031$); indigo v. yellow ($\chi^2 = 14.7308$, df = 1, $P = 0.0001$); indigo v. green ($\chi^2 = 4.5870$, df = 1, $P = 0.0322$)). The trend (from best to worst recall) seen in *Figure 14.1* is: indigo, green, red, and yellow.

EFFECTS OF PYRAZINE ODOUR ON RECALL

Figure 14.2 presents recall curves when bitter taste was associated with the colour red, yellow, green, and indigo, with or without the odour of pyrazine. The results of statistical analyses are presented in *Table 14.2*, and show that addition of pyrazine odour significantly improved recall of bitter taste associated with yellow, green, and red tubes. Indigo was the only coloured tube to which the addition of pyrazine odour did not improve recall.

THE EFFECT OF COLOURS AND PYRAZINE ODOUR ON LEARNING

Table 14.3 summarises the results on the effect of colours, with or without pyrazine odour, on the learning process. Statistical analysis of these data showed that:

(1) there was no significant difference between the mean number of sessions required to reach the learning criterion when the bitter taste of water was associated with different colours;

(2) the addition of pyrazine odour to the association between bitter taste of water and

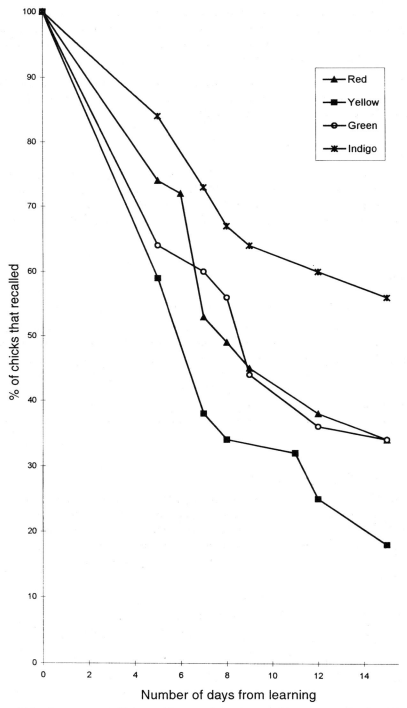

Figure 14.1. Percentage of chicks recalling association between bitter water and colour (▲, red; ■, yellow; ○ green; ✖, indigo) of the water tube, as a function of the number of days from learning.

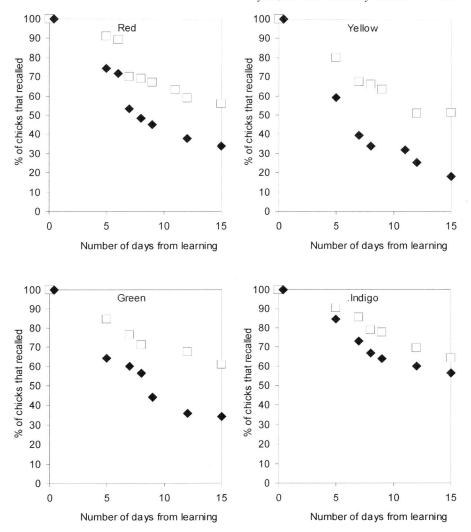

Figure 14.2. Percentage of chicks that recalled association between bitter water and the colour of the water tube, with (□) or without (◆) pyrazine odour, as a function of the number of days from learning.

Table 14.2. Chi-square values for comparisons between different recall curves (see *Figure 14.2*).

Colour of water tube associated with the bitter taste	$\chi^2_{df=1}$ for comparison of the recall curves for this association and the same with pyrazine odour
red	5.54 (*P* = 0.019)
yellow	10.84 (*P* = 0.001)
green	7.68 (*P* = 0.006)
indigo	not significant

Table 14.3.   Mean (± s.e.) to reach the learning criterion (see text) for an association between bitter taste and a colour with/without pyrazine odour.

Colour	Without odour	With pyrazine odour
red	10.0 ± 0.29 (n = 47)	8.8 ± 0.17 (n = 54)
yellow	9.0 ± 0.21 (n = 44)	8.6 ± 0.15 (n = 43)
green	9.7 ± 0.26 (n = 50)	9.0 ± 0.20 (n = 49)
indigo	9.7 ± 0.28 (n = 45)	8.8 ± 0.15 (n = 61)

the colour of the water tube always shortened the learning process, and fewer sessions were then required to reach the learning criterion (*Table 14.3*). Overall, the odour effect was found to be significant (F = 26.95, df = 1,3, P = 0.0001). In order to carry out multiple comparison tests for the different colour–odour combinations and to control for experimental error, we employed Sidak's multiplicative inequality procedure (Sidak, 1967). Accordingly, when α is set to 0.05 and four comparisons are made, α' for each single comparison equals 0.0127. Results of these tests showed that the effect varied as follows: when bitter water was associated with red or indigo tubes, the addition of pyrazine odour shortened the learning process significantly (F = 10.10, df = 1,104, P = 0.002 and F = 11.50, df = 1,99, P = 0.001 respectively); when bitter water was associated with green and yellow tubes, the tested odour had no significant effect on the learning process (F = 4.45, df = 1,97, P = 0.038 and F = 2.23, df = 1,85, P = 0.139 respectively).

Discussion

The chemical defences of insects are amazingly varied (Blum, 1981) but, on the whole, alerting signals are comparatively limited. This is particularly evident in the case of pyrazine odour, which remains dominant and assertive, even when the accompanying visual alerting colours are arranged in completely different patterns. Thus, for example among the tiger moths (Arctiidae), the white ermine moth (*Spilosoma lubricipida*) is snow white, with a few black speckles and a bright yellow abdomen, while the cinnabar moth (*Tyria jacobaeae*) is scarlet, with blue-black forewings, and the cream spot tiger moth (*Arctia villica*) has bright yellow hindwings with black spots, black forewings with large pale yellow blotches, and a scarlet abdomen. Yet all three excrete odoriferous pyrazines (Rothschild, 1961; Moore *et al.*, 1990) (*Plate 2*).

Species of the beetle genus *Pseudolycus* (Oedemeridae) are described by Moore & Brown (1989) as an example of sophisticated mimicry since "they have developed different colour patterns over their extensive ranges ... each to match a particular sympatric model and all smell of methoxy-alkyl pyrazine."

In our laboratory conditions, odours overshadowed or eclipsed colour cues when chicks experienced them together. Our trials with coloured drinking tubes and the odour of pyrazine show this tendency where recall of bitter water is concerned. It was only when pyrazine odour was added to the coloured tubes that recall was significantly improved.

Marples & Roper (1996) found that the odour of almond oil shortened avoidance learning in chicks, thus drawing attention to another effective avian alerting odour. However, they do not state whether or not HCN, which occurs naturally in the almond oil, had been extracted. HCN has been recorded from over a thousand species of plants (Jones, 1979), and it may well rival pyrazine as a common botanical alerting signal. In burnet moths (*Zygaena* spp.), HCN functions as both an alerting and warning signal combined. Curiously enough, burnet moths also excrete pyrazine (Rothschild, 1985; Moore *et al.*, 1990), and they are avoided by almost all bird predators.

Like Roper & Marples (1997b), we also found that coloured drinking tubes were ineffectual in shortening the learning process for quinine-tainted water; but when we offered them to chicks, together with the pyrazine odour, only two colour–odour combinations (red and pyrazine, and indigo and pyrazine) yielded a significant reduction in learning time. Thus, red and pyrazine was the only association which significantly improved both recall and learning of bitter taste. This association is not unexpected from an evolutionary standpoint. Pyrazine (see earlier) has so many features which qualify it for the role of a superior alerting signal, not least of which is its availability. Red as an alerting colour *par excellence,* one assumes, must have evolved in response to the need of a conspicuous colour "in a green world beneath a blue vault". Furthermore, Varela *et al.* (1993) pointed out that, in the various species of birds they examined, the pigment (iodopsin) absorbing at the longest wavelength near 570 mm is by far the most abundant. It appears to be the visual pigment that "dominates photopic spectral sensitivity of the birds". The combination of pyrazines and the colour red as an alerting signal is obviously felicitous.

We have failed to find an explanation for the results of some of our experiments: thus, the chicks' reaction to the indigo coloured tube is baffling. It proved to be the only coloured tube which stimulated recall without the addition of pyrazine and yet, when associated with that odour, unlike the other coloured tubes, it did not improve recall of bitter water. Nevertheless, it elicited shortening of the learning process.

The indigo tube was very dark in shade and tone, considerably darker than the other tubes involved (see earlier), and black paint was certainly mixed most lavishly with indigo to obtain the blue-black coloration characteristic of many aposematic insects. This tube contrasted sharply with the all-white walls of the experimental cage, and possibly we may have unwittingly introduced the potent element of contrast, which affected the chicks' reaction to colours and odours. Authors writing at the turn of the last century (Marshall, 1902; Swynnerton, 1919) noted that ground-foraging birds avoided toxic ground-dwelling beetles which adopted all-black as a warning signal, and exposed themselves against a light background of sandy soil.

A somewhat different puzzle concerns Rowe & Guilford's (1996) finding that pyrazine odour triggered aversion in chicks to food coloured red or yellow, but not green. We could find no support for this observation in our experiments. The discrepancy may solely be due to their use of solids, rather than fluids. It is well known, as we have indicated, that birds react differently to food or drink. Gillette *et al.* (1980) showed that colour aversion in domestic chicks is not easily formed when the colour is associated with fluids, rather than solids. One of us (M. Rothschild, *unpublished*) found that many adult quails developed a permanent aversion to red-coloured water tainted with quinine, but relished various solid food to which it had been added in a visible surface layer.

Great caution is necessary when attempting to interpret such experiments, since bird behaviour is often changeable and capricious, varying from one individual to another, or from brood to brood, and furthermore, bird colour vision is not well understood. Varela *et al.* (1993) pointed out "the tetra or pentachromatic colour space of birds appears to be the most complex in nature, and is likely involved in virtually all areas of the animals' lives", but the authors refer to the different and mysterious functions, the solution of which presents an "appealing facet of natural history".

Nature is, of necessity, parsimonious, opportunistic, and immensely variable. It is, however, suprising for us to find birds, which have hitherto been considered relatively indifferent to smell and dominated by colour cues or the behaviour of potential insect prey, responding so decisively to odours.

Although, at first sight, alerting signals may appear to convey unequivocal messages of 'come' or 'go', on closer examination, they are seen to be more complex. In many plants, and in the melon fly (*Dacus cucurbitae*) (Baker *et al.*, 1982), they convey different messages simultaneously, attracting one species and repelling another. Again, the same signals, used at long intervals, can convey opposite meanings as circumstances dictate. A display of red feathers may be employed to scare off rivals, or attract prospective mates. Moreover, the same message is often conveyed by simultaneous appeal to visual, auditory, or olfactory sensibilities of predators. In mammals, the odour of pyrazines can delay sexual development (Jemiolo & Novotny, 1994), as well as stimulate recall (Kaye *et al.*, 1989), or suppress the development of T cells in mice (Woolfson & Rothschild, 1990). In ants, pyrazine plays a variety of different roles, functioning, for instance, as a trail-laying substance, a defensive spray, or escape pheromone (Brophy, 1989), and in these cases colours are not involved.

It would come as no suprise to find that the type of alerting signal we have investigated is the 'tip of the iceberg', and that pyrazine and many other aromatic substances, with or without a specific colour or sound association, will be found to play a significant role in animal neurophysiology.

Acknowledgements

We thank Yoram Yom-Tov for his support, Yael Alberton for statistical advice and analysis, and Dina Lipkind for technical help. We are also grateful for useful discussions with Robin Aplin, Gunnar Bergström, and Charles Lane. The work was supported by the Ministry of Science, Israel.

References

Baker, R., Herbert, B. H. & Lomer, R. A. (1982) Chemical components of the rectal gland of male *Dacus cucurbitae*, the melon fly. *Experientia* **38**, 232–233.

Blum, M. S. (1981) *Chemical defenses of arthropods.* Academic Press, New York, USA. 562 pp.

Brophy, J. J. (1989) Pyrazines obtained from insects: their source, identification, synthesis and function. In: *Studies in natural products chemistry (structures and elucidation), vol. 5B* (ed. Atta-ur-Rahman), pp. 221–273. Elsevier, Amsterdam, The Netherlands.

Cott, H. B. (1940) *Adaptive coloration in animals.* Methuen, London, UK. 508 pp.

Cox, D. R. & Oakes, D. (1984) Analysis of survival data. In: *Monographs on statistics and applied probability* (eds. D. R. Cox, D. Hinkley, D. Rubin & B. W. Silverman). Chapman & Hall, London, UK.

Gillette, K., Martin, G. M. & Bellingham, W. P. (1980) Differential use of food and water cues in the formation of conditioned aversions by domestic chicks (*Gallus gallus*). *Journal of Experimental Psychology: Animal Behaviour Processes* **6**, 99–111.

Guilford, T., Nicol, C., Rothschild, M. & Moore, B. P. (1987) The biological roles of pyrazines: evidence for a warning odour function. *Biological Journal of the Linnean Society* **31**, 113–128.

Jemiolo, B. & Novotny, M. (1994) Inhibition of sexual maturation in juvenile female and male mice by a chemosignal of female origin. *Psychology and Behaviour* **55**, 519–522.

Jones, D. A. (1979) Chemical defense: primary or secondary function? *American Naturalist* **113**, 445–451.

Kaye, H., Mackintosh, N. J., Rothschild, M. & Moore, B. P. (1989). Odour of pyrazine potentiates an association between environmental cues and unpalatable taste. *Animal Behaviour* **37**, 563–568.

Keppel, G. (1982) *Design and analysis. A researcher's handbook (2nd edition).* Prentice-Hall, Englewood Cliffs, New Jersey, USA. 669 pp.

Marples, N. M. & Roper, T. J. (1996) Effects of novel colour and smell on the response of naïve chicks towards food and water. *Animal Behaviour* **51**, 1417–1424.

Marshall, G. A. K. (1902) Five years' observations and experiments (1896–1901) on the bionomics of South African insects. *Transactions of the Royal Entomological Society of London* **3**, 287–584.

Moore, B. P. & Brown, W. V. (1981) Identification of warning odour components, bitter principles and antifeedants in an aposematic beetle *Metriorrynchus rhipidius* (Coleoptera: Lycidae). *Insect Biochemistry* **11**, 493–499.

Moore, B. P. & Brown, W. V. (1989) Graded levels of chemical defence in mimics of lycid beetles of the genus *Metriorrhynchus* (Coleoptera). *Journal of the Australian Entomological Society* **26**, 229–233.

Moore, B. P., Brown, W. V. & Rothschild, M. (1990) Methylalkylpyrazines in aposematic insects, their host plants and mimics. *Chemoecology* **1**, 43–51.

Penfield, W. & Jasper, H. (1954) *Epilepsy and the functional anatomy of the human brain.* Little, Brown and Co., Boston, USA. 896 pp.

Roper, T. J. & Marples, N. M. (1997a) Colour preferences of domestic chicks in relation to food and water presentation. *Applied Animal Behaviour Science* **54**, 207–213.

Roper, T. J. & Marples, N. M. (1997b) Odour and colour as cues for taste-avoidance learning in domestic chicks. *Animal Behaviour* **53**, 1241–1250.

Rothschild, M. (1961) Defensive odours and Mullerian mimicry among insects. *Transactions of the Royal Entomological Society of London* **113**, 101–121.

Rothschild, M. (1985) British aposematic Lepidoptera. In: *The moths and butterflies of Great Britain, vol. 2* (eds. J. H. Heath & A. M. Emmet), pp. 9–62. Harley Books, Colchester, UK.

Rothschild, M. & Moore, B. P. (1987) Pyrazines as alerting signals in toxic plants and insects. In: *Insects–plants* (eds. G. Labeyrie, G. Fabres & D. Lachaise). *Proceedings of the 6th International Symposium on Insect–Plant Relationships,* pp. 97–101. Junk, Dordrecht, The Netherlands.

Rothschild, M., Moore, B. P. & Brown, W. V. (1984) Pyrazines as warning odour components in the monarch butterfly *Danaus plexippus* and in moths of the genera *Zygaena* and *Amata* (Lepidoptera). *Biological Journal of the Linnean Society* **23**, 375–380.

Rowe, C. & Guilford, T. (1996) Hidden colour aversions in domestic chicks triggered by pyrazine odours of insect warning displays. *Nature, London* **383**, 520–522.

Royal Horticultural Society (1965) *RHS colour chart.* Royal Horticultural Society, London, UK. 4 fans of colour samples.

SAS (1990) *SAS/STAT user's guide, version 6 (4th edition),vols. 1 and 2.* SAS Institute, Cary, USA. 1,848 pp.

Schuler, W. & Roper, T. J. (1992) Responses to warning coloration in avian predators. *Advanced Studies in Behaviour* **21**, 111–146.

Sidak, Z. (1967) Rectangular confidence regions for the means of multivariate normal distributions. *Journal of the American Statistical Association* **62**, 626–633.

Swynnerton, C. F. M. (1919) Experiments and observations bearing on the explanation of form and colouring, 1908–1913. *Journal of the Linnean Society (Zoology)* **33**, 203–385.

Varela, F. J., Palacios, A. G. & Goldsmith, T. H. (1993) Color vision in birds. In: *Vision, brain and behaviour in birds* (eds. H. P. Zeigler & H. J. Bischof), pp. 77–98. Bradford, Cambridge, Massachusetts, USA.

Woolfson, A. & Rothschild, M. (1990) Speculating about pyrazines. *Proceedings of the Royal Society of London, B* **242**, 113–119.

15
A Note on the Odour of Amyl Acetate in the Role of an Alerting Signal

MIRIAM ROTHSCHILD[1] AND ANAT BARNEA[2]

[1]*Ashton Wold, Peterborough, PE8 5LZ, UK and* [2]*Department of Natural and Life Sciences, The Open University of Israel, Ramat-Aviv, Tel Aviv 61392, Israel*

Introduction

Chapter 14 in this book (Barnea *et al.*, 2003) describes studies on the role of pyrazine (2-methoxy-3-isobutyl pyrazine), which can function as an alerting signal that chicks, recalling past experience of disagreeable events, could learn and associate with past events in nature. This alkaloid is widely distributed in toxic insects, such as monarch butterflies, burnet moths, and ladybird beetles, which are avoided by birds. The strong scent of amyl acetate, however, is characteristic of attractive fruit and flowers, and also appears to function as an alerting signal for frugivorous birds and frugivorous insects. In order to investigate the effect of amyl acetate on day-old chicks, we included them in similar experiments as those described by Barnea *et al.* (2003).

Materials

A total of 552 chicks were used for the combined experiments, of which 139 were employed for amyl acetate odour only. In both series of experiments, the birds were offered plain tap water or quinine-tainted tap water in red, yellow, green, indigo, or white drinking tubes associated with or without alerting odours. The amyl acetate was obtained from Sigma (Cat. amyl acetate 628-63-7, FW 130-2, hygroscopic $C_7H_{14}O_2$ diluted 1:1 in tap water) and the pyrazine from Pyrazine Specialities Inc., Georgia, USA. *Tables 15.1, 15.3* and *Figure 15.1* follow those in Chapter 14 (Barnea *et al.*, 2003).

The distribution of amyl acetate (a mixture of isomers) in nature

Amyl acetate is well known for the melliferous smell it imparts to bananas. In the past, this oily extract of the fruit was used as a machine lubricant, and some reputable dictionaries still give 'banana oil' as a synonym for amyl acetate, although some of the other 200 volatiles identified from banana are also present in the extract (Simmonds, 1966; Nursten, 1970) and contribute to the odour of the fruit.

Insect and Bird Interactions
© Intercept Ltd., PO Box 716, Andover, Hampshire, SP10 1YG, UK.

Unlike pyrazines, amyl acetate is rarely found in insects or in the foliage of plants, but has been recorded in the leaves of banana (M. Rothschild, *unpublished*). Isopentyl acetate was noted once in bees by Birch (1974) in a whole extract, and possibly in the heads of two species of New Zealand ant (Brophy, 1989). These two records contrast with the number of those for pyrazines which, for example, have been recorded from no less than 150 species of ant (Brophy, 1989).

Amyl acetate is a characteristic aroma of many fruits, such as pears, apples, oranges, and strawberries, and is recorded from the flowers of *Rosa, Cymbidium, Acrinidia, Cananga, Bubbia, Zygogynum* (Knudsen *et al.*, 1993) and *Musa velutina* (M. Rothschild, *unpublished*).

In the East Indies and Africa, wild bananas are pollinated by sunbirds, bats, and tree shrews (Proctor *et al.*, 1997), and in the New World's cultivated crops by humming birds. In Malaysia, various butterflies are attracted to the copious flow of nectar from the creamy white flowers of the banana. It is tempting to speculate that amyl acetate functions as an alerting signal for these pollinators, and also for the butterflies which feed on ripe fallen apples, plums, and pears in the UK.

Recorded experiments with the odour of amyl acetate and birds

Day-old, male domestic chicks (*Gallus gallus domesticus*) were given coloured beads, accompanied by an odour source of isoamyl acetate, by Burne & Rogers (1996), who recorded a definite dislike of the smell, the chicks responding by a head-shaking reaction on pecking at the bead.

In preliminary experiments carried out to prove that birds responded to odours of various kinds, Wenzel (1967, 1987), in a series of papers, recorded an increase in cardiac and respiration rates in different bird species, including domestic chickens, when exposed to various odours, including that of amyl acetate. Subsequent experiments have shown that the domestic pigeon can not only detect the odour of amyl acetate and butyl acetate (Walker *et al.*, 1979), but can discriminate different intensities of amyl acetate (Schumake *et al.*, 1969). Recording from bundles of the olfactory nerves of about a dozen species of birds, including the domestic pigeon (*Columba livia*), Tucker (1965) and Oley *et al.* (1975) showed that they all responded to the inspiration of the odour of amyl acetate. According to Wenzel (1971), the brown kiwi (*Apteryx australis mantelli*) refused food contaminated with the odour of amyl acetate.

Captive humming birds (*Colibri servicostis*) were found to respond to various odours, including that of amyl acetate, as discriminative stimuli (Ioalè & Papi, 1989). In nature, these birds are principally flower feeders and, according to Grant & Grant (1968), usually select those with little or no scent.

Relatively few experiments have been tried to pit odour against colour as attractants or repellents for birds. Roper & Marples (1997) noted that, when chicks were "trained to avoid quinine solution using a compound visual olfactory cue, the olfactory cue component overshadowed the visual one." On the other hand, Duncan & Slotnick (1985) noted that the food finding of domestic pigeons depended on visual, not olfactory, cues.

The odour of amyl acetate as an alerting signal

Plain tap water or quinine-tainted tap water was offered to the birds in red, yellow,

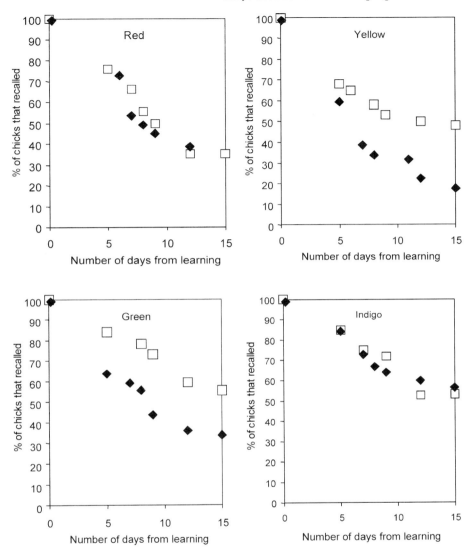

Figure 15.1. Percentage of chicks that recalled association between bitter water and the colour of the water tube, with (□) or without (◆) amyl acetate odour, as a function of the number of days from learning.

green, and indigo coloured drinking tubes, with and without the presence of alerting odours. A synopsis of the results of the comparison with amyl acetate and the pyrazine is given in *Table 15.1*. The experimental design, statistical analysis, general frame of the experiments, and the pyrazine results are described in Chapter 14 (Barnea *et al.*, 2003).

Without either of the alerting signals, there was no significant improvement in the recall or learning process associated with bitter water, except in the case of the indigo tube, which significantly improved recall (*Table 15.1*, *Figure 15.1*) without the addition of an alerting signal. When the odour of amyl acetate was added, recall was

Table 15.1. Reaction of chicks to quinine-tainted water with or without alerting odours offered in coloured drinking tubes.

Alerting odour added	Recall improved	Learning period shortened
amyl acetate	green and yellow tubes	red, green, and indigo tubes
2-methoxy-3-isobutyl pyrazine	red, yellow, and green tubes	red and indigo tubes
no alerting odours added	indigo tube only	no effect

Table 15.2. Chi-square values for comparisons between different recall curves (see *Figure 15.1*).

Colour of water tube associated with the bitter taste	$\chi^2_{df=1}$ for comparison of the recall curves for this association and the same with amyl acetate odour
red	not significant
yellow	6.95 $(P = 0.008)$
green	5.93 $(P = 0.015)$
indigo	not significant

Table 15.3. Mean (± s.e.) to reach the learning criterion (see text) for an association between bitter taste and a colour with/without pyrazine or amyl acetate odour (figures in italics are taken from Barnea *et al.*, 2003).

Colour	Without odour	With pyrazine odour	With amyl acetate odour
red	*10.0 ± 0.29 (n = 47)*	*8.8 ± 0.17 (n = 54)*	8.8 ± 0.21 (n = 40)
yellow	*9.0 ± 0.21 (n = 44)*	*8.6 ± 0.15 (n = 43)*	8.6 ± 0.16 (n = 40)
green	*9.7 ± 0.26 (n = 50)*	*9.0 ± 0.20 (n = 49)*	8.6 ± 0.18 (n = 43)
indigo	*9.7 ± 0.28 (n = 45)*	*8.8 ± 0.15 (n = 61)*	8.8 ± 0.19 (n = 36)

significantly improved if associated with green and yellow tubes, but not red or indigo (*Tables 15.1* and *15.2*), while the learning process significantly improved with red, green, and indigo tubes (*Table 15.3*). When the odour of pyrazine was added, recall was significantly improved with red, yellow, and green tubes, and the learning process with red and indigo tubes, but not green or yellow tubes (*Tables 15.1* and *15.3*). If bitter water was presented to birds in red and indigo tubes in the presence of the odour of amyl acetate or the pyrazine, the learning process was shortened in both cases (F = 7.96, df = 2, P = 0.0005 and F = 6.71, df = 2, P = 0.0017 respectively). If only yellow tubes were used with the odour of amyl acetate or the pyrazine, no significant difference in the learning process was recorded. If only green drinking tubes were employed with both odours, only amyl acetate elicited a significant shortening in the learning process (F = 6.42, df = 2, P = 0.002) (*Tables 15.1* and *15.3*).

These experiments demonstrate certain differences in the reaction of the birds to the coloured tubes associated with the two distinctive alerting signals.

This was perhaps not surprising, since the distribution of the two signals in the animal and plant kingdoms is markedly different, but it is interesting to find that, in our experiments, alerting signals were usually required in order to initiate a specific response to a colour. Marples *et al.* (1994) previously showed that birds trained to avoid toxic ladybird beetles (*Coccinella septempunctata*) needed both their aposematic coloration and the alerting odour of a pyrazine to be presented simultaneously in order to respond as anticipated and show avoidance. Segregated, these cues did not initiate avoidance. This also suggests that alerting signals may exert a subtle influence which is not, in all cases, limited to the arousal of equitable interest, or the anticipation of pleasure or pain.

Discussion

The comparison between the reaction of male, day-old chicks to the pyrazine and amyl acetate, together with their respective distributions in nature, suggests that amyl acetate is often an alerting signal for the presence of edible fruit, or the nectar in flowers – promising an agreeable experience – while in insects, pyrazine frequently functions as a warning signal for a wide range of many toxic or dangerous species of animals and plants. Both substances can, as circumstances dictate, change roles.

In our experiments, amyl acetate was linked with a disagreeable experience, to which signal the chicks responded when it was associated with specific colours, which were sometimes different (see *Table 15.1*) from those which stimulated recall and learning when linked to the pyrazine.

The known distribution of these two alerting signals is, in a sense, inevitably distorted by the interests of the biologists concerned. They are usually found as side effects only of some central theme, and do not form the main objective of the investigation. In the case of pyrazines, trace amounts are so widely distributed in plants, where they enhance and intensify the flavour of other components, that they are generally overlooked, and only noted when their concentrations are greatly increased and function as warning signals, for example in the foliage of *Asclepias* and *Urtica*. Both pyrazines and amyl acetate are often seasonal, and are then consequently not infrequently overlooked. Thus, the nettle (*Urtica urens*) is devoid of the pyrazine until the summer (M. Rothschild, *unpublished*), and amyl acetate is only present when fruit ripens and seeds are ready for dispersal.

We now require experiments in which the chicks are offered water adulterated with an attractant rather than a repellent. It is probable this will reinforce the difference noted here in the birds' reaction when the two alerting signals were linked to different coloured tubes.

Acknowledgements

Our grateful thanks are extended to Gunnar Bergström, Jeffrey Harborne, Tim Roper, and Peter Yeo for their excellent help with the literature. We thank Yoram Yom-Tov and the Zoological Garden at Tel-Aviv University for providing the facilities needed to conduct these experiments, and Yael Alberton for statistical advice and analysis. The work was supported by the Ministry of Science, Israel.

Postscript

The owl butterfly (*Caligo memnon*) was present in numbers free flying in March 1999 in Mr Clive Farrell's Butterfly House at Stratford, UK. They were feeding steadily at 10 a.m. on over-ripe, peeled bananas placed on a table. Transferred to a sheet of cotton wool with a small puddle of 2% amyl acetate in water on the surface, they quickly extended their proboscis and began to imbibe the fluid. Several continued to do so without stopping for 5–32 minutes. Some individuals paused and prodded round the area with the proboscis and either resumed feeding or flew off. They were not attracted to the area from a distance. The behaviour of these butterflies suggested that amyl acetate signals an attraction, and stimulates a linked feeding and searching response.

References

Barnea, A., Gvaryahu, G. & Rothschild, M. (2003) The effect of the odour of pyrazine and colours on recall of past events and learning in domestic chicks (*Gallus gallus domesticus*) (*this volume*).
Birch, M. (1974) Pheromones. In: *Frontiers of biology, vol. 32* (eds. A. Neuberger & E. L. Tatum). North-Holland Publishing Co., Amsterdam, The Netherlands. 495 pp.
Brophy, J. J. (1989) Pyrazines obtained from insects: their source, identification, synthesis and function. In: *Studies in natural products chemistry, vol. 5 (structure and elucidation, B)* (ed. Atta-ur-Rahman), pp. 221–273. Elsevier, Amsterdam,. The Netherlands.
Burne, T. H. J. & Rogers, L. J. (1996) Responses to odorants by the domestic chick. *Physiological Behaviour* **60**, 1441–1447.
Duncan, H. J. & Slotnick, B. M. (1985) Comparison of visual and olfactory stimuli in reversal learning with pigeons. *Chemical Senses* **10**, 409.
Grant, K. A. & Grant, V. (1968) *Hummingbirds and their flowers*. Columbia University Press, New York, USA. 115 pp.
Ioalè, P. & Papi, F. (1989) Olfactory bulb size, odor discrimination and magnetic insensitivity in hummingbirds. *Physiological Behaviour* **45**, 995–999.
Knudsen, J. T., Tollsten, L. & Bergström, G. (1993) Floral scents; a checklist of volatile compounds isolated by head space techniques. *Phytochemistry* **33**, 253–260.
Marples, N. M., van Veelen, W. & Brakefield, P. M. (1994) The relative importance of colour, taste and smell in the protection of an aposematic insect *Coccinella septempunctata*. *Animal Behaviour* **48**, 967–974.
Nursten, H. E. (1970) Volatile compounds: the aroma of fruits. In: *Biochemistry of fruits and their products, vol. 10* (ed. E. D. Hulme), pp. 239–268. Academic Press, London, UK.
Oley, N., DeHan, R. S., Tucker, D., Smith, J. C. & Gradziadei, P. P. C. (1975) Recovery of structure and function following transection of the primary olfactory nerves in pigeons. *Journal of Comparative Physiology* **88**, 477–495.
Proctor, M., Yeo, P. & Lack, A. (1997) Birds, bats and other vertebrates. In: *The natural history of pollination*, pp. 244–247. Harper Collins, New Naturalist Series, London, UK.
Roper, T. J. & Marples, N. M. (1997) Odour and colour as cues for taste-avoidance learning in domestic chicks. *Animal Behaviour* **53**, 1241–1250.
Schumake, S. A., Smith, J. C. & Tucker, D. (1969) Olfactory intensity-difference thresholds in the pigeon. *Journal of Comparative Physiology and Psychology* **67**, 64–69.
Simmonds, N. W. (1966) *Bananas (2nd edition)*. Longmans, London, UK. 512 pp.
Tucker, D. (1965) Electrophysiological evidence for olfactory function in birds. *Nature, London* **207**, 34–36.
Walker, J. C., Tucker, D. & Smith, J. C. (1979) Odour sensitivity mediated by trigeminal nerve in the pigeon. *Chemical Senses and Flavour* **4**, 107–116.
Wenzel, B. M. (1967) Olfactory perception in birds. In: *Olfaction and taste, vol. 2* (ed. T. Hayashi), pp. 203–217. Pergamon Press, Oxford, UK.
Wenzel, B. M. (1971) Olfactory sensation in the kiwi and other birds. *Annals of the New York Academy of Sciences* **188**, 183–193.
Wenzel, B. M. (1987) The olfactory and related systems in birds. *Annals of the New York Academy of Sciences* **519**, 137–149.

16
Measuring Food Availability for an Insectivorous Bird: The Case Study of Wrynecks and Ants

ANNE FREITAG

Museum of Zoology, PO Box 448, CH-1000 Lausanne 17, Switzerland

Introduction

The wryneck (*Jynx torquilla*) is the only migratory bird of the family Picidae. It occurs in Europe during its breeding period, from March to September/October. Until the fifties, it was a common breeding bird living in open habitats, such as orchards or meadows, with scattered trees. However, since that time, its populations have decreased in Switzerland, as well as elsewhere in Europe (Scherner, 1980). Many factors have been put forward to explain this decline (Scherner, 1989), among which modifications of agricultural practices play a crucial role. Not only were the orchards with standard trees, which offered nesting cavities, replaced by intensive orchards, but modern agricultural practices have also greatly affected the food resources. Wrynecks are highly specialised in their diet, feeding almost exclusively on ants and ant brood that they collect directly in ant nests (Bitz & Rohe, 1993; Freitag, 1996). The increasing use of insecticides and herbicides, the disappearance of fallow land, the heavy manuring of pastures and meadows, as well as the intensive use of all arable land, are factors which greatly affect ant communities (Petal, 1974, 1976; Nielsen *et al.*, 1991), and thus the food resources of wrynecks.

To halt the decline of wrynecks and protect them efficiently, it is important to understand what affects the availability of ants. Unfortunately, the impact of modern agricultural practices on ant communities (species diversity, nest density, and population size of ant communities) is not known, and needs thorough study. However, before such studies can be undertaken, it is necessary to know how to characterise the availability of ants for wrynecks. Wolda (1990) defines insect availability for birds as "the abundance of potential prey items in microhabitats used by an insectivore when searching for food". Thus, to characterise ant availability for wrynecks, we first of all need to study what are their potential prey, and where and how they collect food.

In this paper, I present the general process we followed to characterise ant availability for wrynecks, and how we measured the food resources in two breeding sites.

Insect and Bird Interactions
© Intercept Ltd., PO Box 716, Andover, Hampshire, SP10 1YG, UK.

This study was conducted in Switzerland, in the Canton du Valais, in two areas representing two different kinds of habitat where wrynecks are regular breeders. The first site (referred to as 'the Hillside') represents the usual breeding habitat of wrynecks, i.e. orchards with standard trees, not intensively used pastures and meadows, hedgerows, and groves. It is situated on a south exposed slope, 1,000–1,200 m high. The second site (referred to as 'the Plain') is located on the plain along the Rhône river at an altitude of 460 m, and is characterised by intensive agriculture. It mainly consists of intensively managed orchards and market gardens. Nest boxes were installed in both study sites to help us with the location of the breeding wrynecks and to have access to the nestlings.

Characterisation of prey availability

METHODS

Nestling diet analysis

We used two different methods to analyse nestling diet: the ligature method (Johnson et al., 1980; Henry, 1982), and photography of the feedings at nests (Bavoux et al., 1993). The first method was used to assess the composition of the nestling diet, whereas the second one provided information on the feeding behaviour of the adult wrynecks.

Ligature method – The ligatures were made of a cotton wire and held tight with a small elastic band (Strebel, 1991; Freitag, 1996). Collars were placed on nestlings for a period of 60 minutes, during which time the parents normally delivered food. After this time, the nestlings were freed of their collar, and the retained prey were removed from their throat. The food samples were stored in 70% alcohol for later identification. We identified and counted all prey items. The main prey, the ants (Hymenoptera: Formicidae), were identified to species. The developmental stage and the caste of each ant was also recorded. The other prey items were identified to Order. The relative abundance of each ant species in the diet of nestlings was expressed as fresh weight of each ant type (Freitag, 1998).

Photography of feedings – The photography of feedings was performed using a camera fitted with a 105 mm lens and a flash (Freitag, 2000). The photographic device was fixed to a tall stake and installed at 1 m from the nest. Upon entering the nest box, adult wrynecks interrupted an infra-red beam and a picture was taken of the bird's head and bill content. We used ISO 100 colour reversal films to get pictures with a fine resolution. The slides were examined under a stereo microscope to identify the collected food. The ants were identified to genus (Freitag, 2000).

Radio-tracking

To assess the foraging territory of wrynecks and their behaviour during the period they fed the nestlings, we tagged adult wrynecks with a transmitter and followed them during their foraging trips. The transmitter was glued to the interscapular region on the back of the birds (Raim, 1978). The feathers in the mounting area were cut and the transmitter glued with a special surgical glue (Skin-Bond®, Smith & Nephew United,

USA). We used a portable receiver and a hand-held antenna to locate the tagged wrynecks in the field. Each bird was followed 1–2 times a day, for 2–3 hours, for 3–7 days (until the transmitter battery had run down, or the transmitter had fallen off). We located each foraging site with an accuracy of 5–10 m, and recorded the kind of habitat used. If there was visual contact with a wryneck, we observed and recorded its foraging behaviour.

RESULTS

Nestling diet analysis

Ligature method – We obtained 128 food samples from 9 wryneck broods on the Hillside, and 103 food samples from 6 broods on the Plain. Almost all (98.5%) of the 65,000 collected prey items were ants of 11 species (*Table 16.1*). The remaining prey were aphids (Hemiptera: Sternorryncha), found in one food sample only. The nestling diet showed great differences between the two study sites (*Table 16.1*). On the Plain, *Lasius niger* was the most important prey brought to nestlings (71% of the food

Table 16.1.   Nestling diet according to the food collected with the ligature method, expressed in per cent of fresh weight of ants. N, number of food samples.

Ant species	Nestling diet (%)	
	Hillside (n = 128)	Plain (n = 103)
Lasius niger	3	71
Lasius alienus	2	
Lasius flavus	11	
Formica fusca	2	
Formica cunicularia		2
Formica rufibarbis	2	
Formica cinerea		24
Tapinoma erraticum	3	
Tetramorium caespitum	76	3
Myrmica schencki	<1	
Myrmica scabrinodis	<1	

Table 16.2.   Nestling diet according to the photography of feedings: per cent of pictures showing each prey type. N, number of photographs.

Ant species	Nestling diet (% feedings)	
	Hillside (n = 1,308)	Plain (n = 701)
Unidentified ants	1	4
Lasius or *Formica*	3	5
Lasius sp.	65	61
Formica sp.	2	3
Tetramorium or *Tapinoma*	8	7
Tetramorium sp.	15	9
Tapinoma sp.	3	1
Myrmica sp.	2	<1
Other invertebrates	3	10
Unidentified food	1	6

expressed in fresh weight), whereas *Tetramorium caespitum* was the most important prey on the Hillside (76% of the food). All food samples (except the one with aphids) contained ant brood (larvae and/or pupae) mixed with adults (mostly workers). Ant brood constituted most (84% of fresh weight) of the food. A detailed list of prey found in the food samples (castes and developmental stages) is given by Freitag (1996).

Photography of feedings – We obtained 1,308 pictures of feedings from 4 wryneck nests on the Hillside, and 701 pictures of feedings from 3 broods on the Plain. On 95% of the photographs, the prey were ants (*Table 16.2*). On the Hillside, as well as on the Plain, the main prey of the nestlings was *Lasius* or *Tetramorium* ants. Almost all (97%) of the feedings with ants also contained ant brood. In most cases (96% of the feedings), wrynecks brought only one prey type per feeding.

Radio-tracking

We monitored 10 radio-tagged birds, 5 on each study site, and recorded 421 foraging sites. Almost all of them (99%) were located less than 200 m from nesting sites. The mean (± s.d.) distance between the foraging sites and the nest was 115 ± 48 m on the Hillside, and 85 ± 46 m on the Plain.

On the Hillside, wrynecks foraged mostly in moderately grazed pastures, unexploited meadows, uncropped areas, and fallow land. On the Plain, they foraged almost exclusively in intensive orchards, and sometimes on the sandy riverbank. On both study sites, they avoided intensively used meadows, land under cultivation, and forests. When collecting food for their young, adult wrynecks moved on the ground, probing the soil with their bill to find ant nests. They preferentially foraged in areas with low vegetation, and patches of bare soil. When they found an ant nest, they picked up as many ants (brood and adults) as they could, and returned to their nest to feed the nestlings. They mostly exploited only one ant nest on each foraging trip, but sometimes came back two or three times to the same ant nest if brood was still available.

DISCUSSION

During the nestling period, wrynecks have a highly specialised diet. They feed their young almost exclusively on ants. They collect various ant species, but most of these ants show some common characteristics: they are small- to medium-sized species (3–7 mm for the workers), live in the soil or in earth mounds, are common species, and inhabit grassy areas like meadows, pastures, and uncultivated land. The main prey of the nestlings in all studied broods was always *Lasius* species and/or *Tetramorium caespitum*, as has been observed in all previous studies on the diet of wrynecks (König, 1961; Klaver, 1964; Dornbusch, 1968; Bitz & Rohe, 1993).

The omnipresence of ant brood in the food meant for nestlings indicates that wrynecks always collect their prey directly in ant nests. This foraging strategy affords the advantages that the prey are very numerous, with hundreds to tens of thousands of ants per ant nest (Brian & Elmes, 1974; Seifert, 1996), and that wrynecks benefit from the presence of ant brood, which is easy to collect and of a greater nutritional value than adult workers with their heavily sclerotised cuticle. Foraging in habitats with low vegetation and patches of bare soil seems to facilitate the detection of ant nests and the collection of prey. In contrast, areas with high and dense vegetation appear unsuitable for wrynecks.

Thanks to these data on nestling diet and foraging behaviour, we were able to define what we would consider as food available for wrynecks. The wrynecks' potential prey are ants in the nests of *Lasius*, *Formica* (sub-genus *Serviformica* only), *Tapinoma*, *Tetramorium* and *Myrmica* that are located in open habitats situated less than 200 m from the wryneck nest. Aphids and other plant lice (Hemiptera: Sternorryncha), even if sometimes collected by wrynecks, were not considered as potential prey because of their scarcity in the diet of the nestlings. This characterisation of prey availability made it possible for us to quantify the food resources in the foraging territories of the birds (see below).

Measure of prey availability

METHODS

We quantified prey availability for wrynecks in their foraging territories by measuring the abundance of nests of each potential prey ant species. On both study sites, we sampled 4 m² quadrats in habitats considered as suitable foraging sites for wrynecks, i.e. intensive orchards, pastures, uncropped areas, not intensively used meadows, etc. We did not sample forests, land under cultivation, or private gardens, as wrynecks were never seen foraging in such habitats. We carefully inspected each quadrat to find and count all ant nests. We excluded nests situated under heavy stones or in wood, which were considered as being inaccessible to wrynecks. Worker ants were collected for identification.

To estimate the available food resources within a foraging territory, we calculated the approximate total number of available ant nests. The calculation involved multiplying the surface area of all kinds of foraging habitats in a foraging territory by the mean density of ant nests observed in each kind of habitat. The foraging territory was defined as all suitable foraging habitats (as defined in the former paragraph) situated less than 200 m from a nesting site.

RESULTS

We sampled 268 quadrats (138 on the Hillside, and 130 on the Plain), and recorded 959 ant nests (*Table 16.3*). Out of the 24 ant species we recorded, 17 were potential prey. The diversity of potential prey was higher on the Hillside than on the Plain, with respectively 15 and 11 species. About 95% of the ant nests belonged to species which are prey for wrynecks. The most abundant ants were *Lasius* species and *Tetramorium caespitum*. Together, they accounted for more than two-thirds of the ant nests found.

Depending on the study site and the kind of habitat, the mean density of ant nests reached 0.4 to 1.2 nests/m² (*Table 16.4*). We found great differences in the species richness and density of ant nests from one quadrat to another.

The estimated food availability in one foraging territory reached 33,000 ant nests on the Plain, and 68,000 on the Hillside (*Table 16.4*).

DISCUSSION

Having sufficient food resources for breeding birds depends on two main factors: (a)

Table 16.3.   Ant species recorded at the two study sites and the number of ant nests observed. *non-prey species (= never mentioned as prey of wrynecks in the literature). N, number of 4 m² quadrats sampled.

Ant species	Number of ant nests	
	Hillside (n = 138)	Plain (n = 130)
Lasius niger	8	215
Lasius alienus	169	4
Lasius flavus	106	20
Lasius affinis	2	
*Lasius fuliginosus**	1	
*Formica sanguinea**	3	
Formica fusca	1	
Formica cunicularia		2
Formica rufibarbis	8	
Formica cinerea		31
*Camponotus ligniperda**	16	
*Camponotus herculeanus**	2	
Tapinoma erraticum	70	3
Tetramorium caespitum	101	4
Myrmica sabuleti	61	3
Myrmica scabrinodis	57	5
Myrmica schencki	23	
Myrmica rugulosa	4	4
Myrmica laevinodis	1	1
Myrmica ruginodis	1	
*Solenopsis fugax**	7	23
Leptothorax sp.	1	
*Manica rubida**		1
*Ponera coarctata**		1
Total number of ant nests	642	317
Ant nests of prey species	613	292
Total number of species	20	14
Number of prey species	15	11

the prey species must be present in the foraging territory in habitats used by the birds when searching for food; and (b) these prey must be sufficiently abundant (Hutto, 1990). These two conditions seem to be fulfilled for the wrynecks in both study sites, since they succeed in raising their broods.

Table 16.4.   Mean density of ant nests (wrynecks' potential prey species only) according to the study site and the kind of habitat, and estimated prey availability within a wryneck's foraging territory.

Study site	Habitats	Ant nest density (nest/m² ± s.d.)	Mean surface area of each habitat in a foraging territory	Estimated number of ant nests in one foraging territory
Hillside	Pastures	1.0 ± 0.9	29,300 m²	
	Meadows	1.0 ± 1.1	30,700 m²	
	Unexploited meadows	0.6 ± 0.3	3,800 m²	⟹ 68,000 nests
	Fallow land, grassy areas along ways	1.2 ± 1.2	4,700 m²	
Plain	Intensive orchards	0.6 ± 0.6	40,800 m²	
	River bank	0.4 ± 0.3	5,600 m²	⟹ 33,000 nests
	Fallow land, grassy areas along ways	0.6 ± 0.9	10,200 m²	

However, the two study sites show very different patterns in ant nest availability. On the Hillside, the estimated food resources per foraging territory are twice as high as those on the Plain (68,000 ant nests versus 33,000). Interestingly, from the wrynecks' point of view, ant availability seems to be lower on the Hillside than on the Plain. The mean brood size on the Hillside is 5.3 nestlings/brood, whereas it reaches 6.7 nestlings/brood on the Plain (Freitag, 1998), and the wrynecks on the Hillside seem to experience greater difficulties in finding food, as shown by their longer foraging trips. This may suggest that the food resources available to wrynecks are, in fact, lower on the Hillside than on the Plain.

These contradictory observations indicate that our method of measuring food availability is not entirely reliable. We have measured the number of ant nests available, but we have not taken into account three important factors: (a) the size of the ant societies; (b) the ant nests' accessibility; and (c) the possible preference or avoidance of some ant species by wrynecks. For this last factor, the wrynecks do not seem to be very selective between the ant species on which they prey. The main prey in the wrynecks' diet (*Lasius* and/or *Tetramorium* according to the study site) are also the more abundant ant species in the foraging territories (*Tables 16.1, 16.2* and *16.3*). The size of the ant societies plays a more important role in the food availability. This factor was not taken into account because of the difficulty of appraising population size, but a small ant nest with 100 workers is not equivalent to a large one with 50,000 individuals. Not only does the larger one contain much more food, but also the probability that a wryneck will detect such an ant nest is higher. The third factor we have not taken into account is the accessibility of the ant nest. The wrynecks not only need an abundant supply of ant nests, but also need the soil and vegetation structure to be such that these are readily available (Hutto, 1990). This could explain why the wrynecks seem to experience higher food availability on the Plain than on the Hillside. In the intensive orchards, the vegetation is very sparse (because of the use of herbicides), making it easy for the birds to walk on the ground, and the muddy soil of the Rhône Plain is easy to dig. In such areas, ant brood is more accessible than in pastures with low, but dense vegetation and hard soil, or in meadows where ant nests are hidden in tall plant tufts. On the Plain, the lower prey diversity and density seem to be compensated by the higher accessibility of the ant nests.

Thus, our previous definition of food availability, based solely on the presence of prey species in foraging habitats, must be extended. Food accessibility (i.e. the possibility for wrynecks to find and access the ant nests, and to collect ant brood) and the size of ant societies must also be taken into account. For wrynecks, prey diversity is not a restrictive factor for food availability. *Lasius*, *Tetramorium*, *Formica*, and *Myrmica* ants are present and dominant in almost all open habitats (Seifert, 1986, 1996). Thus, the prey species will be present in almost all potential breeding habitats of wrynecks. And a low prey density, as observed on the Plain, does not seem to affect the feeding of the nestlings. The ant nests' accessibility is more crucial.

In further studies, it will be a rewarding challenge to find a way to quantify prey accessibility besides ant nest abundance. This will be decisive in understanding how agricultural practices affect food availability for wrynecks.

Acknowledgements

This research was financially supported by the Centre Suisse de Cartographie de la Faune (Neuchâtel), the Société Académique Vaudoise, the Murithienne (Société Valaisanne des Sciences Naturelles) and the Société Vaudoise d'Entomologie. I thank Professor Daniel Cherix for his great help during this work, and Arnaud Maeder for assistance with fieldwork. Use of the ligature method and tagging of wrynecks were approved by the Swiss Ornithological Institute and the concerned services. I am grateful to Professor Helmut van Emden for critically reading the manuscript and correcting my English.

References

Bavoux, C., Burneleau, G., Juillard, M. & Nicolau-Guillaumet, P. (1993) Le hibou petit-duc *Otus scops* sur l'île d'Oléron (France). Régime alimentaire des poussins. *Nos Oiseaux* **42**, 159–170.

Bitz, A. & Rohe, W. (1993) Nahrungsökologische Untersuchungen am Wendehals (*Jynx torquilla*) in Rheinland-Pfalz. *Beihefte zu den Veröffentlichungen für Naturschutz und Landschaftspflege in Baden-Württemberg* **67**, 83–100.

Brian, M. V. & Elmes, G. W. (1974) Production by the ant *Tetramorium caespitum* in a southern English heath. *Journal of Animal Ecology* **43**, 889–903.

Dornbusch, M. (1968) Zur Nestlingsnahrung des Wendehalses. *Der Falke* **15**, 130–131.

Freitag, A. (1996) Le régime alimentaire du torcol fourmilier (*Jynx torquilla*) en Valais (Suisse). *Nos Oiseaux* **43**, 497–512.

Freitag, A. (1998) Analyse de la disponibilité spatio-temporelle des fourmis et des stratégies de fourragement du torcol fourmilier (*Jynx torquilla* L.). PhD Thesis, University of Lausanne, Switzerland.

Freitag, A. (2000) La photographie des nourrissages: une technique originale d'étude du régime alimentaire des jeunes torcols fourmiliers *Jynx torquilla*. *Alauda* **68**, 81–93.

Henry, C. (1982) Etude du régime alimentaire des passereaux par la méthode des colliers. *Alauda* **50**, 92–107.

Hutto, R. L. (1990) Measuring the availability of food resources. *Studies in Avian Biology* **13**, 20–28.

Johnson, E. J., Best, L. B. & Heagy, P. A. (1980) Food sampling biases associated with the 'ligature method'. *The Condor* **82**, 186–192.

Klaver, A. (1964) Waarnemingen over de biologie van de Draaihals. *Limosa* **37**, 221–231.

König, C. (1961) Beobachtungen an einer Brut des Wendehalses (*Jynx torquilla* L.). *Anzeiger der Ornithologischen Gesellschaft in Bayern* **6**, 81–83.

Nielsen, M. G., Peakin, G. J. & Wright, P. J. (1991) The effect of perturbation on *Lasius flavus* communities in grassland ecosystems. In: *Terrestrial and aquatic ecosystems: perturbation and recovery* (ed. O. Ravero), pp. 284–288. Ellis Horwood, London, UK.

Petal, J. (1974) Analysis of a sheep pasture ecosystem in the Pieniny Mountains (The Carpathians). XV. The effect of pasture management on ant population. *Ekologia Polska* **22**, 679–692.

Petal, J. (1976) The effect of mineral fertilization on ant populations in meadows. *Polish Ecological Studies* **2**, 209–218.

Raim, A. (1978) A radio transmitter attachment for small passerine birds. *Bird Banding* **49**, 326–332.

Scherner, E. R. (1980) *Jynx torquilla* Linnaeus 1758 – Wendehals. In: *Handbuch der Vögel Mitteleuropas* (eds. U. N. Glutz von Blotzheim & K. M. Bauer), pp. 881–916. Akademische Verlagsgesellschaft, Wiesbaden, Germany.

Scherner, E. R. (1989) Wendehals und Populationsbiologie: der "Vogel des Jahres 1988" und die Pflicht zur Forschung. *Laufener Seminarbeiträge* **3**, 24–39.

Seifert, B. (1986) Vergleichende Untersuchungen zur Habitatswahl von Ameisen (Hymenoptera: Formicidae) im mittleren und südlichen Teil der DDR. *Abhandlungen und Berichte des Naturkundemuseum Görlitz* **59**, 1–124.

Seifert, B. (1996) *Ameisen. Beobachten, bestimmen.* Naturbuch Verlag, Augsburg, Germany. 351 pp.

Strebel, S. (1991) Bruterfolg und Nahrungsökologie der Dohle *Corvus monedula* im Schloss Murten FR. *Der Ornithologische Beobachter* **88**, 217–242.

Wolda, H. (1990) Food availability for an insectivore and how to measure it. *Studies in Avian Biology* **13**, 38–43.

17
Relationship Between Chick Diet and Breeding Performance of Great Tits in a Caterpillar-Poor Environment

EMILIO BARBA, JOSÉ A. GIL-DELGADO AND JUAN S. MONRÓS

Instituto 'Cavanilles' de Biodiversidad y Biología Evolutiva & Unidad de Ecología, Universidad de Valencia, E-46100 Burjassot, Valencia, Spain

Introduction

It is well known that great tits (*Parus major*) feed their chicks preferentially with caterpillars (e.g. Perrins, 1979). This preference is so marked that the breeding season of the species is timed so that the peak in food demand by the chicks roughly coincides with the peak of caterpillar availability (van Balen, 1973; Perrins, 1991). Moreover, broods raised when caterpillars are scarce have a low breeding success (Lack, 1958; van Balen, 1973), and those populations that breed in habitats with low caterpillar abundance, such as cities (Berressem *et al.*, 1983), suburban gardens (Cowie & Hinsley, 1988), or evergreen forests (Belda *et al.*, 1998), are less productive than those breeding in broad-leaved forests, which are caterpillar-rich habitats.

Using 'caterpillar richness' in the chick diet as a criterion, orange plantations might be considered a suboptimal habitat for great tits, since caterpillars amount only to 26% of the chick food in this habitat (Barba & Gil-Delgado, 1990). Our aim here is to look in detail at the composition of the diet of particular nests in orange plantations, and relate this with the fate of the breeding attempt. The expectation is that the breeding success of particular nests will decrease as the proportion of caterpillars in the chick diet decreases.

Study area and methods

The study area was located within an extensive orange monoculture near Sagunto, eastern Spain (39°42'N, 0°15'W, 30 m above sea level). The area was composed of small groves separated by irrigation canals and narrow paths. The tree distribution and plant species in the field layer are given by Gil-Delgado *et al.* (1979). Wooden nesting boxes have been available in this area since 1986, and great tits occupy most of them each year (e.g. Barba *et al.*, 1995).

Insect and Bird Interactions
© Intercept Ltd., PO Box 716, Andover, Hampshire, SP10 1YG, UK.

During 1988, neck collars were fitted to the chicks of some nests to obtain food samples (Barba & Gil-Delgado, 1990). Collars were fitted to 2- to 13-day-old chicks, and simultaneously to all the chicks present in the nest (*Plate 3*). No nest was sampled more than once a day. The collars were typically applied for two-hour periods, though one-hour periods were chosen in very young broods, or under unfavourable weather conditions. The food collected in each nest (*Plates 4* and *5*) was kept in individual vials containing 70% alcohol. Nest box number and sampling date were noted on each vial.

For this study, we used 17 nests, 13 first and 4 second broods, in which at least 14 prey items were collected (range 14–59). The average number of days of sampling in these nests was 6.3 (s.d. = 1.4, range 4–10), so it is not possible that the prey of any one nest belonged only to one or two 'runs' brought consecutively to the nest from one place by the parents (see e.g. Cowie & Hinsley, 1988). Laying and hatching dates, clutch size, and the number of hatchlings and fledglings were recorded for each nest in routine visits.

Adult Lepidoptera (hereafter referred to as 'moths') were identified to species, where possible, using the genitalia and following Pierce (1942, 1967) and Calle (1982). Since *Peridroma saucia* and *Noctua pronuba* were the most abundant moth species, these were treated separately in the analyses, while the other 10 species (*Agrotis segetum, A. exclamationis, A. ipsilon, Xestia c-nigrum, Mythimna unipuncta, M. loreyi, Spodoptera littoralis, Athetis hospes, Autographa gamma* and *Hoplodrina ambigua*) were grouped as 'other moths'. Other prey were grouped less specifically into 'caterpillars', Lepidoptera pupae ('pupae' hereafter), 'spiders' and 'other prey' (including Hymenoptera, Coleoptera, Orthoptera, egg cocoons, and some pieces of oranges). Snail shells and sand brought to the chicks were not included in the results presented here. More details are available in Barba & Gil-Delgado (1990). Each of these seven prey classes amounted to at least 5% of the chick diet.

According to the number of prey of each of the seven classes, the nests were clustered by group average, using the chord distance as the index of resemblance (see e.g. Ludwig & Reynolds, 1988).

As many prey classes as we could identify as different (n = 28) were used to calculate the diversity of the diet (Shannon index, H'). Differences between diversity indices were tested following Magurran (1988, pp. 35–36).

Results

The dendrogram resulting from the cluster analysis (*Figure 17.1*) allowed us to differentiate three groups of nests, the food and breeding data of which are summarised in *Table 17.1*. For each of the three groups, there was a major prey type which represented more than 40% of all prey items delivered to the nests, while the second in abundance never reached 25%. The Kolmogorov–Smirnov test was used to determine the significance of differences in the major prey brought to each group of nests. 'Caterpillars' were significantly more abundant than *P. saucia* or 'pupae' in the nests of Group I (P <0.05), *P. saucia* was more abundant than 'other moths' in Group II (P <0.01), and 'other moths' were more abundant than *P. saucia* or 'pupae' in Group III (P <0.05; Kolmogorov–Smirnov two sample test, one-tailed). Thus, the major prey for each group as a whole was consistently the most abundant in the nests

Table 17.1.   Breeding and food data of the three groups of great tits resulting from the cluster analysis. Numbers in brackets are standard deviations.

	Group I	Group II	Group III
Breeding data			
Number of nests	7	6	4
Mean laying date	14.57 (3.05)	27.33 (10.88)	53.75 (3.50)
Mean clutch size	8.29 (1.11)	7.83 (0.98)	7.00 (1.15)
Mean number of hatchlings	7.43 (1.62)	6.50 (1.64)	6.25 (2.36)
Mean number of fledglings	6.86 (1.46)	6.00 (1.67)	2.25 (0.50)
Breeding success (%)	82.8	76.6	32.1
Partial losses (%)	9.4	8.0	64.0
Food data			
% Caterpillars	42.9	12.9	5.5
% Pupae	14.1	14.4	15.6
% *Peridromia saucia*	14.1	43.2	15.6
% *Noctua pronuba*	7.3	3.8	4.4
% Other moths	8.8	21.2	41.1
% Spiders	6.3	1.5	7.8
% Other prey	6.3	3.0	10.0
Total prey	204	129	86
Diversity (H')	1.89	1.91	2.63

belonging to that group. This allowed us to consider the nests of each group as sufficiently homogeneous in relation to the major prey type brought to the chicks, with the major prey being different in the three groups.

Groups I and II were respectively composed of early and late first clutches, while the four second clutches formed Group III (*Table 17.1*). We tested for differences in partial losses (hatchlings that died in the nest) and breeding success (proportion of eggs producing fledglings) among the three groups using 2 × 2 contingency tables for the number of eggs (for breeding success) or hatchlings (for partial losses) which succeeded or failed in producing fledglings. There were no significant differences either for partial losses ($\chi^2 = 0.01$, n.s.) or breeding success ($\chi^2 = 0.62$, n.s.) between Groups I and II. By contrast, partial losses were significantly higher, and breeding success significantly lower, in Group III than in the other two (partial losses, I–III,

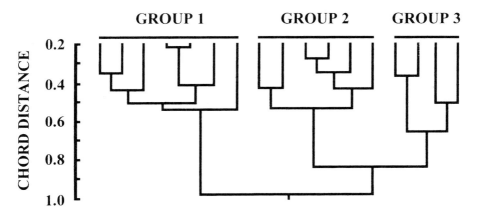

Figure 17.1.   Dendrogram showing the similarity among the nestling diet of 17 different great tit broods.

$\chi^2 = 27.84$; II–III, $\chi^2 = 23.14$; breeding success, I–III, $\chi^2 = 21.65$; II–III, $\chi^2 = 14.55$; $P < 0.001$ in all cases).

The abundance of caterpillars in the chick diet decreased from Group I to Group III. The proportion of moths (including *P. saucia*, *N. pronuba*, and 'other moths') increased from 30% in Group I to more than 60% in Groups II and III. In spite of the differences in the main prey consumed by chicks of Groups I and II, the diversity of the diet did not differ ($t_{260} = 0.15$, n.s.). The diet of Group III was significantly more diverse than that of Group I ($t_{258} = 6.7$, $P < 0.001$) and Group II ($t_{209} = 5.4$, $P < 0.001$). This is so because the main prey type in Group III ('other moths') was composed of at least ten moth species.

Discussion

Initially, the large contribution of moths to the chick diet was considered as a probable cause for the low clutch sizes of the great tit population studied here when compared with central and northern European populations (Barba & Gil-Delgado, 1990). We thought that orange plantations were a caterpillar-poor environment, and this restricted the number of chicks that a pair could raise in any breeding attempt, and consequently led to smaller clutch sizes. However, further information from other populations has shown that the clutch size in Sagunto is similar, or even higher than that of other Mediterranean populations (Belda *et al.*, 1998). Therefore, the lower clutch size observed is unlikely to be a consequence of the chick diet, but seems to be characteristic of populations living in Mediterranean areas. The present paper examined whether differences between nests in the nestling diet within a population affect the reproductive success and, specifically, whether chicks mostly fed with moths survived as well as those mostly fed with caterpillars.

In the population studied, some great tit pairs had to raise their first brood when caterpillars contributed as few as 13% to the chick diet (Group II). They did this quite successfully, as only 8% of the chicks starved. We think that this is explained by the high consumption of *Peridroma saucia* (43% of the prey brought). Adult Lepidoptera provide more energy per wet weight than caterpillars (Eguchi, 1980; Sage, 1982), but they are usually much less abundant and probably need a longer handling time, since wings and legs are usually removed (Barba *et al.*, 1996). This trade-off seems balanced only by a short search time, in other words, by concentrating predation mainly on one, probably very abundant, species. When caterpillars became less abundant in the diet, and moth diversity increased, mortality of the young also increased (Group III). Moths are preyed upon when they rest during the day, and the resting place is more or less species-specific, or even morph-specific (Boardman *et al.*, 1974; Kettlewell & Conn, 1977). The increase in the number of places to search, and the probable low densities of other species, seems to become uneconomic for the birds, as has been shown by Török & Ludvig (1988) for blackbirds (*Turdus merula*) (see also Royama, 1970; Blondel *et al.*, 1991). That chick mortality increases when the diversity of prey brought increases has also been noticed by Michelland (1982) with tit populations.

We do not know to what extent the relationship between chick diet and chick survival is causal. However, it is worth remembering that Group III was composed of second clutches, that is, these pairs had successfully raised a first brood. Thus, differences in the foraging ability of the parents, or the 'quality' of the territory *per se* are unlikely explanations for the differences in chick mortality. Therefore, the increase in chick mortality, caused by starvation, may be a consequence of a change in the pool of available prey species, probably accompanied by an absolute decrease in prey numbers.

Acknowledgements

We would like to thank J. A. López for helping in the field and laboratory work, and J. Baixeras for his advice on the determination of Lepidoptera.

References

Barba, E. & Gil-Delgado, J. A. (1990) Seasonal variation in the nestling diet of the great tit *Parus major* in orange groves in eastern Spain. *Ornis Scandinavica* **21**, 296–298.

Barba, E., Gil-Delgado, J. A. & Monrós, J. S. (1995) The costs of being late: consequences of delaying great tit *Parus major* first clutches. *Journal of Animal Ecology* **64**, 642–651.

Barba, E., López, J. A. & Gil-Delgado, J. A. (1996) Prey preparation by adult great tits *Parus major* feeding nestlings. *Ibis* **138**, 532–538.

Belda, E. J., Barba, E., Gil-Delgado, J. A., Iglesias, D. J., López, G. M. & Monrós, J. S. (1998) Laying date and clutch size of great tits (*Parus major*) in the Mediterranean region: a comparison of four habitat types. *Journal für Ornithologie* **139**, 269–276.

Berressem, K. G., Berressem, H. & Smith, K.-H. (1983) Vergleich der Brutbiologie von Höhlenbrütern in innerstädtischen und stadtfernen Biotopen. *Journal für Ornithologie* **124**, 431–445.

Blondel, J., Dervieux, A., Maistre, M. & Perret, P. (1991) Feeding ecology and life history variation of the blue tit in Mediterranean deciduous and sclerophyllous habitats. *Oecologia* **88**, 9–14.

Boardman, M., Askew, R. R. & Cook, L. (1974) Experiments on resting site selection by nocturnal moths. *Journal of Zoology* **172**, 343–355.

Calle, J. A. (1982) *Noctuidos Españoles*. Ministerio de Agricultura, Pesca y Alimentación, Madrid, Spain. 430 pp.

Cowie, R. J. & Hinsley, S. A. (1988) Feeding ecology of great tits (*Parus major*) and blue tits (*Parus caeruleus*) breeding in suburban gardens. *Journal of Animal Ecology* **57**, 611–626.

Eguchi, K. (1980) The feeding ecology of the nestling great tit *Parus major minor* in the temperate ever-green broad-leaved forest. I. Food consumption and maintenance cost. *Journal of the Yamashina Institute of Ornithology* **12**, 10–20.

Gil-Delgado, J. A., Pardo, R., Bellot, J. & Lucas, I. (1979) Avifauna del naranjal valenciano. II. El gorrión común (*Passer domesticus* L.). *Mediterránea Serie de Estudios Biologicos* **3**, 69–99.

Kettlewell, H. B. D. & Conn, D. L. T. (1977) Further background-choice experiments on cryptic Lepidoptera. *Journal of Zoology* **181**, 371–376.

Lack, D. (1958) A quantitative breeding study of British tits. *Ardea* **46**, 91–124.

Ludwig, J. A. & Reynolds, J. F. (1988) *Statistical ecology*. Wiley, New York, USA. 337 pp.

Magurran, A. E. (1988) *Ecological diversity and its measurement*. Croom Helm, London, UK. 179 pp.

Michelland, D. (1982) Survie en millieu insulaire: quelle stratégie? Le cas des Mésanges en Corse. *Revue d'Ecologie (La Terre et la Vie)* **36**, 187–210.

Perrins, C. M. (1979) *British tits*. Collins, London, UK. 304 pp.

Perrins, C. M. (1991) Tits and their caterpillar food supply. *Ibis* **133** (supplement 1), 49–54.

Pierce, F. N. (1942) *The genitalia of the group Noctuidae of the Lepidoptera of the British Islands (female)*. Classey, Twickenham, UK. 62 pp. + 15 plates.

Pierce, F. N. (1967) *The genitalia of the group Noctuidae of the Lepidoptera of the British Islands (male)*. Classey, Twickenham, UK. 88 pp. + 32 plates.

Royama, T. (1970) Factors governing the hunting behaviour and selection of food by the great tit (*Parus major* L.). *Journal of Animal Ecology* **39**, 619–668.

Sage, R. D. (1982) Wet- and dry-weight estimates of insects and spiders based on the length. *American Midland Nauralist* **108**, 407–411.

Török, J. & Ludvig, E. (1988) Seasonal changes in foraging strategies of nesting blackbirds (*Turdus merula* L.). *Behavioral Ecology and Sociobiology* **22**, 329–333.

van Balen, J. H. (1973) A comparative study of the breeding ecology of the great tit *Parus major* in different habitats. *Ardea* **61**, 1–93.

PART 4

Ectofauna

18
Avian Defences Against Ectoparasites

BRETT R. MOYER AND DALE H. CLAYTON

Department of Biology, University of Utah, Salt Lake City, UT 84112, USA

Introduction

Birds are infested by a variety of detrimental ectoparasites (Janovy, 1997). A good deal of information concerning the impact of ectoparasites on birds has been summarised in several recent reviews (Loye & Zuk, 1991; Lehmann, 1993; Clayton & Moore, 1997). These reviews show that ectoparasites can have severe effects on birds, selecting for efficient host defences. Unfortunately, the literature on avian defences against ectoparasites is scattered. The first goal of this chapter is to provide a comprehensive catalogue of these defences. The second goal is to critically evaluate the evidence supporting the defensive function of each trait. The third and final goal is to argue that future research should investigate how defensive traits interact, rather than continuing to focus on each defensive trait in isolation. Studying defences in isolation can lead to misleading interpretations regarding their adaptive function.

Birds are exploited by a diverse community of ectoparasites, including insects such as lice (Phthiraptera), fleas (Siphonaptera), true bugs (Hemiptera), and flies (Diptera) (Marshall, 1981), as well as ticks and other mites (Acari) (Walter & Proctor, 1999). These taxa vary in the resources they exploit (skin, feathers, blood, etc.), and in the intimacy of their association. For example, feather lice (Phthiraptera: Ischnocera) specialise on feathers and spend their entire life cycle on the bird. Other taxa, like flat flies (Diptera: Hippoboscidae) and swallow bugs (Hemiptera: Cimicidae), feed on blood and are more transient parasites. In this review, we focus on more 'permanent' parasites, such as lice. Generally speaking, defences against permanent parasites are likely also to help defend against more ephemeral parasites.

Many recent studies have demonstrated detrimental effects of ectoparasites on avian fitness (Møller et al., 1990; Loye & Zuk, 1991; Lehmann, 1993; Clayton & Moore, 1997). Ectoparasites can influence several components of host reproductive success. These components include nest desertion (Moss & Camin, 1970; Duffy, 1983; Oppliger et al., 1994), as well as reductions in mating success (Andersson, 1994; Hillgarth & Wingfield, 1997), clutch size (Møller, 1993), hatching success (Oppliger et al., 1994; Clayton & Tompkins, 1995), nestling survival (Møller, 1987; Shields & Crook, 1987; Richner et al., 1993), and fledgling success (Clayton & Tompkins, 1995). Adult survival can also be reduced by ectoparasites (Brown et al.,

1995; Clayton *et al.*, 1999). Owing to these detrimental effects, avian traits that minimise infestations have a selective advantage.

Birds combat ectoparasites in a variety of ways. All else being equal, the simplest way to minimise the detrimental effects of ectoparasites is to avoid them in the first place. Once infested, however, various defensive mechanisms exist to minimise the detrimental impact. These mechanisms include morphological barriers, immune responses, and behavioural defences. This review will focus mainly on morphological and behavioural defences against ectoparasites. Wikel (1996) and Wakelin & Apanius (1997) provide recent reviews of immune defences in birds.

Habitat choice

One defence against parasites is to occupy environments that are relatively free from parasites. For example, avian blood parasites (Haematozoa) are scarce on open tundra (Greiner *et al.*, 1975; Bennett *et al.*, 1992), in arid regions (Little & Earle, 1995; Tella, 1996), and at high elevations (van Riper III *et al.*, 1986). The main reason for the paucity of Haematozoa in these habitats is low abundance of vectors, such as mosquitoes.

Another defence may be to occupy environments too extreme for the survival of the parasites themselves. For instance, the abundance of fleas on dogs in Egypt is positively correlated with the relative humidity of the environment (Amin, 1966). Similarly, in a comparison of the feather lice on birds in habitats ranging from desert to rainforest, Moyer *et al.* (*in press*) found a positive correlation between louse prevalence (% of individuals infested) and relative humidity (*Figure 18.1*). Inca doves (*Columbina inca*) and mourning doves (*Zenaida macroura*) were sampled from the Sonoran Desert (Arizona, USA) and from the more humid Rio Grande Valley (Texas, USA). Only 3% of Arizona birds (n = 346) had lice, compared to 80% of the Texas birds (n = 62) (Moyer *et al.*, 2002). Moyer *et al.* (2002) further showed experimentally that lice cannot survive on birds kept at low relative humidity (<35% RH).

Avoidance of parasites also operates on a more local scale. For example, quite a few studies have now shown that birds avoid nesting or roosting in sites that are infested with ectoparasites (Christe *et al.*, 1994; Oppliger *et al.*, 1994, Merilä & Allandar, 1995; Merino & Potti, 1995; Hart, 1997; Loye & Carroll, 1998; Rytkonen *et al.*, 1998; Thompson, 1999). Recent evidence further demonstrates that the microclimate of nest cavities influences colonisation by ectoparasites. Heeb *et al.* (2000) manipulated the humidity of uninfested nest boxes of blue tits (*Parus caeruleus*). More hen fleas (*Ceratophyllus gallinae*) subsequently colonized the dry nests than humid nests.

Mate choice

Another anti-parasite defence is for the members of one sex (often females) to choose mates that are parasite-free. Since the publication of Hamilton & Zuk's (1982) seminal paper, parasite-mediated mate choice has been a topic of widespread interest and research (Hillgarth & Wingfield, 1997). Hamilton & Zuk argued that parasite-free males are more likely to carry genes for resistance to parasites, genes that will be passed on to a choosy female's offspring. According to the Hamilton–Zuk hypothesis,

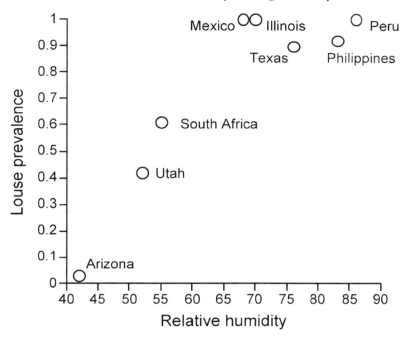

Figure 18.1.   Louse prevalence (% birds infested) plotted against the average annual relative humidity near the site of capture (n = 1,295 birds). Sampling localities are as follows: Tucson, Arizona, USA; Salt Lake City, Utah, USA; Free State and Mpumalanga, South Africa; Campeche, Mexico; Manteno, Illinois, USA; Weslaco, Texas, USA; Cagayan de Oro City, Philippines; and near Manu, Peru (Moyer *et al.*, *in press*).

females choose resistant males on the basis of secondary sexual traits whose full expression depends on health and vigour. Examples of such traits are brightly coloured plumage that is subject to fading, or vigorous courtship displays that parasitised individuals cannot perform. Hamilton & Zuk argued that, over time, sexual selection will lead to the elaboration of parasite-indicative traits that improve the ability of females to identify resistant males.

The Hamilton–Zuk hypothesis is a 'good-genes' model of sexual selection which assumes that choice of resistant males benefits females indirectly through the inheritance of resistance by offspring. Parasite-mediated mate choice could also yield more direct fitness benefits. For example, females might choose unparasitised males simply to avoid the direct transmission of parasites to themselves and/or their offspring (Able, 1996). Females might also benefit directly from the choice of unparasitised males if they require a healthy mate to provide resources such as parental care (Milinski & Bakker, 1990). Further information on this intriguing subject, including work involving birds and ectoparasites, can be found in recent reviews by Andersson (1994) and Hillgarth & Wingfield (1997).

Plumage as a barrier

FEATHER TOUGHNESS

A tough integument could conceivably be another defence against ectoparasites. Literally 'having a thick skin' might deter blood-feeding ectoparasites, although we know of no data relevant to this hypothesis. Tough plumage could also deter feather-feeding ectoparasites, analogous to foliage containing cellulose which helps deter feeding by herbivores (Howe & Westley, 1988). Some recent work suggests that feather toughness may be an important defence against ectoparasites.

Feathers that contain melanin are known to be more resistant to mechanical abrasion than feathers without this pigment (Burtt, 1986; Bonser, 1995). Two recent studies suggest that melanin may also limit damage by feather-feeding lice (Kose & Møller, 1999; Kose et al., 1999). Kose and colleagues studied the interaction between the barn swallow (*Hirundo rustica*) and its louse (*Hirundoecus malleus*), which chews holes in the host's tail feathers (Møller, 1991). The authors showed that this louse feeds more on white than dark regions of the tail, resulting in more extensive damage to white regions. Interestingly, this damage may be used as a cue by female swallows for detecting and avoiding lousy males as mates. The results of these studies are consistent with the hypothesis that melanin discourages feeding by lice. However, a direct test of this hypothesis has yet to be conducted.

Further tests of the role of feather toughness as a defence against ectoparasites are needed. It should be relatively easy to test the influence of feather pigments on the feeding and population ecology of ectoparasites. It is intriguing to speculate on the generality of possible relationships between feather-feeding ectoparasites and the plumage colours of birds. For example, how widespread is parasite-indicative plumage, such as the white tail spots of barn swallows, that reveal louse damage to discriminating mates? Are some species of birds black because they are parasitised by more species of feather-feeding ectoparasites than are white species? Are other features of feather structure, including pigments other than melanin, important in preventing ectoparasite damage? These and other questions relevant to feather toughness have received little attention.

FEATHER TOXICITY

Somewhat more attention has been devoted to the possibility that the toxic feathers or skin (Dumbacher & Pruett-Jones, 1996) of some birds may protect them against ectoparasites. For example, the feathers and skin of several species in the genus *Pitohui* contain homobatrachotoxin, the neurotoxin found in the skin of poison dart frogs (Dumbacher et al., 1992). Although this toxin probably plays some role in deterring predators of *Pitohui*, recent evidence suggests that it deters ectoparasites also (Mouritsen & Madsen, 1994; Poulsen, 1994; Dumbacher, 1999). Dumbacher (1999) conducted a series of Petri dish trials in which he exposed feather lice from a variety of bird species to *Pitohui* feathers and the feathers of non-toxic birds. He found that, given a choice, lice avoided feeding or resting on *Pitohui* feathers. Furthermore, lice on *Pitohui* feathers showed higher mortality than lice on non-toxic feathers. Since homobatrachotoxin affects a wide range of invertebrates (Dumbacher, 1999), it may deter a range of ectoparasites.

Body maintenance behaviour

Avian body maintenance includes grooming, dusting, sunning, and anting (Cotgreave & Clayton, 1994). Grooming behaviour, defined as preening and scratching combined

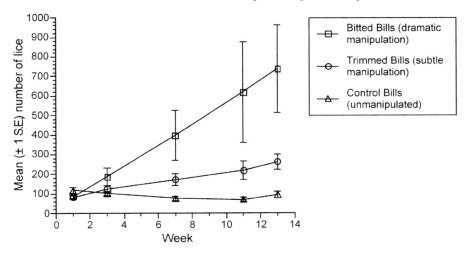

Figure 18.2.   Change-over time in the number of lice on feral pigeons with bitted bills (n = 18), the bill overhang trimmed (n = 19), or unmanipulated bills (n = 18). Birds with bitted bills were fitted with small, C-shaped pieces of plastic that are inserted between the mandibles and crimped slightly in the nostrils to prevent dislodging, but without piercing the tissue. Bits create a 1.0–3.0 mm gap between the mandibles that impairs preening, resulting in direct increases in louse load (Clayton, 1990, 1991; Booth *et al.*, 1993; Clayton & Tompkins, 1995; Clayton *et al.*, 1999). Although bits interfere with preening, they do not interfere with feeding since feral pigeons feed on grain that is easy to pick up, despite the mandibular gap created by bits. In an experimental test for possible side effects of bits, Clayton & Tompkins (1995) found that bits did not significantly alter the body mass or reproductive success of (parasite-free) feral pigeons. Birds with trimmed bills had the 1–2 mm maxillary overhang removed. This procedure is harmless to the bird and was repeated each week because the overhang grows back rapidly, much like a fingernail. Birds with control bills were not manipulated.

Lice were estimated using regression models that predict the total population size ($r^2 = 0.82$) from timed visual counts of lice on various body regions (Clayton, 1991). A comparison of the (log-transformed) number of lice on bitted, trimmed, and control birds over the course of the experiment revealed a significant effect of treatment on the change in louse population size (ANOVA, $F_{(2,52)} = 31.4$, $P < 0.0001$). Lice increased significantly faster on bitted and trimmed birds than on controls (Tukey HSD, $P < 0.05$). Furthermore, lice increased significantly faster on bitted birds than on trimmed birds (Tukey HSD, $P < 0.05$).

The effect of bill treatment on lice was not due to a difference in the amount of preening: time devoted to preening during the experiment did not differ significantly among treatments (repeated measures ANOVA: treatment, $F_{(2,52)} = 1.0$, $P = 0.39$; time, $F_{(1,52)} = 14.3$, $P < 0.001$; time × treatment, $F_{(2,52)} = 0.3$, $P = 0.74$). The effect of bill treatment on lice was the result of an apparent difference in the efficiency of preening. This experiment confirms earlier work showing that efficient preening is critical for controlling lice (see text). It further shows that the maxillary overhang is an important component of efficient preening. However, the more rapid increase in lice on bitted birds than on trimmed birds indicates that the overhang is not the only component of efficient preening. Bitting prevents contact of the upper and lower mandibles along their entire lengths, which is a more dramatic manipulation than merely trimming the maxillary overhang.

(Clayton & Cotgreave, 1994), is critical for defence against ectoparasites (Marshall, 1981; Hart, 1997). Preening is of two types: self-preening and allopreening – the latter when one individual preens another. Dusting, sunning, and anting may also play a role in ectoparasite defence, but have received less attention than grooming, as outlined below. Other behavioural defences, such as fly-repelling behaviour, are important for defence against ephemeral parasites (reviewed by Lehane, 1991; Hart, 1997).

GROOMING: SELF-PREENING

A substantial body of work shows that preening is a major defence against ectoparasites, and that bill morphology is an important component of preening efficiency. Numerous anecdotal reports document that wild birds with deformed bills have elevated ectoparasite loads (Rothschild & Clay, 1952; Ash, 1960; Pomeroy, 1962; Ledger, 1969; Marshall, 1981). Controlled experiments, in which bill morphology was dramatically manipulated (reviewed by Clayton, 1991; Hart, 1997), triggered rapid increases in ectoparasite load. The results of these studies clearly show the importance of preening, and normal bill morphology, for controlling ectoparasites.

Recent work demonstrates that even subtle features of bill morphology are critical for controlling ectoparasites. A comparative analysis of 52 species of neotropical birds revealed a significant negative correlation between length of the maxillary overhang of the bill and the mean number of lice on a given species (Clayton & Walther, 2001). The maxillary overhang is the distal portion of the upper mandible (maxilla) that curves over the lower mandible. The negative correlation between length of the overhang and louse load suggests that the overhang is important for controlling lice during preening. This functional hypothesis was tested through a series of manipulative experiments in which the 1–2 mm overhang was trimmed from feral pigeons (*Columba livia*) (Moyer *et al.*, *in press*). Removal of this small overhang caused louse load to triple in just three months (*Figure 18.2*). Preening is a complex behaviour in which the bill is used in a variety of ways (Simmons, 1985a). Different components of bill morphology may be important to different aspects of preening. Unfortunately, preening behaviour has not been studied in detail for many species of birds.

A major conclusion from these studies is that the evolution of bill morphology is probably influenced by the need for efficient preening, in addition to the more generally recognised need for efficient foraging. Indeed, preening and foraging could conceivably represent opposing selective forces shaping bill morphology. To date, avian biologists have focused almost exclusively on bills as tools for feeding. However, the critical importance of efficient preening for ectoparasite control suggests that the adaptive radiation of bill morphology may need to be re-interpreted with both preening and feeding firmly in mind.

GROOMING: ALLOPREENING

Allopreening may help to reduce ectoparasite loads, particularly on the head and neck, which are difficult or impossible to self-preen, and which are the sites of most allopreening (Harrison, 1965). Allopreening is a widespread behaviour which has been observed in many species and higher taxa of birds (Harrison, 1965). It is most common between courting individuals, mates, and between parents and their offspring. Harrison (1965) argued that allopreening serves mainly a social function, such as re-inforcement of pair bonds, and is of little or no importance for ectoparasite control. However, several more recent studies have implicated allopreening in the removal of ectoparasites (Fraga, 1984; Brooke, 1985; Murray, 1990; Wernham-Calladine, 1995).

For example, in a study of breeding penguins, Brooke (1985) showed that allopreening (paired) individuals had significantly fewer ticks than unpaired individuals, which could only self-preen. Unfortunately, Brooke was unable to control for possible co-variates of tick load, such as genetic resistance. Such resistance, if present, could have been responsible for the low tick loads of some individuals, as well as for their ability to attract mates (see 'mate choice' above). This would lead to a spurious (inverse) correlation between tick load and allopreening. A more rigorous test of the role of allopreening requires analysis of co-variation between allopreening and parasite load (c.f. Mooring, 1995), or experimental manipulation of allopreening and its subsequent effect on ectoparasites.

GROOMING: SCRATCHING

Although the role of allopreening remains unclear, scratching with the feet definitely controls ectoparasites on inaccessible regions, such as the head. Birds with a deformed or missing leg often have large numbers of ectoparasites (and their eggs) concentrated around the head and neck (Clayton, 1991). The obvious explanation is that, although a bird can preen itself while standing on one leg, it is unable to scratch itself. Head-scratching is known to kill or damage fleas on chickens (Suter, 1964, cited in Marshall, 1981, p. 107).

Some birds may use scratching as compensation for the absence of other methods of ectoparasite control. The unpaired penguins in Brooke's (1985) study spent significantly more time scratching than did the paired individuals with access to allopreening. Scratching may also compensate for inefficient preening in species of birds with unwieldy bills. In a phylogenetically-controlled comparative study, Clayton & Cotgreave (1994) reported that long-billed species average 16.2% of their grooming time scratching, compared to only 2.3% in short-billed species. In a series of paired taxonomic comparisons, long-billed species scratched significantly more than related short-billed taxa (Clayton & Cotgreave, 1994).

The efficiency of scratching for ectoparasite control may be enhanced by the presence of a pectinate claw on the middle toenail (Brewer, 1839; Brauner, 1953; Clay, 1957). A recent survey documented this curious feature in dozens of species of birds from 17 families representing eight orders (Moyer *et al.*, *in press*). The serrations on some pectinate claws are similar to the teeth of combs designed to rid humans of head lice, suggesting that the claw may help in removing lice and other ectoparasites during scratching. This hypothesis, which has not been tested, could be explored by comparing the ectoparasite loads of birds with trimmed claws to those with untrimmed controls.

Although experimental data do not exist, a recent comparative study tested Clay's (1957) assertion that bird species with pectinate claws have fewer species of head lice than clawless species. In a series of paired taxonomic comparisons, Moyer *et al.* (*in press*) found no significant difference in the number of species of head lice on birds with and without claws. However, it is unclear an inverse correlation between claw presence and parasite species richness should be expected anyway. If parasite richness decreases on birds that evolve pectinate claws, then selection maintaining the claw will be relaxed, leading to disappearance of the structure (assuming it is costly to maintain). Thus, the results of the comparative analysis of Moyer *et al.* (*in press*) should not be viewed as a conclusive test of the hypothesis that pectinate claws help to control ectoparasites.

DUSTING

Many species of birds (*Table 18.1*) engage in dusting, a behaviour in which they

Table 18.1.   Examples of birds known to dust (compiled from Simmons, 1985b; with Latin names where given). Classification based on Howard & Moore (1991).

STRUTHIONIFORMES	**STRIGIFORMES**
STRUTHIONIDAE	STRIGIDAE
Ostrich (*Struthio*)	Owl
RHEIFORMES	**CAPRIMULGIFORMES**
RHEIDAE	CAPRIMULGIDAE
Rhea (*Rhea*)	Nightjar
FALCONIFORMES	**COLIIFORMES**
ACCIPITRIDAE	COLIIDAE
Hawk	Mousebird
FALCONIDAE	**CORACIIFORMES**
Falcon	MOMOTIDAE
GALLIFORMES	Motmot
PHASIANIDAE	MEROPIDAE
Grouse (*Lagopus*)	Bee-eater
Bobwhite (*Colinus*)	CORACIIDAE
Fowl (*Gallus*)	Roller
Quail (*Coturnix*)	UPUPIDAE
Partridge (*Alectoris*)	Hoopoe
Pheasant (*Chrysolophus*, *Phasianus*)	BUCEROTIDAE
GRUIFORMES	Hornbill
TURNICIDAE	**PASSERIFORMES**
Buttonquail	ALAUDIDAE
CARIAMIDAE	Lark
Seriema	TROGLODYTIDAE
OTIDIDAE	Wren
Bustard	TIMALIIDAE
CHARADRIIFORMES	Wrentit (*Chamaea*)
THINOCORIDAE	EMBERIZIDAE
Seedsnipe	Sparrow (*Spizella*, *Pooecetes*)
COLUMBIFORMES	ICTERIDAE
COLUMBIDAE	Grackle (*Quiscalus*)
Dove	PLOCEIDAE
PTEROCLIDIDAE	Sparrow (*Passer*, *Petronia*, *Montifringilla*)
Sandgrouse	GRALLINIDAE
	Chough (*Corcorax*)

'ruffle' fine earth or sand through the plumage (Simmons, 1985b). Several authors have suggested that dusting may help control ectoparasites through dislodgement (Hoyle, 1938), abrasion of the cuticle leading to desiccation (Murray, 1990; Hendricks & Hendricks, 1995), plugging of the spiracles leading to poor respiration, or by reducing feather lipids upon which some ectoparasites feed (Borchelt & Duncan, 1974). Dusting is known to remove excess feather oil that can lead to matting of the plumage (Healy & Thomas, 1973; Borchelt & Duncan, 1974; van Liere, 1992), but no direct test of the impact of dusting on ectoparasites has been carried out.

SUNNING

Sunning, a behaviour in which birds expose themselves to solar radiation while adopting a stereotyped posture, has been recorded for over 170 species in nearly 50 families (Kennedy, 1969; Simmons, 1986). Sunning may control ectoparasites by killing them directly through overheating (Moyer & Wagenbach, 1995), or indirectly by increasing their vulnerability to preening as they try to escape the heat (Simmons, 1986). Sunning also appears to be an adaptation for conserving energy, since birds sun in cool temperatures to warm themselves and reduce metabolic expenditure (Morton, 1967; Ohmart & Lasiewski, 1971; Simmons 1986). However, sunning birds often show signs of heat stress, such as panting (Simmons, 1986). Black noddies (*Anous minutus*) sun more frequently in periods of high, rather than low, temperature (Moyer & Wagenbach, 1995). Blem & Blem (1993) observed swallows "panting markedly", while sunning on substrates in excess of 50°C.

No direct test of the impact of sunning on ectoparasites has been carried out. However, two recent studies provide evidence that is consistent with the ectoparasite control hypothesis. Blem & Blem (1993) compared the rate of sunning in swallows that were fumigated to remove ectoparasites, with the rate of sunning by non-fumigated controls. They found that fumigated birds sunned less frequently than controls (Blem & Blem, 1993), suggesting that the need for sunning decreases with a reduction in ectoparasite load. Moyer & Wagenbach (1995) exposed model black noddy wings to sun and shade. The duration of exposure was typical of a sunning bout, and the temperature of model wings did not exceed that of sunning noddies. The mortality of feather lice placed on model wings in the sun was higher than that of lice placed on wings in the shade, suggesting that sunning may help noddies combat lice. Additional research is needed to test the impact of sunning on the ectoparasites of live birds. Since birds sun readily in captivity (Simmons, 1986), it should be possible to use captive birds for rigorous tests of the impact of sunning on ectoparasites.

ANTING

Another oft mentioned defence against parasites is 'anting' behaviour, in which birds crush and rub ants on their feathers (active anting), or allow ants to crawl through the plumage (passive anting). The fact that birds ant exclusively with ants that secrete acid or other pungent fluids suggests that anting may kill or deter ectoparasites. Although anting has been observed in over 200 bird species, its function remains controversial (Simmons, 1986; Clayton & Wolfe, 1993; Hart, 1997). At present, no study has provided convincing evidence that anting combats ectoparasites.

Birds also 'ant' with items such as fruit peel, flowers, mothballs, and other substances, many of which have anti-parasite properties (Clark *et al.*, 1990). After observing a common grackle (*Quiscalus quiscula*) anting with a hemisphere of lime, Clayton & Vernon (1993) showed that lime oil vapour rapidly kills feather lice in Petri dishes. The impact of anting with such substances on ectoparasites has not been tested *in situ*.

Nest maintenance behaviour

An important component of ectoparasite defence is a suite of behaviours that control parasites in nests, such as nest 'sanitation'. For example, male house wrens (*Troglodytes aedon*) remove old nest material from their nest cavities prior to each nesting bout. Pacejka *et al.* (1996) showed that removal of the old material dramatically reduces the number of parasitic mites in the nest cavity (*Figure 18.3*). The authors argued that this reduction delays mite population growth, allowing young birds to leave the nest before mites reach a detrimental level.

The hole-nesting great tit (*Parus major*) also engages in a form of nest sanitation.

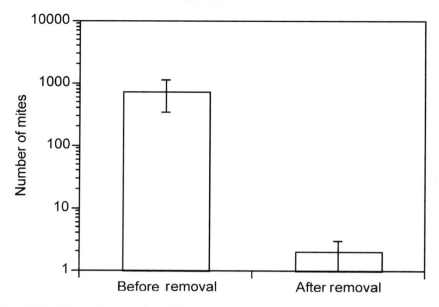

Figure 18.3.   Mean (± 1 s.e.) number of mites in house wren nest boxes sampled before (n = 6 boxes) and after (n = 6) the male wren removed nest material remaining from the previous breeding season. Redrawn from Pacejka *et al.* (1996).

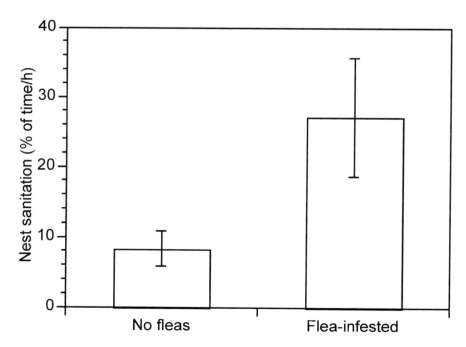

Figure 18.4.   Mean (± 95% CI) percentage of a night hour taken up by nest sanitation behaviour for female great tits in nests without fleas (n = 15) and with fleas (n = 14). Redrawn from Christe *et al.* (1996).

Christe *et al.* (1996) described it as "a period of active search with the head dug into the nest material." Although the precise effect of this behaviour is unclear, the authors showed that females devote significantly more time to sanitation in flea-infested nests than in uninfested nests (*Figure 18.4*). The increased sanitation behaviour in infested nests comes at the expense of sleep, suggesting that the behaviour is costly to the birds (Christe *et al.*, 1996).

Another behaviour demonstrated to control nest parasites is the insertion of green vegetation into nests (reviewed by Clark, 1991; Clayton & Wolfe, 1993; Dumbacher & Pruett-Jones, 1996; Hart, 1997). Clark & Mason (1985) showed that European starlings (*Sturnus vulgaris*) select species of plants that contain volatile chemicals with antibacterial, insecticidal, or miticidal properties. The same authors later showed that nests containing such herbs have lower infestations of blood-sucking mites (Clark & Mason, 1988). More recent research suggests that the addition of herbs to the nest does not necessarily serve to reduce ectoparasite loads, but may help nestlings cope with the detrimental effects of the ectoparasites. Gwinner *et al.* (2000) manipulated green vegetation in 148 starling nests. They found no difference in the ectoparasite loads (mites, lice, fleas) of nests with and without the herbs starlings normally choose to insert. However, nestlings from nests with herbs had higher red blood cell counts and body masses than nestlings from nests without herbs. Gwinner *et al.* (2000) argued that herbs may stimulate the immune system of nestlings such that they can better deal with the detrimental effects of blood-feeding ectoparasites.

Plants apparently may not be the only biological control agents used by birds to control parasites in the nest. Eastern screech owls (*Otus asio*) deposit live blind snakes (*Leptotyphlops dulcis*) in their nests. Gehlbach & Baldridge (1987) found that nestlings from nests with live snakes grew significantly faster than nestlings from nests without snakes. Since the snakes consume soft-bodied insect larvae from the nests, the authors suggested that the snakes might reduce larval parasitism on owl nestlings. But, as the presence of snakes was not experimentally manipulated in this study, there may be other factors that co-vary with the presence of snakes (e.g. hunting ability of parents) that might contribute to the difference in nestling growth rate.

While owls may exploit snakes to reduce nest ectoparasites, oropendolas and caciques reportedly exploit cowbirds (Smith, 1968). Smith claimed that adult oropendola and cacique tolerate brood parasitism by cowbirds when bot fly parasitism is likely, because cowbird nestlings preen the host nestlings of their bot fly larvae. Presumably owing to this attention, host broods with cowbirds produced more nestlings than broods without cowbirds (Smith, 1968). The use of snakes and cowbirds to control nest ectoparasites are intriguing accounts that warrant further study and replication.

Conclusions

In this chapter, we have attempted to review some of the ways in which birds combat ectoparasites. Most of the research in this area has tended to focus on how single defences combat single types of ectoparasites. Future research should address how birds use suites of defences to control ectoparasites. This approach is important because individual defences can interact in at least two fundamental ways. First, defences can be complementary. They could target different types of ectoparasites, or ectoparasites on different host body regions. For example, we have mentioned that preening can control ectoparasites on the wing, while scratching can control ectoparasites on the head.

Second, defences can interact synergistically – where their combined effect is greater than the sum of their individual effects. One example regards the possible synergistic interaction of preening and sunning. In the case of a single defence, an ectoparasite can exploit a refuge to escape that defence. For instance, some wing lice are dorsoventrally flattened so that they can slide between the barbs of flight feathers and escape the preening bill. Sunning, however, heats the flight feathers such that lice flee the interbarb refuge and move down the feather towards the body (B. R. Moyer, *unpublished*). Thus, lice may not have a refuge from preening in the presence of sunning, and their mortality may increase synergistically. In short, in addition to conducting needed tests of additional candidate defences (e.g. feather toughness and dusting), future research should address how defences interact.

An intriguing pattern arising from the basic research on avian defences is that different birds employ rather different suites of defences. Several factors might help to explain the variation in how different birds combat ectoparasites. First, the nature and intensity of selection by ectoparasites varies among different birds. Second, environmental and phylogenetic constraints may influence the defences available to different birds. Third, ectoparasite defence is just one component of avian life history demands; accordingly, the optimum defensive strategy will depend on a bird's life history trade-offs.

Ectoparasite populations and communities vary greatly among different birds, and may thereby influence defensive strategies. As we discussed earlier (under 'habitat choice'), ectoparasite loads can vary by more than an order of magnitude in different environments. These differences may cause variation in the intensity of selection for defence among birds. Just as the size of an ectoparasite population can vary, so can the diversity of the ectoparasite community. For instance, a single species of tinamou can be infested by a dozen species of lice, while ostriches are only ever infested by one species (Marshall, 1981). The optimum defensive strategy against a single ecto-parasite is likely to differ from that against a diverse community of ectoparasites. Cotgreave & Clayton (1994) found that bird species infested with more species of lice devoted more time to maintenance behaviour than bird species with fewer species of lice.

Even if ectoparasite pressure is uniform among birds, environmental and phylogenetic constraints can limit the defences available to different birds. Some defences may not be effective in some habitats. For example, dusting may not be an option in a marshy habitat. Similarly, insertion of green vegetation in nests may not be an option on barren oceanic islands. The phylogenetic history of a bird can also place constraints on defence. For example, scratching as a means of controlling ectoparasites appears to be constrained by foot morphology; species belonging to web-footed families scratch very little, regardless of other factors (Clayton & Cotgreave, 1994)

Parasite defence is just one of many life history demands. Birds must balance limited resources among competing life history traits, forcing trade-offs. Accord-ingly, variation in parasite defence might be governed partly by variation in other life history parameters. For instance, we have described how having a specialized bill for feeding can influence parasite defence. Birds with unwieldy bills compensate for inefficient preening by scratching more. The composition of grooming behaviour is thus dictated partly by constraints related to foraging ecology. Likewise, as we discussed earlier, the evolution of bill morphology is probably influenced by the need for efficient preening, in addition to the more generally recognised need for efficient foraging. To what extent do foraging, breeding strategy, migration, and other life history components influence the composition of ectoparasite defence in different bird species (see Piersma, 1997)?

Just as variation in general life history demands may help to explain variation in ectoparasite defence, the converse is also true. We may be better able to understand variation in general life history traits by understanding how the demands for ecto-parasite defence vary among different bird taxa. A species living in an environment with few ectoparasites should be released, to some extent, from constraints imposed by ectoparasite defence on other life history traits.

In conclusion, birds have a variety of defences against ectoparasites. For a more complete understanding of how birds combat ectoparasites, future research should investigate how these different defences interact, and why the composition of the defensive arsenal differs among bird taxa. This variation might be explained by considering how avian defences are influenced by (1) parasite communities, (2) environmental and phylogenetic constraints on the host, and (3) broader life history trade-offs. An appreciation of variation in ectoparasite pressure and the consequent investment in avian defence may also shed reciprocal light on other parameters influencing the evolution of avian life histories.

Acknowledgements

This research was supported by an NSF CAREER award (DEB-9703003) to DHC, by a University of Utah Research Awards Committee grant to DHC, and grants to BRM from Sigma Xi and the Frank M. Chapman Memorial Fund of the American Museum of Natural History.

References

Able, D. J. (1996) The contagion indicator hypothesis for parasite-mediated sexual selection. *Proceedings of the National Academy of Sciences of the United States of America* **93**, 2229–2233.

Amin, O. M. (1966) The fleas (Siphonaptera) of Egypt: distribution and seasonal dynamics of fleas infesting dogs in the Nile Valley and Delta. *Journal of Medical Entomology* **3**, 293–298.

Andersson, M. (1994) *Sexual selection*. Princeton University Press, Princeton, USA. 599 pp.

Ash, J. S. (1960) A study of the Mallophaga of birds with particular reference to their ecology. *Ibis* **102**, 93–110.

Bennett, G. F., Montgomerie, R. & Seutin, G. (1992) Scarcity of Haematozoa in birds breeding on the arctic tundra of North America. *Condor* **94**, 289–292.

Blem, C. R. & Blem, L. B. (1993) Do swallows sunbathe to control ectoparasites? An experimental test. *Condor* **95**, 728–730.

Bonser, R. H. C. (1995) Melanin and the abrasion resistance of feathers. *Condor* **97**, 590–591.

Booth, D. T., Clayton, D. H. & Block, B. A. (1993) Experimental demonstration of the energetic cost of parasitism in free-ranging hosts. *Proceedings of the Royal Society of London, B* **253**, 125–129.

Borchelt, P. L. & Duncan, L. (1974) Dustbathing and feather lipid in Bobwhite (*Colinus virginianus*). *Condor* **76**, 471–472.

Brauner, J. (1953) Observations on the behavior of a captive poor-will. *Condor* **55**, 68–74.

Brewer, T. M. (1839) *Wilson's American ornithology*. Magagnos Company, New York, USA. 380 pp.

Brooke, M. de L. (1985) The effect of allopreening on tick burdens of molting eudyptid penguins. *Auk* **102**, 893–895.

Brown, C. R., Brown, M. B. & Rannala, B. (1995) Ectoparasites reduce long-term survival of their avian host. *Proceedings of the Royal Society of London, B* **262**, 313–319.

Burtt, E. H. (1986) An analysis of physical, physiological, and optical aspects of avian coloration with emphasis on wood-warblers. *Ornithological Monographs* **38**, 1–125.

Christe, P., Oppliger, A. & Richner, H. (1994) Ectoparasite affects choice and use of roost sites in the great tit *Parus major*. *Animal Behaviour* **47**, 895–898.

Christe, P., Richner, A. & Oppliger, A. (1996) Of great tits and fleas: sleep baby sleep ... *Animal Behaviour* **52**, 1087–1092.

Clark, C. C., Clark, L. & Clark, L. (1990) 'Anting' behavior by common grackles and European starlings. *Wilson Bulletin* **102**, 167–169.

Clark, L. (1991) The nest protection hypothesis: the adaptive use of plant secondary compounds by European starlings. In: *Bird–parasite interactions: ecology, evolution, and behaviour* (eds. J. E. Loye & M. Zuk), pp. 205–221. Oxford University Press, Oxford, UK.

Clark, L. & Mason, J. R. (1985) Use of nest material as insecticidal and anti-pathogenic agents by the European starling. *Oecologia* **67**, 169–176.

Clark, L. & Mason, J. R. (1988) Effect of biologically active plants used as nest material and the derived benefit to starling nestlings. *Oecologia* **77**, 174–180.

Clay, T. (1957) The Mallophaga of birds. In: *First symposium on host specificity among parasites of vertebrates*, pp. 120–155. Universite de Neuchatel, Imprimerie Paul Attinger S. A., Neuchatel, Switzerland.

Clayton, D. H. (1990) Mate choice in experimentally parasitized rock doves: lousy males lose. *American Zoologist* **30**, 251–262.

Clayton, D. H. (1991) Coevolution of avian grooming and ectoparasite avoidance. In: *Bird–parasite interactions: ecology, evolution, and behaviour* (eds. J. E. Loye & M. Zuk), pp. 258–289. Oxford University Press, Oxford, UK.

Clayton, D. H. & Cotgreave, P. (1994) Relationship of bill morphology to grooming behaviour in birds. *Animal Behaviour* **47**, 195–201.

Clayton, D. H. & Moore, J. (1997) *Host–parasite evolution: general principles and avian models*. Oxford University Press, Oxford, UK. 473 pp.

Clayton, D. H. & Tompkins, D. M. (1995) Comparative effects of mites and lice on the reproductive success of rock doves (*Columba livia*). *Parasitology* **110**, 195–206.

Clayton, D. H. & Vernon, J. G. (1993) Common grackles anting with lime fruit and its effect on ectoparasites. *Auk* **110**, 951–952.

Clayton, D. H. & Walther, B. A. (2001) Influence of host ecology and morphology on the diversity of neotropical bird lice. *Oikos* **94**, 455–467.

Clayton, D. H. & Wolfe, N. D. (1993) The adaptive significance of self-medication. *Trends in Ecology and Evolution* **8**, 60–63.

Clayton, D. H., Lee, P. L. M., Tompkins, D. M. & Brodie, E. D. (1999) Reciprocal natural selection on host–parasite phenotypes. *American Naturalist* **154**, 261–270.

Cotgreave, P. & Clayton, D. H. (1994) Comparative analysis of time spent grooming by birds in relation to parasite load. *Behaviour* **131**, 171–187.

Duffy, D. C. (1983) The ecology of tick parasitism on densely nesting Peruvian seabirds. *Ecology* **64**, 110–119.

Dumbacher, J. P. (1999) Evolution of toxicity in pitohuis: I. Effects of homobatrachotoxin on chewing lice (order Phthiraptera). *Auk* **116**, 957–963.

Dumbacher, J. P. & Pruett-Jones, S. (1996) Avian chemical defense. In: *Current ornithology* (eds. V. Nolan & E. D. Ketterson), pp. 137–174. Plenum Press, New York, USA.

Dumbacher, J. P., Beehler, B. M., Spande, T. F., Garraffo, H. M. & Daly, J. W. (1992) Homobatrachotoxin in the genus *Pitohui:* chemical defense in birds? *Science, New York* **258**, 799–801.

Fraga, R. M. (1984) Bay-winged cowbirds (*Molothrus badius*) remove ectoparasites from their brood parasites, the screaming cowbirds (*M. rufoaxillaris*). *Biotropica* **16**, 223–226.

Gehlbach, F. R. & Baldridge, R. S. (1987) Live blind snakes (*Leptotyphlops dulcis*) in eastern screech owl (*Otus asio*) nests: a novel commensalism. *Oecologia* **71**, 560–563.

Greiner, E. C., Bennett, G. F., White, E. M. & Coombs, R. F. (1975) Distribution of the avian Hematozoa of North America. *Canadian Journal of Zoology* **53**, 1762–1787.

Gwinner, H., Oltrogge, M., Trost, L. & Nienaber, U. (2000) Green plants in starling nests: effects on nestlings. *Animal Behaviour* **59**, 301–309.

Hamilton, W. D. & Zuk, M. (1982) Heritable true fitness and bright birds: a role for parasites? *Science, New York* **218**, 84–387.

Harrison, C. J. O. (1965) Allopreening as agonistic behaviour. *Behaviour* **24**, 161–209.

Hart, B. L. (1997) Behavioural defence. In: *Host–parasite evolution: general principles and avian models* (eds. D. H. Clayton & J. Moore), pp. 59–77. Oxford University Press, New York, USA.

Healy, W. M. & Thomas, J. W. (1973) Effects of dusting on plumage of Japanese quail. *Wilson Bulletin* **85**, 442–448.

Heeb, P., Kolliker, M. & Richner, H. (2000) Bird-ectoparasite interactions, nest humidity, and ectoparasite community structure. *Ecology* **81**, 958–968.

Hendricks, P. & Hendricks, L. N. (1995) Behavior and interaction of Bewick's and house wrens at a common dusting site, with comments on the utility of dusting. *Journal of Field Ornithology* **66**, 492–496.

Hillgarth, N. & Wingfield, J. C. (1997) Parasite-mediated sexual selection: endocrine aspects. In: *Host–parasite evolution: general principles and avian models* (eds. D. H. Clayton & J. Moore), pp. 78–104. Oxford University Press, New York, USA.

Howard, R. & Moore, A. (1991) *A complete checklist of the birds of the world.* Academic Press, San Diego, USA. 622 pp.

Howe, H. F. & Westley, L. C. (1988) *Ecological relationships of plants and animals.* Oxford University Press, Oxford, UK. 273 pp.

Hoyle, W. L. (1938) Transmission of poultry parasites by birds with special reference to the 'English' or house sparrow and chickens. *Transactions of the Kansas Academy of Science* **41**, 379–384.

Janovy, J. (1997) Protozoa, helminths, and arthropods of birds. In: *Host–parasite evolution: general principles and avian models* (eds. D. H. Clayton & J. Moore), pp. 303–337. Oxford University Press, New York, USA.

Kennedy, R. J. (1969) Sunbathing behaviour of birds. *British Birds* **62**, 249–258.

Kose, M. & Møller, A. P. (1999) Sexual selection, feather breakage and parasites: the importance of white spots in the tail of the barn swallow (*Hirundo rustica*). *Behavioural Ecology and Sociobiology* **45**, 430–436.

Kose, M., Mand, R. & Møller, A. P. (1999) Sexual selection for white tail spots in the barn swallow in relation to habitat choice by feather lice. *Animal Behaviour* **58**, 1201–1205.

Ledger, J. A. (1969) Ectoparasite load in a laughing dove with a deformed mandible. *Ostrich* **41**, 191–194.

Lehane, M. J. (1991) *Biology of blood-sucking insects.* Harper Collins, London, UK. 288 pp.

Lehmann, T. (1993) Ectoparasites: direct impact on host fitness. *Parasitology Today* **9**, 8–13.

Little, R. M. & Earle, R. A. (1995) Sandgrouse (Pterocleidae) and sociable weavers *Philetarius socius* lack avian Haematozoa in semi-arid regions of South Africa. *Journal of Arid Environments* **30**, 367–370.

Loye, J. E. & Carroll, S. P. (1998) Ectoparasite behavior and its effects on avian nest site selection. *Annals of the Entomological Society of America* **91**, 159–163.

Loye, J. E. & Zuk, M. (eds.) (1991) *Bird–parasite interactions: ecology, evolution, and behaviour.* Oxford University Press, Oxford, UK. 406 pp.

Marshall, A. G. (1981) *The ecology of ectoparasitic insects.* Academic Press, London, UK. 459 pp.

Merilä, J. & Allandar, K. (1995) Do great tits prefer ectoparasite-free roost sites? An experiment. *Ethology* **99**, 53–60.

Merino, S. & Potti, J. (1995) Pied flycatchers prefer to nest in clean nest boxes in an area with detrimental nest ectoparasites. *Condor* **97**, 828–831.

Milinski, M. & Bakker, T. C. M. (1990) Female sticklebacks use male coloration in mate choice and hence avoid parasitized males. *Nature, London* **344**, 330–333.

Møller, A. P. (1987) Advantages and disadvantages of coloniality in the swallow *Hirundo rustica. Animal Behaviour* **35**, 819–832.

Møller, A. P. (1991) Parasites, sexual ornaments, and mate choice in the barn swallow. In: *Bird–parasite interactions: ecology, evolution, and behaviour* (eds. J. E. Loye & M. Zuk), pp. 328–343. Oxford University Press, Oxford, UK.

Møller, A. P. (1993) Ectoparasites increase the cost of reproduction in their hosts. *Journal of Animal Ecology* **62**, 309–322.

Møller, A. P., Allander, K. & Dufva, R. (1990) Fitness effects of parasites on passerine birds: a review. In: *Population biology of passerine birds* (eds. J. Blondel, A. Gosler, J. Lebreton & R. H. McCleery), pp. 269–280. Springer, Berlin, Germany.

Mooring, M. S. (1995) The effect of tick challenge on grooming rate by impala. *Animal Behaviour* **50**, 377–392.

Morton, M. L. (1967) The effects of insolation on the diurnal feeding pattern of white-crowned sparrows (*Zonotrichia leucophrys gambelii*). *Ecology* **48**, 690–694.

Moss, W. W. & Camin, J. H. (1970) Nest parasitism, productivity, and clutch size in purple martins. *Science, New York* **168**, 1000–1003.

Mouritsen, K. N. & Madsen, J. (1994) Toxic birds: defence against parasites? *Oikos* **69**, 357–358.

Moyer, B. R. & Wagenbach, G. E. (1995) Sunning by black noddies (*Anous minutus*) may kill chewing lice (*Quadraceps hopkinsi*). *Auk* **112**, 1073–1077.

Moyer, B. R., Drown, D. M. & Clayton, D. H. (2002) Low humidity reduces ectoparasite pressure: implications for host life history evolution. *Oikos* **97**, 223–228.

Moyer, B. R., Pacejka, A. J. & Clayton, D. H. (2003) How birds combat ectoparasites. *Current Ornithology* **17** (in press).

Murray, M. D. (1990) Influence of host behaviour on some ectoparasites of birds and mammals. In: *Parasitism and host behaviour* (eds. C. J. Barnard & J. M. Behnke), pp. 290–315. Taylor & Francis, New York, USA.

Ohmart, R. D. & Lasiewski, R. C. (1971) Roadrunners: energy conservation by hypothermia and absorption of sunlight. *Science, New York* **172**, 67–69.

Oppliger, A., Richner, H. & Christe, P. (1994) Effect of an ectoparasite on lay date, nest-site choice, desertion, and hatching success in the great tit (*Parus major*). *Behavioural Ecology* **5**, 130–134.

Pacejka, A. J., Santana, E., Harper, R. G. & Thompson, C. F. (1996) House wrens *Troglodytes aedon* and nest-dwelling ectoparasites: mite population growth and feeding patterns. *Journal of Avian Biology* **27**, 273–278.

Piersma, T. (1997) Do global patterns of habitat use and migration strategies co-evolve with relative investments in immunocompetence due to spatial variation in parasite pressure? *Oikos* **80**, 623–631.

Pomeroy, D. E. (1962) Birds with abnormal bills. *British Birds* **55**, 49–72.

Poulsen, B. O. (1994) Poison in *Pitohui* birds: against predators or ectoparasites? *Emu* **94**, 128–129.

Richner, H., Oppliger, A. & Christe, P. (1993) Effect of an ectoparasite on reproduction in great tits. *Journal of Animal Ecology* **62**, 703–710.

Rothschild, M. & Clay, T. (1952) *Fleas, flukes and cuckoos*. Collins, London, UK. 304 pp.

Rytkonen, S., Lehtonen, R. & Orell, M. (1998) Breeding great tits *Parus major* avoid nestboxes infested with fleas. *Ibis* **140**, 687–690.

Shields, W. M. & Crook, J. R. (1987) Barn swallow coloniality: a net cost for group breeding in the Adirondacks? *Ecology* **68**, 1373–1386.

Simmons, K. E. L. (1985a) Comfort behaviour. In: *A dictionary of birds* (eds. B. Campbell & E. Lack), pp. 101–104. Buteo Books, Vermillion, South Dakota, USA.

Simmons, K. E. L. (1985b) Dusting. In: *A dictionary of birds* (eds. B. Campbell & E. Lack), pp. 161–162. Buteo Books, Vermillion, South Dakota, USA.

Simmons, K. E. L. (1986) *The sunning behaviour of birds.* Bristol Ornithological Club, Bristol, UK. 119 pp.

Smith, N. G. (1968) The advantage of being parasitized. *Nature, London* **219**, 690–694.

Suter, P. R. (1964) Biologie von *Echidnophaga gallinacea* (Westw.) und Vergleich mit anderen Verhaltenstypen bie Flöhen. *Acta Tropica* **21**, 193–238.

Tella, J. L. (1996) Absence of blood-parasitization effects on lesser kestrel fitness. *Auk* **113**, 253–256.

Thompson, C. F. (1999) Ectoparasite behavior and its effects on avian nest-site selection: corrections and comment. *Annals of the Entomological Society of America* **92**, 108–109.

van Liere, D. W. (1992) The significance of fowl's bathing in dust. *Animal Welfare* **1**, 187–202.

van Riper III, C., van Riper, S. G., Goff, M. L. & Laird, M. (1986) The epizootiology and ecological significance of malaria in Hawaiian land birds. *Ecological Monographs* **56**, 27–344.

Wakelin, D. & Apanius, V. (1997) Immune defence: genetic control. In: *Host–parasite evolution: general principles and avian models* (eds. D. H. Clayton & J. Moore), pp. 30–58. Oxford University Press, New York, USA.

Walter, D. E. & Proctor, H. C. (1999) *Mites: ecology, evolution, and behaviour.* CABI, New York, USA. 322 pp.

Wernham-Calladine, C. V. (1995) Guillemot preening activity in relation to tick infestation. *Bulletin of the British Ecological Society* **26**, 187–195.

Wikel, S. K. (1996) *Immunology of host–ectoparasitic arthropod relationships.* CABI, Wallingford, UK. 331 pp.

19
The Acquisition of Host-Specific Feather Lice by Common Cuckoos (*Cuculus canorus*)

M. de L. BROOKE[1] AND HIROSHI NAKAMURA[2]

[1]*Department of Zoology, University of Cambridge, Downing Street, Cambridge, CB2 3EJ, UK and* [2]*Faculty of Education, Shinshu University, Nishinagano, Nagano 380, Japan*

Introduction

Feather lice (Phthiraptera) are flightless obligate ectoparasitic insects which spend their entire lives adhering to the plumage or skin of their avian hosts. The majority of feather lice species are specific to a particular host species or genus (Rothschild & Clay, 1952; Marshall, 1981). The transmission of feather lice from one generation of brood parasitic birds to the next generation therefore poses a puzzle. If the feather lice of the parasite and of the host that raises it belong to different species, the parasite chick cannot receive its lice from its host. In the absence of any contact with its parents, how then does the young parasite acquire its lice?

Since cuckoos throughout the world harbour three genera of lice, namely *Cuculoecus, Cuculicola*, and *Cuculiphilus*, the question has been posed in respect of cuckoos by a number of authors (Rothschild & Clay, 1952; Dogiel, 1964; Marshall, 1981), but not answered. The young cuckoo, in the order Cuculiformes, is raised by passerine birds in the order Passeriformes, which harbour different genera of lice. As far as is known, neither the female cuckoo nor her mate visit the host nest after laying. There is therefore no apparent opportunity for transfer of cuculiform lice between parent and young cuckoo. It could be that the cuckoo lice leave the female cuckoo while she is laying and then linger in the host nest or on the host until the young cuckoo has grown feathers to which the lice could attach. This explanation has problems. The lice would have to detect the exact period, of only about 10 seconds duration (Wyllie, 1981), when the female cuckoo was laying. They would then have to survive off their usual host until the young cuckoo had hatched and grown, a period of 3–4 weeks, whereas, as far as is known, feather lice do not normally survive away from their hosts for more than a matter of days (Ledger, 1980; Marshall, 1981; Lehane, 1991).

Reproduced from the *Journal of Zoology (London)* **244**, 167–173 (1998), © The Zoological Society of London published by Cambridge University Press. Reprinted with permission.
Insect and Bird Interactions
© Intercept Ltd., PO Box 716, Andover, Hampshire, SP10 1YG, UK.

Two alternative explanations have been proposed. The first is that transfer of feather lice occurs during mating. Under this scenario, the young cuckoo would leave the nest free of cuckoo lice and would remain free for its first year of life (assuming first breeding at one year old). When the first-year cuckoo mated with infested, older birds, it would acquire lice (Hillgarth, 1996). This mode of transmission yields testable predictions. Cuckoos in the nest and in their first year would be louse-free (excluding possible temporary infestation with host louse species). Then, during the course of the first breeding season, a steadily increasing proportion of these birds would become infested, but louse loads would possibly be lower than those of older, adult birds which had been infested for at least a year. On the basis of finding fledgling cuckoos to be louse-free, and mature cuckoos shot the following summer to be lousy, Dogiel (1936, 1964) has championed this explanation. However, his work did not demonstrate that first-year cuckoos, which had not mated, arrived louse-free at the breeding grounds. The second explanation involves phoresy, whereby the lice would attach to flat flies (Hippoboscidae), as is not infrequently observed (Clay & Meinertzhagen, 1943; Rothschild & Clay, 1952; Keirans, 1975). By securing a flight aboard a flat fly from an infested cuckoo to an uninfested cuckoo, the lice could potentially reach new hosts. If this mode of transmission operated, it might be that cuckoo nestlings would occasionally become infested. But most infestation would presumably occur post-fledging, and it would not then be distinguishable from infestation occurring by direct bird–to–bird contact.

This study reports observations on common cuckoos (*Cuculus canorus*) designed to distinguish between these various alternatives. The three louse species found on this bird are *Cuculoecus latifrons*, *Cuculicola latirostris* (both belonging to the suborder Ischnocera), and *Cuculiphilus fasciatus* (suborder Amblycera). *Cuculoecus* is a louse primarily found on the head and neck (Rothschild & Clay, 1952), while the other two lice occur more widely over the body.

Study areas and methods

The methods used in our study of common cuckoos relate to the various transmission possibilities mentioned above.

DO CUCKOO NESTLINGS LEAVE THE NEST WITHOUT CUCKOO LICE?

To assess this, 21 nestling cuckoos were checked for feather lice using the method of Fowler & Cohen (1983). Briefly, the bird, its head held in an airtight collar, is suspended for 10 minutes over a container filled with chloroform vapour into which ectoparasites fall. They are collected for mounting and microscopic examination. The nestlings (13 in reed warbler (*Acrocephalus scirpaceus*) nests in Oxfordshire and Cambridgeshire, England (1994–1996), 7 in great reed warbler (*A. arundinaceus*) nests in Nagano Prefecture, Japan (1995), and 1 in an azure-winged magpie (*Cyanopica cyana*) nest in Japan (1995)) were checked at roughly the largest size they could be handled without prompting premature fledging. Thus, primaries were showing beyond the sheaths and wing length ranged from 77–128 mm (n = 20).

To assess whether cuckoo nestlings were more or less likely to acquire the host-specific lice than the hosts' own young, nestling reed warblers (1–3 per nest) were deloused in Cambridgeshire in 1994, and nestling great reed warblers (2 per nest) were deloused in Japan in 1995. As with the cuckoo nestlings, the young warblers were examined at roughly the largest size they could be handled without prompting premature fledging. Mist-netted adult and juvenile reed warblers were also examined for lice in July and August 1994. The method of obtaining lice from warblers was as described for young cuckoos, except treatment with chloroform lasted only 5 minutes.

Since it is probably true that Fowler & Cohen's method does not secure all ectoparasites on a bird (Clayton & Walther, 1997), a failure to find lice on nestling cuckoos could represent a 'false negative' owing to imperfect sampling of low louse infestation levels. To check this, and the possibility that lice leave the laying female cuckoo to linger in the host nest in anticipation of a young cuckoo, we also checked for feather lice in the actual nests where cuckoos had laid. This examination was conducted on 19 reed warbler nests obtained from Oxfordshire and Cambridgeshire in 1996. As soon as practicable after the nest had failed at the egg stage (n = 1) or at the chick stage (n = 10), or after the cuckoo chick had fledged (n = 8), the entire nest was cut out of the reeds and brought into the laboratory. It was then examined carefully under a binocular microscope for 10 minutes. This examination was done immediately, and repeated 1 day later, 2 days later, and 5 days later. During this period in the laboratory, 13 of the nests were kept in polythene bags, while 6 were suspended on a metal grille under a 60 W light and over water in a Berlese funnel (Southwood, 1966).

ARE LICE ACQUIRED DURING MATING?

To explore this possibility, free-flying cuckoos were caught during the breeding season in canopy-level nets in *Acacia* groves along the Chikuma River, Nagano, Japan, where the principal host is the great reed warbler (Nakamura, 1990). The birds were separated into those in their first year (i.e. those hatching the previous calendar year), and those that were older by the presence in the former group of variable numbers of juvenile secondaries and coverts (Baker, 1993). The birds were then ringed to avoid repeat delousing, and deloused for 10 minutes using Fowler & Cohen's (1983) method. While the bird was suspended over chloroform, its head was examined, and any *Cuculoecus* found removed with forceps. The feather lice were initially preserved in alcohol. Subsequently, they were mounted, identified by reference to the collections of the Natural History Museum, London, and sexed by examination of the genitalia. Nymphs were differentiated from adult lice by smaller head capsules (measured under the microscope) and incomplete development of the genitalia.

The main sample comprised 35 birds caught by the authors between 27th May and 2nd July, 1995, a period stretching from a few days before the start of cuckoo laying until towards the end of the cuckoo laying period. Detailed analysis of the ectoparasites is restricted to the lice from these birds, since the delousing treatment was standard. In addition, 6 birds (3 at Wicken Fen, Cambridgeshire, 1 at Knettishall, Suffolk and 2 at Dungeness, Kent) were caught in England in 1994, and another 3 at Wicken Fen in 1995.

ARE LICE ACQUIRED OUTSIDE THE BREEDING SEASON VIA PHORESY AND/OR BY
DIRECT CONTACT BETWEEN BIRDS?

Since cuckoos are not readily captured alive outside the breeding season, and since
we did not wish to collect birds, we could only sample louse loads via museum skins.
We visited the Natural History Museum, Tring (NHM), and combed cuckoo skins
with a nit comb. Each skin was combed for about 5 minutes and any lice obtained
were identified. Although not without drawbacks, for instance lice may abandon a
dead bird before its prepared skin reaches the museum, and dead lice may be
inadvertently exchanged between skins in the same museum drawer, the method did
yield lice (the maximum number obtained from one bird was 3) and, as will emerge
below, the results were compatible with the field studies.

Results

DO CUCKOO NESTLINGS LEAVE THE NEST WITHOUT CUCKOO LICE?

Neither warbler-specific nor cuckoo-specific feather lice were obtained from any of
the 20 cuckoo nestlings examined from the nests of *Acrocephalus* warblers. However, warbler-specific lice were found on warbler chicks; *Menacanthus* sp.
(Amblycera) occurred on 5 of 57 reed warbler nestlings examined from 24 nests, and
Bruelia sp. (Ischnocera) occurred on 4 of 52 great reed warbler nestlings examined
from 26 nests. Thus, the prevalence of warbler-specific lice (9/109) on warbler chicks
was higher, but not significantly higher, than on cuckoo chicks in warbler nests (0/
20). The prevalence on reed warbler nestlings was similar to that on fledged juvenile
(3/26) and adult (1/18) reed warblers mist-netted in July and August, suggesting that,
for the warblers, contact within the nest is a primary means of louse transmission (Lee
& Clayton, 1995).

The single cuckoo nestling sampled from an azure-winged magpie nest yielded one
specimen of *Philopterus* sp. (Ischnocera), belonging to a louse genus often associated
with corvids (Hopkins & Clay, 1952).

No feather lice whatsoever were found in the 19 reed warbler nests examined.

ARE LICE ACQUIRED DURING MATING?

The Japanese sample and its bearing on louse transfer during cuckoo mating

The principal sample of 35 cuckoos caught in Nagano Prefecture, Japan, comprised
15 first-year and 20 older cuckoos. All the cuckoos were found to carry cuckoo-
specific feather lice (*Table 19.1*), and the likelihood that any one of the three louse
species would be found on a particular cuckoo was not significantly affected by the
age of the cuckoo (*Table 19.1a*). There was also no difference between louse species
in prevalence, either on first-year or older cuckoos.

While all cuckoos bore lice, first-year cuckoos were somewhat more likely to have
only one species of louse, while older cuckoos were more likely to be infested with
two species. There was no difference between cuckoo age classes in the proportion of
birds infested with all three louse species (*Table 19.1b*).

Table 19.1. The prevalence of feather lice on common cuckoos caught in Nagano Prefecture, Japan, in spring 1995.

| | Cuckoos | | Significance |
	First year	Older	(χ^2 test, 1 d.f.)
a) Proportion carrying			
Cuculicola	10/15	17/20	n.s.
Cuculoecus	12/15	18/20	n.s.
Cuculiphilus	10/15	13/20	n.s.
b) Proportion of cuckoos with			
One louse species	8/15	3/20	<0.05
Two louse species	2/15	12/20	<0.05
Three louse species	5/15	5/20	n.s.
At least one louse species	15/15	20/20	n.s.

Table 19.2. Numbers of feather lice (mean ± s.d.; range) on common cuckoos caught in Nagano Prefecture, Japan, in spring 1995. No differences between first-year and older cuckoos (i.e. within rows) were significant (2-tailed Mann–Whitney tests).

| | | Cuckoos | |
		First year (n = 15)	Older (n = 20)
Louse species			
Cuculicola	Adults	7.6 ± 13.75 (0–47)	14.3 ± 25.88 (0–113)
	Nymphs	4.3 ± 7.96 (0–30)	20.6 ± 48.79 (0–113)
Cuculoecus	Adults	5.9 ± 13.87 (0–57)	3.3 ± 3.13 (0–12)
	Nymphs	2.9 ± 4.54 (0–14)	1.7 ± 2.03 (0–5)
Cuculiphilus	Adults	1.4 ± 2.87 (0–11)	1.8 ± 3.55 (0–15)
	Nymphs	4.4 ± 8.50 (0–33)	6.9 ± 19.64 (0–91)

For none of the three louse species was there a significant difference in the number of lice, either adult or nymphs, retrieved from first-year and older cuckoos (*Table 19.2*).

There is thus no evidence that first-year cuckoos differed from older birds in the prevalence of louse infestation or in the number of lice harboured, or in adult/nymph ratios which might reflect the recent breeding history of the louse population. Some such differences would be anticipated if the first-year birds had acquired their cuckoo-specific lice by mating with older birds during the sampling period, 27th May to 2nd July.

If mating were a primary route of louse transfer, then we predicted that first-year birds in the early part of the breeding season would be less heavily infested than those caught in the latter part. We cannot use prevalence as a measure, since all first-year birds, whenever caught, were infested (*Table 19.1*). We have, however, divided the 15 first-year cuckoos into the eight caught before 22nd June (actually 4th–21st June) and the seven caught after 23rd June (24th June–2nd July). The former group did not have significantly fewer lice (*Table 19.3*).

Table 19.3.   Numbers of feather lice (mean ± s.d. for adults and nymphs combined) on first-year common cuckoos caught in Nagano Prefecture, Japan, in spring 1995. No differences between cuckoos in the two time periods were significant (2-tailed Mann–Whitney tests).

| | Cuckoos caught | |
	Before 22nd June (n = 8)	After 23rd June (n = 7)
Louse species		
Cuculicola	11.1 ± 15.71	12.9 ± 26.32
Cuculoecus	13.8 ± 22.24	3.1 ± 4.94
Cuculiphilus	8.0 ± 13.80	3.3 ± 6.50

Other comments on the Japanese sample

From *Table 19.2*, adult/nymph ratios generally appear higher in the head louse, *Cuculoecus*, the largest of the species, than in the other two species. Among 30 birds carrying *Cuculoecus*, adult/nymph ratios exceeded unity in 20 cases. The corresponding values for *Cuculicola* and *Cuculiphila* were 13/27 (v 20/30; χ^2 = 1.31, 1 d.f., NS) and 5/23 (χ^2 = 8.81, 1 d.f., P <0.01), respectively.

As has been recorded in a variety of studies of avian lice (Marshall, 1981; Clayton *et al.*, 1992; Rózsa *et al.*, 1996), the sex ratio of the three louse species found on cuckoos was skewed in favour of females (*Table 19.4*). In a majority of birds, a majority that was significant for *Cuculicola* and *Cuculoecus* (P <0.01, sign test), female lice outnumbered male lice. For each species of louse, the total number of female and male lice deviated significantly from parity (P <0.01), and the ratio of females to males was fairly similar in the three louse species ranging from 2.11 in *Cuculiphilus* to 2.56 in *Cuculoecus* (χ^2 = 0.603, 2 d.f., P >0.1).

Twenty-seven of the Japanese cuckoos could be sexed confidently by their plumage and, more conclusively, by their call when released. There were no significant differences in the numbers of lice on the 16 males and 11 females.

British samples

The details of the lice obtained from the cuckoos trapped in England are given in *Table 19.5*. Although the data are sparse, some interesting points emerge. First, the first-year bird caught in May at Dungeness, a bird observatory on the south coast of England, was almost certainly on migration, and therefore had not mated (assuming

Table 19.4.   Sex ratio of adult feather lice obtained from common cuckoos caught in Nagano Prefecture, Japan, in spring 1995. Note that the number of infested cuckoos (first three numerical columns) is slightly lower than in *Table 19.1* because of the small number of birds that had only immature lice of a particular species.

| | No. of cuckoos with | | | Total no. of lice | |
	More female than male lice	Equal numbers	More male than female lice	Female	Male
Cuculicola	20	2	4	275	125
Cuculoecus	22	1	6	115	45
Cuculiphilus	11	3	1	38	18

Table 19.5.   Feather lice recorded on nine common cuckoos trapped in England. Cuckoos classified below as 'older' were more than one year old.

Date	Place	Age of cuckoo	Louse species found
15th May, 1994	Wicken Fen	Older	None
23rd May, 1994	Dungeness	First year	*Cuculicola*
1st June, 1994	Wicken Fen	Older	*Cuculicola Cuculiphilus*
1st June, 1994	Wicken Fen	First year	*Cuculicola*
2nd June, 1994	Knettishall	Older	*Cuculicola*
15th July, 1994	Dungeness	Juvenile	None
22nd May, 1995	Wicken Fen	First year	*Cuculicola*
22nd May, 1995	Wicken Fen	First year	*Cuculicola*
22nd May, 1995	Wicken Fen	First year	*Cuculicola Cuculiphilus*

that cuckoos do not copulate away from the breeding grounds). It was, nevertheless, lousy. Furthermore, all four first-year cuckoos caught at Wicken Fen carried lice, although the dates of capture were at the very start of the nesting season (first cuckoo egg dates at Wicken Fen normally in the last week of May; *personal observation*). Second, the Dungeness juvenile caught in mid-July did not yield lice. This is compatible with the failure to find cuckoo-specific lice on any cuckoo nestlings. Third, the failure to record *Cuculoecus* was probably because of inconsistent searching of the birds' heads during the delousing treatment. There is no reason to suppose this louse species does not infest British cuckoos, since it was obtained during the museum combing of cuckoos shot in the breeding season in Britain (see below).

ARE LICE ACQUIRED OUTSIDE THE BREEDING SEASON VIA PHORESY AND/OR BY
DIRECT CONTACT BETWEEN BIRDS?

We combed the skins of 50 fledged juvenile cuckoos from Great Britain (>80% of the holding in the NHM) and another six from continental Europe (the entire holding). Only one skin yielded a louse, identified as *Cuculicola*. This bird was collected in 1935 at Lochgilphead, Argyllshire, Scotland, on the exceptionally late date of 10th November. Nevertheless, the record was accepted by Witherby *et al.* (1938). Six juvenile cuckoos collected between September and November in Saudi Arabia, where they were likely to be on migration (Moreau, 1972), yielded no lice. The NHM's entire holding of first-year common cuckoos obtained in winter quarters in Africa (n = 37) was combed, and lice were found on four skins. The proportion with lice was as follows: July/Aug. 0/1; Sept./Oct. 1/12; Nov./Dec. 2/14; Jan./Feb. 0/6; March–May 1/4. The earliest louse found was a specimen of *Cuculiphilus* on a young cuckoo obtained in Somalia on 18th September. All three cuckoo lice species were represented on the birds obtained in Africa. Finally, all British common cuckoos (n = 42), and a further seven from continental Europe, all obtained in the breeding season as adults (late April–end June), were combed. Again, all three louse species were represented. The overall prevalence of lice in this combined sample of first-year and older birds was 6/49, and therefore not significantly different from the prevalence of 4/37 on first-year birds wintering in Africa. Thus, the prevalence of lice on first winter and breeding season cuckoos was 10/86, higher (χ^2 = 3.33, 1 d.f., 1-tailed P <0.05) than the prevalence, 1/56, on juvenile cuckoos still in the breeding area or on their first migration.

Despite the admitted problems associated with combing skins (see 'Study areas and methods'), the pattern of occurrence revealed by skins is compatible with the field observations. In the early autumn, few, if any, juvenile cuckoos are lousy. As the autumn progresses, lice are acquired. This acquisition can apparently occur either in Africa or, occasionally, in Europe. By the time birds leave Africa to begin spring migration, louse prevalence is probably similar to that observed on the breeding grounds, nearly 100%, judging by the observations of mist-netted birds in England and Japan.

Discussion

Our study of cuckoo nestlings and of nests in which cuckoos had laid essentially confirmed previous findings (Dogiel, 1936, 1964; Clay & Meinertzhagen, 1943) in establishing that the young cuckoo leaves the nest without cuckoo lice. Compatible with these observations is the fact that the skins of young, but fledged cuckoos combed in the Natural History Museum were louse-free, with the single exception of a bird shot unusually late, in November, in Scotland. Also louse-free was a single, free-flying juvenile sampled in July. Thus, the conclusion that acquisition of lice does not occur within the host nest seems secure.

Moreover, only one young cuckoo, sampled from an azure-winged magpie nest, carried a host louse (see also Dubinin, 1951; Dogiel, 1964). Although we cannot know how long this louse would have survived, the fact that no free-flying cuckoos sampled in our study harboured non-cuckoo feather lice suggests that any such host lice usually do not persist for long. However, Lindholm et al. (1998) have recently discovered that host lice can be found on diederik cuckoos (*Chrysococcyx caprius*) at least one year old. Since the cuckoos sampled were male, making any contact with the hosts unlikely, it seems the lice had remained on the birds since they fledged.

While acquisition of cuckoo lice has often been assumed to occur during mating (Dogiel, 1964; Marshall, 1981; Page et al., 1996), the present results refute that assumption. All first-year common cuckoos carried cuckoo lice, including one likely to be on migration and those sampled early in the breeding season on the nesting grounds. We did not detect any difference in the numbers or age distribution of lice carried by first-year and older cuckoos. The only significant difference between the two classes was that first-year birds more commonly had one species of louse and, more rarely, had two species of louse than older birds. This difference may simply reflect the lesser age of the first-year birds, and therefore the shorter time available to acquire the full set of louse species. While lice may transfer between birds during mating (Hillgarth, 1996), our field data do not argue for contact during copulation as the principal means by which the cuckoo lice reach new cuckoo hosts. In nearly 100% of cases, lice have reached the cuckoo by the start of the breeding season, and therefore, we presume, before mating.

This scenario is supported by the results of combing skins. Lice can occasionally be found on the skins of young birds obtained on the breeding grounds in late autumn, and more frequently on the skins of young birds from Africa. By the time the birds are ready to return on migration to the breeding grounds, skins obtained from Africa are apparently as likely to yield lice as skins obtained on the Palaearctic breeding grounds. This picture is identical for all three species of cuckoo lice.

Another point of identity is that the sex ratios among adult lice of the three species are significantly female biased, but to a similar degree. Sex ratio bias, mostly towards females, has been noted in a proportion of feather louse populations studied (Marshall, 1981; Clayton *et al.*, 1992). Female bias has been interpreted as a consequence of local mate competition (Hamilton, 1967), arising because the louse populations on individual birds are more or less isolated from other such populations, and therefore inbred. This reasoning leads to the prediction that the greater the degree of isolation, the greater the female bias, a suggestion supported by a comparison between bird species (Rózsa *et al.*, 1996). Since the sex ratio biases of all three species of cuckoo lice are similar, it may be that the three species transfer by similar methods, so that populations on different cuckoos are isolated to a similar degree. Neither the sex ratio results, nor indeed any other part of the data, suggest major differences in transmission between the two Ischnoceran lice and the one Amblyceran species (Keirans, 1975; Clayton *et al.*, 1992).

If young cuckoos leave the nest louse-free and return to the breeding grounds carrying lice, the timing of louse acquisition has been narrowed to a period of approximately nine months. Two non-exclusive possibilities for the method of acquisition are direct contact and phoresy. Both methods would lead to a progressive build-up in louse infestation, as hinted by the combing studies. Both routes would presumably be available only infrequently, and so cause inbreeding, and hence female bias, in the louse populations on individual hosts. How likely are these methods?

Although we ourselves have no experience of common cuckoos in the Asiatic and African wintering grounds, the consensus of bird-watching colleagues is that the birds are not often seen, and then usually singly. However, there are records both from Africa (Moreau, 1972) and from migration points (Christie, 1979; Hurrell, 1980; Rogers, 1980; Cramp, 1985) of aggregations of tens of birds gathering excitedly at caterpillar outbreaks. D. A. Christie (*personal communication*), amplifying his published observations, writes "I do certainly recall seeing three [cuckoos] perch for some 10 seconds in very close contact, two of these definitely in body contact; and another sub-group of five or six perched for no more than a couple of seconds, with at least three of these birds so close together that full body contact was inevitable". Such occasions can therefore involve the contact needed for louse transfer. In addition to any transfer at daytime feeding aggregations of cuckoos, there is the possibility of transfer at communal dustbaths, or at roosts if ever the birds roost communally (Marshall, 1981).

Clay & Meinertzhagen (1943) report examining five young cuckoos in Britain and Estonia in late summer. None had Mallophaga. However, one had nine, two had five, one had two, and one had no hippoboscids. One of the hippoboscids, *Ornithomyia avicularis*, carried a female of the louse *Bruelia merulensis*, a species which is a parasite of the blackbird (*Turdus merula*). Since Meinertzhagen's veracity as an observer has been questioned (Knox, 1993), these records merit cautious assessment. Even if they do suggest regular use of cuckoos by hippoboscids, they do not provide an instance of a hippoboscid carrying a cuckoo louse (see also Eichler, 1939). To our knowledge, there are no such records. Even if there were, hippoboscids are generally less specific in their host preferences than are feather lice (Maa, 1966; Marshall, 1981). It would therefore be surprising if cuckoo lice used hippoboscids as the sole means of transferring to new hosts.

Records of phoresy generally involve Ischnoceran feather lice, while Amblyceran species may crawl between hosts (Keirans, 1975). The fact that our data on prevalence and sex ratios suggest no difference between the transmission methods of the two Ischnoceran and one Amblyceran species infesting cuckoos is a further argument that phoresy is not the primary means of transmission. But, we certainly cannot exclude the possibility of phoresy.

We conclude that transfer of the feather lice of the common cuckoo does mostly occur horizontally, either by direct contact between birds and/or by phoresy before the breeding season. It will be difficult to devise experiments to distinguish the relative importance of these alternatives.

A final point is that this evidence of horizontal transmission by the lice of a bird species which is not generally gregarious implies that horizontal transmission could also occur in other bird species, especially those that are more sociable than cuckoos. Thus, the presumption that feather lice primarily move between hosts by vertical transmission may be premature. If lice, indeed, commonly move by both vertical and horizontal routes, then the theoretically-based belief that vertically-transmitted lice are less virulent than other, horizontally-transmitted ectoparasites may need to be refashioned (Ewald, 1993; Clayton & Tompkins, 1994).

Acknowledgements

It is a pleasure to thank David Walker of Dungeness Bird Observatory, plus Ian Hartley, Sue McRae, and the Wicken Fen Ringing Group for catching English cuckoos, and Fugo Takasu and Takashi Funakoshi for help catching Japanese cuckoos. Nick Davies, David Noble, and Mike Bayliss harvested 'cuckoo' nests, while Mike was also a stalwart finder of cuckoo nestlings. Chris Lyal of the Natural History Museum (London) kindly confirmed louse identifications. We are also grateful to the curators of the Natural History Museum (Tring) for access to their collections. The manuscript benefitted from the comments of Anna Lindholm and Dale Clayton. MdeLB's visit to Japan was made possible by a grant from the Royal Society, and the fact that Nick Davies nobly marked extra exam scripts. HN's research is supported by a Grant-in-Aid for Scientific Research (B) from the Japanese Ministry of Education.

References

Baker, K. (1993) *Identification guide to European non-passerines*. BTO Guide no. 24, Thetford, Norfolk, UK. 175 pp.

Christie, D. A. (1979) Large gatherings of cuckoos. *British Birds* 72, 552.

Clay, T. & Meinertzhagen, R. (1943) The relationship between Mallophaga and hippoboscid flies. *Parasitology* 35, 11–16.

Clayton, D. H. & Tompkins, D. M. (1994) Ectoparasite virulence is linked to mode of transmission. *Proceedings of the Parasitological Society of London, B* 256, 211–217.

Clayton, D. H. & Walther, B. A. (1997) Collection and quantification of arthropod parasites of birds. In: *Host–parasite evolution: general principles and avian models* (eds. D. H. Clayton & J. Moore), pp. 419–440. Oxford University Press, Oxford, UK.

Clayton, D. H., Gregory, R. D. & Price, R. D. (1992) Comparative ecology of neotropical bird lice (Insecta: Phthiraptera). *Journal of Animal Ecology* 61, 781–795.

Cramp, S. (ed.) (1985) *The birds of the Western Palaearctic IV*. Oxford University Press, Oxford, UK. 960 pp.

Dogiel, V. A. (1936) The details of infestation of the cuckoo with the mallophagan parasites [in Russian]. *Priroda* **8**, 113–114.

Dogiel, V. A. (1964) *General parasitology*. Oliver & Boyd, Edinburgh & London, UK. 516 pp.

Dubinin, V. B. (1951) Feather mites (Analgesoidea) Part 1. Introduction to their study. *Fauna SSSR Paukoobraznye* **6**, 1–363.

Eichler, W. (1939) Deutsche Lausfliegen, ihre Lebenweise und ihre hygienische Bedeutung. *Zeitschrift für hygienische Zoologie und Schädlingsbekämpfung* **31**, 210–226.

Ewald, P. W. (1993) The evolution of virulence. *Scientific American*, *April 1993*, 56–62.

Fowler, J. A. & Cohen, S. (1983) A method for the quantitative collection of ectoparasites from birds. *Ringing and Migration* **4**, 185–189.

Hamilton, W. D. (1967) Extraordinary sex ratios. *Science, New York* **156**, 477–488.

Hillgarth, N. (1996) Ectoparasite transfer during mating in ring-necked pheasant *Phasianus colchicus*. *Journal of Avian Biology* **27**, 260–262.

Hopkins, G. H. E. & Clay, T. (1952) *A check-list of the genera and species of Mallophaga*. British Museum (Natural History), London, UK. 362 pp.

Hurrell, H. G. (1980) Large gatherings of cuckoos. *British Birds* **73**, 412–413.

Keirans, J. E. (1975) A review of the phoretic relationship between Mallophaga (Phthiraptera: Insecta) and Hippoboscidae (Diptera: Insecta). *Journal of Medical Entomology* **12**, 71–76.

Knox, A. G. (1993) Richard Meinertzhagen – a case of fraud examined. *Ibis* **135**, 320–325.

Ledger, J. A. (1980) *The arthropod parasites of vertebrates in Africa south of the Sahara, vol. 4. Phthiraptera (Insecta)*. South African Institute for Medical Research, Johannesburg, South Africa. 327 pp.

Lee, P. L. M. & Clayton, D. H. (1995) Population biology of swift (*Apus apus*) ectoparasites in relation to host reproductive success. *Ecological Entomology* **20**, 43–50.

Lehane, M. J. (1991) *Biology of blood-sucking insects*. Harper Collins, London, UK. 288 pp.

Lindholm, A. K., Venter, G. J. & Ueckermann, E. A. (1998) Persistence of passerine ectoparasites on the diederik cuckoo *Chrysococcyx caprius*. *Journal of Zoology (London)* **244**, 145–153.

Maa, T. C. (1966) Studies in Hippoboscidae (Diptera). The genus *Ornithoica* Rondani. *Pacific Insects Monographs* **10**, 10–124.

Marshall, A. G. (1981) *The ecology of ectoparasitic insects*. Academic Press, London, UK. 459 pp.

Moreau, R. E. (1972) *The Palaearctic–African bird migration systems*. Academic Press, London, UK. 384 pp.

Nakamura, H. (1990) Brood parasitism by the cuckoo *Cuculus canorus* in Japan and the start of new parasitism on the azure-winged magpie *Cynopica cyana*. *Japanese Journal of Ornithology* **39**, 1–18.

Page, R. D. M., Clayton, D. H. & Paterson, A. M. (1996) Lice and co-speciation: a response to Barker. *International Journal of Parasitology* **26**, 213–218.

Rogers, M. J. (1980) Large gatherings of cuckoos. *British Birds* **73**, 413–414.

Rothschild, M. & Clay, T. (1952) *Fleas, flukes and cuckoos*. Collins New Naturalist Series, London, UK. 304 pp.

Rózsa, L., Rékási, J. & Reiczigel, J. (1996) Relationship of host coloniality to the population ecology of avian lice (Insecta: Phthiraptera). *Journal of Animal Ecology* **65**, 242–248.

Southwood, T. R. E. (1966) *Ecological methods*. Methuen, London, UK. 391 pp.

Witherby, H. F., Jourdain, F. C. R., Ticehurst, N. F. & Tucker, B. W. (1938). *The handbook of British birds, vol. 2*. Witherby, London, UK. 352 pp.

Wyllie, I. (1981) *The cuckoo*. Batsford, London, UK. 176 pp.

20

Moth and Bird Interactions: Guano, Feathers, and Detritophagous Caterpillars (Lepidoptera: Tineidae)

GADEN S. ROBINSON

Department of Entomology, The Natural History Museum, Cromwell Road, London, SW7 5BD, UK

Introduction

Our generalised impression of the Lepidoptera, the butterflies and moths, is that of a group of highly evolved, highly diverse insects with dispersive nectar-feeding adults and with foliage-feeding larvae. Indeed, these 'typical' species comprise the vast bulk of Lepidoptera diversity – there are more than 135,000 known species, and they form an important component of the diet of insectivorous birds.

But not all Lepidoptera larvae eat green plants. Here and there, groups have evolved that are, instead, detritophagous. And detritophagy – feeding on dead plant material – has opened an avenue at the end of which lie extraordinary larval biologies. Amongst these, we find the traditional bird-eats-insect story reversed – it is the insects that eat the birds, or rather food substrates that are produced by birds. These substrates may contain keratin (the cross-linked polypeptide that occurs in nature as feathers, fur, skin, and horn), or chitin (the tanned polysaccharide that forms the insect exoskeleton). The substrates are: raptor pellets (compacted bone, feather, skin, and fur), guano (compacted and finely divided chitin, together with faecal and nitrogenous waste), and feathers (fine-grained keratin that has been shed or pulled out for nesting material, or which is part of a dead bird). So, moth larvae that can cope with digesting keratin and chitin are associated either with nests lined with feathers, bird corpses, or roost areas where guano, feathers, or pellets are shed. Detritophagous moths capable of dealing with keratin and chitin feed also in mammal nests and on mammalian corpses and, if capable of coping with the extreme conditions, impinge on man as clothes moths.

Detritophagous moths without specialist digestive abilities may also be involved in the decomposition of mouldy plant material in old bird nests. Two examples of these, the brown house moth and white-shouldered house moth (*Hofmannophila pseudospretella* and *Endrosis sarcitrella*), also enter man's world as common domestic inhabitants and occasional pests of stored food products (just another form of plant

Insect and Bird Interactions
© Intercept Ltd., PO Box 716, Andover, Hampshire, SP10 1YG, UK.

detritus). Rotting seaweed in the crude nests of seabirds on St. Paul's Rocks provides the larval food of possibly the most isolated insect population in the world (Robinson, 1983).

That some moth larvae are habitual inhabitants of bird nests has been known for a long time. Cornelius (1869) and Morris (1870) record species of *Tinea* from nests, but concerted efforts to document the fauna of nests were not made until the mid-20th century. The work of Woodroffe and Southgate is notable in this respect, with useful reviews by Hicks and Petersen (Woodroffe & Southgate, 1950, 1951; Southgate & Woodroffe, 1951; Woodroffe, 1953; Hicks, 1959, 1962, 1971; Petersen, 1963).

The phylogenetic context of trophic specialisation

EARLY LEPIDOPTERA LINEAGES

The phylogeny of the 'primitive' Lepidoptera is well established (Davis, 1999; Kristensen, 1999a,b; Kristensen & Skalski, 1999). The most species-rich lineages are cladistically the most subordinate. The phylogenetic pattern exhibited by the early lineages of the Lepidoptera is the cited classic example of the 'Hennigian comb', with the first diverged extant lineages exhibiting step-by-step acquisition of the apomorphies, which characterise the most subordinate and successful groups (Kristensen & Skalski, 1999). It is among the early lineages that most of the detritophagous species are found. But even amongst these, detritophagy – as opposed to foliage-feeding by larvae – is the exception rather than the rule (*Figure 20.1*).

Kristensen (1997) has reviewed the early evolution and larval feeding modes of the Lepidoptera–Trichoptera lineage (the Amphiesmenoptera) and suggested that soil-dwelling is the original lifestyle of this clade. It has been retained by the Micropterigoidea, the earliest diverging lineage of the Lepidoptera, but subsequently diverging lineages of moths have arboreal larvae. Kristensen has suggested that parsimony favours the notion that the last common ancestor of the non-micropterigoid Lepidoptera had canopy-living and canopy-ovipositing adults, with the obvious inference that the shift came about through the utilisation of arboreal pollen sources by adults of early moths with soil-dwelling larvae.

The first fifteen clades of non-micropterigoid Lepidoptera to diverge (Robinson & Nielsen, 1993; Kristensen, 1997; Kristensen & Skalski, 1999) (*Figure 20.1*) show, with a few exceptions, a clear pattern of lifestyle. The larvae are arboreal, most are endophagous, and in about one-third of the groups, the ovipositor of the adult female is adapted for insertion of the egg into plant tissue. Only the last two clades (Yponomeutoidea and Gelechioidea) exhibit any diversity of exophagous larval feeding, but this is the norm in the Apoditrysia, the remaining monophylum which encompasses all the remaining Lepidoptera:

- **Agathiphagoidea**: endophagous seed-feeders in *Agathis*; non-piercing ovipositor;
- **Heterobathmioidea**: leaf-miners in *Nothofagus*; non-piercing ovipositor;
- **Eriocranioidea**: leaf-miners in Fagales and Rosales; piercing ovipositor;
- **Acanthopteroctetoidea**: leaf-miners in Rhamnaceae; piercing ovipositor;
- **Lophocoronoidea**: lifestyle unknown; piercing ovipositor;
- **Neopseustoidea**: lifestyle unknown; piercing ovipositor;

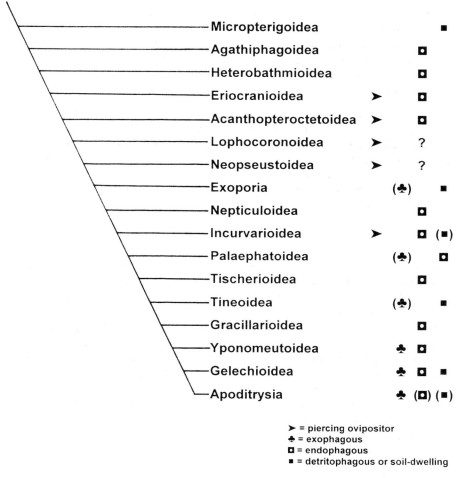

> = piercing ovipositor
♣ = exophagous
◻ = endophagous
■ = detritophagous or soil-dwelling

Figure 20.1.   Phylogenetic and biological summary of the basal lineages of the Lepidoptera; the apoditrysian Lepidoptera comprise more than 85% of the species diversity of the order but contribute only a handful of detritophagous species. Cladogram after Kristensen & Skalski, 1999 and Robinson & Nielsen, 1993.

- **Exoporia**: soil-dwelling, fungivory widespread, some root/trunk/branch borers; transition from fungivory to phytophagy recorded in some species; non-piercing ovipositor;
- **Nepticuloidea**: leaf- or cambium-miners; non-piercing ovipositor;
- **Incurvarioidea**: leaf-miners, later instars ground-dwelling detritophages (fungivores); all instars ground-dwelling in a few species; endophytic oviposition with piercing ovipositor by *all* species;
- **Palaephatoidea**: early instars leaf-miners in Proteaceae, later instars feeding concealed in folded leaves; non-piercing ovipositor;
- **Tischerioidea**: leaf-miners; non-piercing ovipositor;
- **Tineoidea**: all detritophagous/fungivorous except exophagous arboreal-feeding 'higher' Psychidae; non-piercing ovipositor in practically all species;
- **Gracillarioidea**: leaf- or cambium-miners; non-piercing ovipositor;

- **Yponomeutoidea**: leaf-miners, stem-borers or exophagous, partially concealed by webbing; non-piercing ovipositor;
- **Gelechioidea**: predominantly concealed but exophagous (e.g. Oecophoridae: Stenomatinae; most Depressariidae), but includes leaf-miners (e.g. Elachistidae, Coleophoridae, Cosmopterigidae) and detritophages (many Oecophorinae, Batrachedridae, Lecithoceridae); non-piercing ovipositor.

So, detritophagy is a departure from the original lifestyle of the non-micropterigoid Lepidoptera, but one that has arisen on several occasions (Rawlins, 1984). It is within the Tineoidea that the most intriguing caterpillar–bird interactions occur. Here, detritophagy has progressed beyond the ability to feed on a diet that comprises effectively a matrix of decaying plant material permeated by fungal mycelia to a diet that may be predominantly fungal, and thence to the chemically difficult substrates of chitin and keratin.

THE TINEOIDEA

The superfamily Tineoidea comprises four families – Tineidae, Eriocottidae, Acrolophidae, and Psychidae. A fifth family, Arrhenophanidae, is widely recognised (e.g. by Davis & Robinson, 1999), but its separation from the Psychidae cannot presently be justified on phylogenetic criteria. A phylogeny of the four (Robinson, 1988b; Robinson & Nielsen, 1993) has the Acrolophidae and Psychidae as sister-groups, with Eriocottidae as their sister-group; the Tineidae is the sister-group of the other three families (*Figure 20.2*), but there is only sparse evidence for the monophyly of the Tineidae, and the group may be a grade and not a clade (Robinson & Nielsen, 1993; Davis & Robinson, 1999). However, Minet (*unpublished*) has recently suggested that the dorsoventrally compressed terminal segment of the labial palpus in adult Tineidae may be evidence for their monophyly. The biologies of the four families are similar:

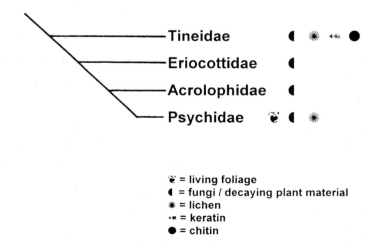

    ẽ = living foliage
    ◖ = fungi / decaying plant material
    ✸ = lichen
    ◂◂ = keratin
    ● = chitin

Figure 20.2. Phylogenetic and biological summary of the Tineoidea; Arrhenophanidae are included with Psychidae. Cladogram after Robinson & Nielsen, 1993.

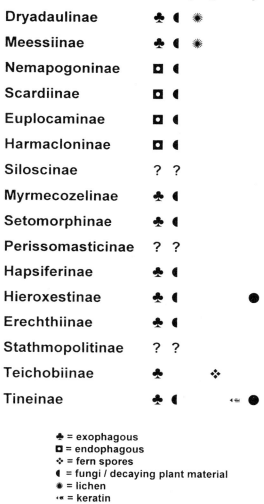

Dryadaulinae	♣ ◖ ✳
Meessiinae	♣ ◖ ✳
Nemapogoninae	◻ ◖
Scardiinae	◻ ◖
Euplocaminae	◻ ◖
Harmacloninae	◻ ◖
Siloscinae	? ?
Myrmecozelinae	♣ ◖
Setomorphinae	♣ ◖
Perissomasticinae	? ?
Hapsiferinae	♣ ◖
Hieroxestinae	♣ ◖ ●
Erechthiinae	♣ ◖
Stathmopolitinae	? ?
Teichobiinae	♣ ❖
Tineinae	♣ ◖ ◂ ●

♣ = exophagous
◻ = endophagous
❖ = fern spores
◖ = fungi / decaying plant material
✳ = lichen
◂ = keratin
● = chitin

Figure 20.3. Biological summary of the subfamilies of the Tineidae; no phylogeny has been resolved. Chitin digestion in Hieroxestinae is unproven, but several species of *Wegneria* are associated with cave guano.

- **Tineidae:** primarily fungivorous – food includes mouldy plant material, lichen, fungi, rotten wood, and (usually mouldy) keratin and chitin;
- **Eriocottidae:** soil-dwelling, feeding on mouldy plant detritus;
- **Acrolophidae:** soil-dwelling, on roots and plant detritus;
- **Psychidae:** 'primitive taxa' lichenophagous; 'derived groups' phytophagous on angiosperm foliage.

Parsimony in the context of this phylogeny suggests that foliage-feeding, while a typical lepidopteran trait, is a derived, and thus a secondary, feature among the so-called derived groups of Psychidae (bagworm moths).

THE TINEIDAE

Within the Tineidae there is a clear division of feeding habit among the sixteen subfamilies currently recognised (*Figure 20.3*), although there is no phylogeny resolved for them; two subfamilies (Meessiinae and Myrmecozelinae) are thought to be paraphyletic.

Robinson & Nielsen (1993) point out that "the association of Tineidae with fungi is all-pervading", and that the universality of fungus in tineid larval diets is a key feature of the tineid lifestyle. In one subfamily, Tineinae, detritophagy encompasses animal, as well as plant, remains, and it is here that the only known examples of keratophagy and chitinophagy within the Lepidoptera occur, though one possible exception may occur in the subfamily Hieroxestinae (in which some species of *Wegneria* are associated with guano). Tineine larvae utilise keratin sources such as wool, fur, feathers, and horn. Most do so in environmental conditions in which that source may provide a substrate for keratinolytic fungi of various kinds, and most species feed on sources that have already been colonised by such fungi (Zagulajev, 1960; Robinson, 1988b, 1990; Robinson & Nielsen, 1993). The tentative suggestion has been made (Robinson, 1990) that, in the case of most Tineinae, fungal mediation is necessary for keratin digestion. Such mediation may involve utilisation of fungal enzymes, or the utilisation of symbiotic micro-organisms in the gut (although there is no evidence for the latter). Such fungal mediation may also be necessary in the case of plant-detritophagous and chitinophagous species. Indeed, it is unlikely that these insects can avoid the ingestion of substantial quantities of fungal mycelium. The question has not yet been answered as to whether these species really make any use of the plant or insect detritus they ingest, or whether it is merely incidental to a diet of fungal mycelium.

However, a few species are capable of feeding and surviving in conditions in which the humidity is too low to permit significant fungal growth. These are the horn moths (*Ceratophaga*) (Gozmány & Vári, 1973) and the clothes moths (*Tinea pellionella*-group and *Tineola bisselliella*); the latter (reviewed by Robinson, 1979) are domestic pests, and appear capable of digesting keratin without fungal mediation. An intermediate situation occurs in the case of *Tinea pallescentella*, in which first-instar larvae graze fungal mycelium from the surface of hairs, whereas later instar larvae feed on the hairs themselves (Zagulajev, 1960).

The route to keratophagy

The phylogenies described above permit cautious inferences about the evolutionary route to keratin feeding. Keratophagy occurs as an isolated phenomenon within a single subfamily of tineid moths; the remaining members of the family are fungivores in some way or another, and keratin digestion mediated by fungi is seen as an intermediate step (*Figure 20.4*). Very few other insects (perhaps just feather lice and some dermestid beetles) are capable of digesting keratin.

ADAPTATIONS FOR KERATIN DIGESTION

The large body of work on the physiology of Tineidae is almost entirely restricted to

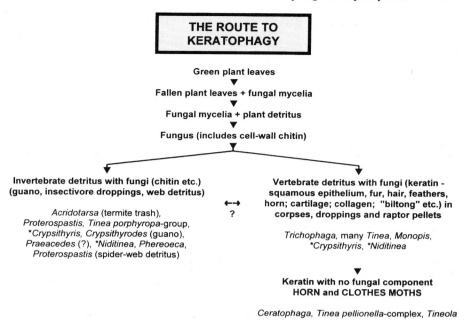

Figure 20.4. Hypothetical evolution of trophic types among Tineidae.

studies of larvae of *Tineola*, the webbing clothes moth. So, it is not always possible to discern which features are distinctive aspects of keratin digestion, since comparative data for non-keratophagous species are not available. Early studies have been summarised by Hinton (1956) and Waterhouse (1957, 1958). Baker (1986) and Baker & Bry (1987) have contributed more recent work on enzymes (reviewed by Robinson & Nielsen, 1993). The keys to keratin digestion by *Tineola* are the high pH and strongly reducing conditions maintained in the midgut; the practically anaerobic gut is held at a very low redox potential of –280 mV, and is able to withstand the considerable production of free hydrogen sulphide as the cross-links of the keratin molecules are severed. While high pH is by no means unusual in the insect midgut, strong reducing conditions are. An inversion of redox potential in the hind gut (to +250 mV) permits the breakdown of cysteine to cystine and water; such metabolic water has been suggested by Hinton (1956) to be crucial to the ability of *Tineola* to survive dry conditions.

While micro-organisms utilise keratinases to digest keratin, the process in *Tineola* is primarily due to proteinase activity (Baker & Bry, 1987), and no less than 29 different proteolytic enzymes appear to be involved.

The parameters of the physiological borderline beyond which keratin digestion is possible, are yet to be defined; the phylogenetic position of this borderline appears to delimit the Tineinae, but whether the strongly reducing conditions of the *Tineola* midgut are unique to that genus or are developed progressively within the group is not known. The question is begged whether extreme gut conditions are more general in

Table 20.1.   Tineine genera with species restricted to bat- or bird-guano in caves (chitinophages).

Genus	No. of species (genus total in brackets)	Region
Crypsithyrodes	3 (3)	Pantropical
Crypsithyris	1 (32)	Old World
Kangerosithyris	1 (1)	Oriental
Tinea	5 (67)	Oriental
Tetrapalpus	1 (1)	Neotropical
Proterospastis	1 (26)	Pacific

Table 20.2.   Tineine genera associated with compacted keratin sources (raptor pellets, carnivore faeces, horn) (keratinophages).

Genus	No. of species	Region
Trichophaga	8	Old World
Monopis	79	Old World, present but depauperate in New World
Ceratophaga	16	Africa, Oriental
Tinea	67	Worldwide [occasional association only]

the Tineidae, and whether those conditions may predispose larvae to be able to tackle other polymers, such as polysaccharides. The most obvious polysaccharide in tineine diets is chitin, in the form of fungal cell walls and arthropod remains.

CHITINOPHAGY AND GUANO-FEEDING

A hen-and-egg question hangs over the subject of chitin- or guano-feeding. Was it a corollary to keratin-digestion – a niche that was opened simply by dint of the fact that once a larva can digest keratin, it can digest anything? Certainly, the diversity of guano-feeding tineine Lepidoptera is very low, and encompasses five separate monophyletic entities (*Table 20.1*), whereas the diversity and 'phylogenetic compactness' of species that feed on keratin sources – raptor pellets, weathered carnivore droppings, horn, fur, and feathers – is much more marked (*Table 20.2*). It should be noted, however, that several species of *Wegneria* (Tineidae: Hieroxestinae) are associated with guano in caves, and these may have developed independently the ability to digest chitin (Robinson, 1980; Robinson & Tuck, 1997).

KERATOPHAGY: PELLET- AND FEATHER-FEEDING

Whereas *Trichophaga* and *Monopis* utilise keratin in a loose form (fur, feathers) or 'packaged' (as carnivore droppings or raptor pellets), *Ceratophaga* specialises in probably the most compact and 'difficult' keratin food source imaginable – the horns of antelope and buffalo. This is the only organism capable of breaking down dry horn and, it is rumoured, digesting the skin and foot-pads of dead elephants (Coe, 1978).

Most species of keratophagous moths are associated with keratin in the finely divided form of fur, hair, and feathers. The commonest, and apparently most frequent, occurrence of this resource is in the form of the feather lining of bird nests, and this food source may be used by, worldwide, over 100 species (*Table 20.3*).

Table 20.3. Tineine genera associated with birds' nests containing feathers (keratinophages).

Genus	No. of species	Region
Monopis	79	Old World, present but depauperate in New World
Niditinea	13	Worldwide
Tinea	67	Worldwide

Table 20.4. Segregation of tineid groups by physical characteristics of nest type.

Concealed (tree-hole) nests (and nest boxes)		Exposed (e.g. tree or hedge) nests	
Damp	Dry	Damp	Dry
Niditinea piercella-group (worldwide) (*Figure 20.5*)		*Niditinea fuscella*-group (worldwide) (*Figure 20.6*)	*Niditinea fuscella*-group (worldwide)
	Tinea corynephora-group (Australia)	*Tinea trinotella, semifulvella* etc. (Eurasia) (*Figures 20.13 and 20.14*)	*Tinea columbariella*-group (Eurasia)
		Monopis congestella-group (Oriental) (*Figures 20.8 and 20.9*)	*Monopis monachella*-group (Asia) (*Figure 20.7*)
Monopis fenestratella (Europe)	*Monopis argillacea*-group (Australia)	*Monopis impressella*-group (Old World tropics) (*Figure 20.11*)	*Monopis laevigella*-group (Europe, Africa) (*Figure 20.12*)
		Tinea bivirgella-group (Andes) (?) (*Figures 20.15 and 20.16*)	

We are able, in some cases, to differentiate clearly between different nest 'qualities', and to associate different species or genera with different nest types or different microclimates. It seems likely that there is differentiation of the bird nest niche into different physical types by sympatric keratophagous species (*Table 20.4*).

Niditinea provides a graphic example of how two species-groups within one genus have 'partitioned' nest types between them. While members of the *Niditinea fuscella*-group (*Figure 20.6*) are common inhabitants of both sheltered and damp exposed nests, members of the *Niditinea piercella*-group (*Figure 20.5*) have only ever been found as larvae in nests in holes in trees, and this appears to be their exclusive habitat.

ADAPTATIONS TO NEST DWELLING AND LARVIPARITY

Specific adaptations of Tineidae that are nest dwelling, excluding the robust physiological adaptations that permit them to digest keratin, are, for the most part, subtle. In adults, these adaptations include an erratic negative phototropism – many nest-inhabiting species are only rarely encountered in light-traps, for example – and a 'panic' response that may take one of two forms. Some *Monopis* species drop and sham death on disturbance, while others crash-dive into the nest and burrow into the nesting material.

The ability of adults to locate sources of keratin must also be an adaptation. How

Figures 20.5–20.10.   Tineid moths associated with birds' nests: **20.5**, *Niditinea piercella*, Britain (note pale head) – tree-hole nests, west Palaearctic; **20.6**, *Niditinea fuscella*, USA – exposed nests, Holarctic; **20.7**, *Monopis longella*, Thailand – exposed nests, south-east Asia; **20.8**, *Monopis congestella* Walker, Sulawesi (larviparous) – exposed nests, Indo-Australian region; **20.9**, *Monopis icterogastra*, Australia (larviparous) – tree-hole nests (?); **20.10**, *Monopis* sp., Australia (oviparous) [not to scale; wingspan of all specimens 14–18 mm].

they do it is uncertain, although Robinson & Nielsen (1993) suggest that it may be an olfactory response to low levels of hydrogen sulphide from decomposing keratin. However, more recently, Traynier & Schumacher (1995) have found that *Tinea pellionella* and *Tinea translucens* prefer to oviposit on wool that is associated with yeast odours. This gives rise to the suspicion that a volatile associated with B-complex vitamins (or their decomposition) may be involved. This notion may be supported by the observation that clothes moth attacks on garments tend to be concentrated in regions contaminated by sweat or urine. Hinton (1956) noted that B-complex vitamins are a growth stimulant for *Tineola*. The technique for deploying artificial nests (described below) involves the contamination of the nest using beer, dissolved vitamin B-complex tablets, or urine. Observations suggest that uncontaminated artificial nests are far less attractive.

The larval case-making habit, widespread among Tineidae, is facultative in *Monopis longella* in south-east Asia (Robinson & Nielsen, 1993), and in *Monopis icterogastra* and an unnamed related species in Australia (Gilbert & Horak, *unpublished*), and may be so in many other species. In a nest of loose composition, the larva makes a permanent, portable case and feeds externally but, if the material of the nest is compacted, no case is made and the larva burrows in the substrate.

Within the genus *Monopis* we find a remarkable adaptation: a large species-group, the *congestella*-group, is larviparous. Instead of laying eggs, the female lays fully-developed first instar larvae. This phenomenon was first described in *Monopis icterogastra* (*Figure 20.9*) by Scott (1863) in New South Wales, and has since been noted in numerous species of the *congestella*-group in south-east Asia and Australasia (*Figures 20.8* and *20.9*). But all is not straightforward: *Monopis icterogastra* has a sibling species (*Figure 20.10*), to which it is practically identical in external features, which is conventionally oviparous. It seems intuitively unlikely that 'species X' is the sister of the remaining *congestella*-group, all members of which are larviparous. If it is not, then there is some mechanism of homoplasy that permits a precise reversal to the oviparous condition that involves *all* associated morphological modifications.

The reasons for the evolution of larviparity among nest-dwelling Tineidae are unclear. It may be that the mortality of eggs or newly-hatched larvae in bird nests may be unusually high, and it thus makes sense to deposit offspring with some defensive capability. But what are the potential predators? Mites have been tentatively suggested, but evidence is needed. It may be that a female rarely finds more than one oviposition site; her search time would permit maturation of larvae, and reduced fecundity would be neutralised by the low 'carrying capacity' of the feather content of a single nest.

The costs of larviparity are in fecundity. A larviparous species can produce up to about 60 larvae (the norm appears to be about 40), whereas an oviparous species may lay more than 120 eggs. The morphological modifications that a species undergoes to achieve larviparity appear to be common to all larviparous Tineidae. The common oviduct is grossly enlarged and folded back on itself, and the ovipositor is considerably reduced in length. The *apophyses anteriores,* muscle attachments used to manoeuvre the apex of the abdomen and ovipositor, are reduced, and the eighth abdominal segment is shortened and broadened.

Perhaps most surprisingly, this modification has evolved not once but independently at least twice in the nest-dwelling Tineidae. It has been discovered recently that the Andean representatives of the genus *Tinea* – the apparently monophyletic *Tinea bivirgella*-group (*Figures 20.15* and *20.16*) – are also larviparous.

What is clear is that it is a most unusual modification, and one that is quite atypical within the biology of the Lepidoptera. A very small number of species – fewer than

11.

12.

13.

14.

15.

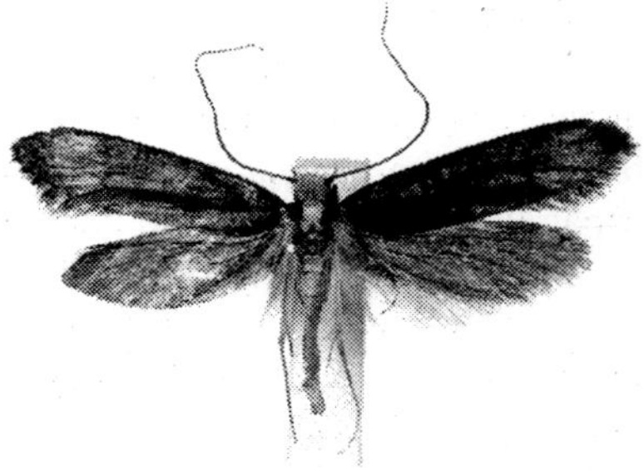

16.

Figures 20.11–20.16.   Tineid moths associated with birds' nests: **20.11**, *Monopis ministrans*, Cameroun – exposed nests, West Africa; **20.12**, *Monopis laevigella*, Britain – exposed nests, Palaearctic; **20.13**, *Tinea trinotella*, Britain – exposed nests, west Palaearctic; **20.14**, *Tinea semifulvella*, Britain – exposed nests, west Palaearctic; **20.15**, *Tinea* sp. (*bivirgella*-group), Peru (larviparous) – ? nests, montane Neotropical; **20.16**, *Tinea* (?) *montezuma*, Peru (larviparous) – ? nests, montane Neotropical [not to scale; specimens in *Figures 20.11–20.13* with ~14 mm wingspan, in *Figures 20.14–20.16* with 19, 31 and 20 mm wingspans respectively].

ten – from other families are thought to be larviparous. Four or five species of Coleophoridae (*Coleophora*) are apparently larviparous (Toll, 1952, 1962), together with two or three species of Walshiidae (*Ascalenia*) (Kasy, 1969). Other than these known cases, Hodges (1964) suspected larviparity in two genera of Cosmopterigidae (*Aeaea* and *Synploca*). Evidence for larviparity in all of these is the finding of a single larva in the oviduct of more than one dissected female, with one spectacular exception: *Coleophora albella* (previously known as *C. leucapennella*) is known to mature multiple larvae – it has a monstrously enlarged common oviduct, similar to those of larviparous Tineidae. Otherwise, larviparity, if it does occur in these species, involves the maturing of one larva at a time.

Single larvae being found in the oviduct of a dissected female is a phenomenon that has been reported from other groups, but only as a single observation, where it was probably the result of post-mortem development.

The occurrence and biology of the so-called larviparous non-tineid species is otherwise very different from that of the larviparous Tineidae – they are, to all intents and purposes, typical Lepidoptera with no features of their biology that could conceivably be interpreted as favouring the evolution of larviparity.

The phenomenon of larviparity in nest-dwelling Lepidoptera is one that begs further investigation and explanation.

Studying nest-dwelling Lepidoptera – artificial nests

Much of the foregoing has been derived and interpreted primarily from taxonomic research. Field research on nest-dwelling tineid moths is surprisingly sparse. One reason may be that it is difficult to collect abandoned nests in quantity, and in some areas it may be illegal, despite the lack of occupants.

Artificial nests were developed as oviposition (and larviposition) baits for keratophagous Tineidae (Robinson, 1988a, 1990; Robinson & Nielsen, 1993), primarily for use in the wet tropics, where the problems of recovery of real nests are compounded by the height of the forest canopy, but they have also been used successfully in temperate regions (Jensen, 1989). Artificial nests comprise a handful of feathers (usually from domestic chickens or pluckings from game) enclosed in rot-resistant fabric netting (knotted sections of black fishnet stocking were used in most trials) to form a ball. 'Nests' are deployed in trees and bushes, or under the eaves of houses, and are soaked with either beer, solution of vitamin B tablets, or urine, to enhance the vitamin B content (and possibly to provide olfactory clues for location – see above) prior to deployment. Cholesterol was added as a food supplement to nests deployed in trials in Australia (Horak, *unpublished*).

Artificial nests are exposed for anything from two to fifteen weeks (and longer in temperate climates), and locations are usually chosen to provide a range of microhabitats, from those fully exposed to rain to completely sheltered.

Colonisation of artificial nests in the Malay Peninsula and Borneo has occurred within a few days of their deposition. However, in seasonally dry areas of Thailand, colonisation in the dry season has taken much longer. So, moisture needs to be present for some species to oviposit – or larviposit. We found that artificial nests exposed for about ten weeks typically produced fifty to several hundred adult Tineidae, and that in a controlled environment chamber it was easy to rear some species through several generations together with their parasitoids (which also infest the nest in the wild). However, at least one species required a larger space than that available in the culture boxes for a successful courtship flight. This species reared itself successfully through three generations by escaping, mating, and depositing larvae through the mesh of rearing boxes, and causing confusion and mayhem by cross-infestation.

Artificial nest exposures in West Africa were similarly successful, resulting in the rearing of two species that had previously been known from only single individuals. Artificial nests have been used in Australia in a very extensive transect survey to establish whether there are 'wild' reservoirs of clothes moths: preliminary results suggest that pest species are reared always from 'nests' placed in close proximity to human habitation (Horak, *unpublished*).

Several baseline experiments are still to be carried out – artificial nest exposures have not been attempted in the New World to establish the identity of nest-inhabitors; assumptions about the North American and Andean Tineidae are based on museum specimens. We have not yet developed an artificial nest-in-a-hole to see whether the clear faunal differentiation that we see between exposed nest faunas and hole-nest faunas in the Holarctic region is repeated elsewhere. These are future projects, planned to yield predominantly morphological and taxonomic data.

However, the artificial nest also offers itself as a novel sampling tool that an imaginative ecologist might turn to his advantage, and I should be very interested to hear of proposals involving this technique.

References

Baker, J. E. (1986) Amylase/proteinase ratios in larval midguts of ten stored-product insects. *Entomologia Experimentalis et Applicata* **4**, 41–46.

Baker, J. E. & Bry, R. E. (1987) Nutritional ecology of wool- and fur-feeding insects. In: *Nutritional ecology of insects, mites, spiders and related invertebrates* (eds. F. Slansky & J. G. Rodriguez), pp. 971–992. Wiley, New York, USA.

Coe, M. (1978) The decomposition of elephant carcases in the Tsavo (East) National Park, Kenya. *Journal of Arid Environments* **1**, 71–86.

Cornelius, C. (1869) Vogelnester und Insekten. *Stettiner Entomologische Zeitung* **30**, 407–410.

Davis, D. R. (1999) The monotrysian Heteroneura. In: *Lepidoptera, moths and butterflies. 1. Evolution, systematics and biogeography. Handbook of zoology, vol. 4, part 35, Lepidoptera* (ed. N. P. Kristensen), pp. 65–90. de Gruyter, Berlin, Germany.

Davis, D. R. & Robinson, G. S. (1999) The Tineoidea and Gracillarioidea. In: *Lepidoptera, moths and butterflies. 1. Evolution, systematics and biogeography. Handbook of zoology, vol. 4, part 35, Lepidoptera* (ed. N. P. Kristensen), pp. 91–117. de Gruyter, Berlin, Germany.

Gozmány, L. A. & Vári, L. (1973) The Tineidae of the Ethiopian Region. *Transvaal Museum Memoir* **18**, 1–238.

Hicks, E. A. (1959) *Check-list and bibliography on the occurrence of insects in birds' nests.* Ames, Iowa, USA. 681 pp.

Hicks, E. A. (1962) Check-list and bibliography on the occurrence of insects in birds' nests. Supplement 1. *Iowa State Journal of Science* **36**, 233–344.

Hicks, E. A. (1971) Check-list and bibliography on the occurrence of insects in birds' nests. Supplement 2. *Iowa State Journal of Science* **46**, 123–338.

Hinton, H. E. (1956) The larvae of the species of Tineidae of economic importance. *Bulletin of Entomological Research* **47**, 251–346.

Hodges, R. W. (1964) A review of the North American moths of the family Walshiidae (Lepidoptera: Gelechioidea). *Proceedings of the US National Museum* **115**, 289–330.

Jensen, H. K. (1989) En ny handtering til fangst af nogle sjaeldne mol. *Lepidoptera, København* **5**, 239–245.

Kasy, F. (1969) Vorläufige Revision der Gattung *Ascalenia* Wocke (Lepidoptera, Walshiidae). *Annalen des Naturhistorischen Museums, Wien* **73**, 339–375.

Kristensen, N. P. (1997) Early evolution of the Lepidoptera + Trichoptera lineage: phylogeny and the ecological scenario. *Mémoires du Muséum National d'Histoire Naturelle, Paris* **173**, 253–271.

Kristensen, N. P. (1999a) The non-glossatan moths. In: *Lepidoptera, moths and butterflies. 1. Evolution, systematics and biogeography. Handbook of zoology, vol. 4, part 35, Lepidoptera* (ed. N. P. Kristensen), pp. 41–49. de Gruyter, Berlin, Germany.

Kristensen, N. P. (1999b) The homoneurous Glossata. In: *Lepidoptera, moths and butterflies. 1. Evolution, systematics and biogeography. Handbook of zoology, vol. 4, part 35, Lepidoptera* (ed. N. P. Kristensen), pp. 51–63. de Gruyter, Berlin, Germany.

Kristensen, N. P. & Skalski, A. W. (1999) Phylogeny and palaeontology. In: *Lepidoptera, moths and butterflies. 1. Evolution, systematics and biogeography. Handbook of zoology, vol. 4, part 35, Lepidoptera* (ed. N. P. Kristensen), pp. 7–25. de Gruyter, Berlin, Germany.

Morris, F. O. (1870) *A natural history of British moths, vol. 4*. Longmans, Green, Reader & Dyer, London, UK. 304 pp.

Petersen, G. (1963) Tineiden als Bestandteil der Nidicolenfauna. *Beiträge zur Entomologie* **13**, 411–427.

Rawlins, J. E. (1984) Mycophagy in Lepidoptera. In: *Fungus–insect relationships* (eds. Q. Wheeler & M. Blackwell), pp. 382–423. Columbia University Press, New York, USA.

Robinson, G. S. (1979) Clothes moths of the *Tinea pellionella* complex: a revision of the world's species (Lepidoptera: Tineidae). *Bulletin of the British Museum (Natural History), Entomology* **38**, 57–128.

Robinson, G. S. (1980) Cave-dwelling tineid moths: a taxonomic review of the world species (Lepidoptera: Tineidae). *Transactions of the British Cave Research Association* **7**, 83–120.

Robinson, G. S. (1983) Darwin's moth from St. Paul's Rocks: a new species of *Erechthias* (Tineidae). *Systematic Entomology* **8**, 303–311.

Robinson, G. S. (1988a) Keratophagous moths in tropical forests – investigations using artificial birds' nests. *Entomologist* **107**, 34–45.

Robinson, G. S. (1988b) A phylogeny for the Tineoidea (Lepidoptera). *Entomologica Scandinavica* **19**, 117–129.

Robinson, G. S. (1990) Clothes moths and keratin in Asian rainforest. In: *Insects and the rainforests of south-east Asia (Wallacea)* (eds. W. J. Knight & J. D. Holloway), pp. 305–308. Royal Entomological Society of London, London, UK.

Robinson, G. S. & Nielsen, E. S. (1993) Tineid genera of Australia (Lepidoptera). *Monographs on Australian Lepidoptera, 2*. CSIRO, Melbourne, Australia. 343 pp.

Robinson, G. S. & Tuck, K. R. (1997) Phylogeny and composition of the Hieroxestinae (Lepidoptera: Tineidae). *Systematic Entomology* **22**, 363–396.

Scott, A. W. (1863) Description of an ovo-viviparous moth belonging to the genus *Tinea*. *Transactions of the Entomological Society of New South Wales* **1**, 33–36.

Southgate, B. J. & Woodroffe, G. E. (1951) The insect fauna of birds' nests. *Proceedings and Transactions of the South London Entomological and Natural History Society, 1950–1951*, 44–45.

Toll, S. (1952) Eupistidae (Coleophoridae) of Poland. *Documenta Physiographica Poloniae, Kraków* **32**, 1–292.

Toll, S. (1962) Materialien zur Kenntnis der paläarktischen Arten der Familie Coleophoridae (Lepidoptera). *Acta Zoologica Cracoviensia* **7**, 577–720.

Traynier, R. M. M. & Schumacher, R. K. (1995) Yeast odours localize oviposition by *Tinea translucens* and *T. pellionella* (Lepidoptera: Tineidae). *Journal of Stored Products Research* **31**, 301–305.

Waterhouse, D. F. (1957) Digestion in insects. *Annual Review of Entomology* **2**, 1–18.

Waterhouse, D. F. (1958) Wool digestion and mothproofing. In: *Advances in pest control research, vol. 2* (ed. R. L. Metcalf), pp. 207–262. Interscience, New York, USA.

Woodroffe, G. E. (1953) An ecological study of the insects and mites in the nests of certain birds in Britain. *Bulletin of Entomological Research* **44**, 739–772.

Woodroffe, G. E. & Southgate, B. J. (1950) Notes on the insect fauna of birds' nests. *Middle Thames Nauralist* **3**, 28–31.

Woodroffe, G. E. & Southgate, B. J. (1951) Birds' nests as a source of domestic pests. *Proceedings of the Zoological Society of London* **121**, 55–62.

Zagulajev, A. K. (1960) Tineidae; part 3 – subfamily Tineinae [In Russian]. *Fauna SSSR* **78**, 1–267 [Translation, 1975, New Delhi, India].

Index

Page numbers which appear in **bold** represent a substantial section of text devoted to the topic